Progress in Medicinal Chemistry 24

Progress in Medicinal Chemistry 24

Edited by

G.P. ELLIS, D.SC., PH.D., F.R.S.C.

Department of Applied Chemistry,
University of Wales Institute of Science and Technology,
P.O. Box 13,
Cardiff, CF1 3XF,
United Kingdom

and

G.B. WEST, B. PHARM., D.SC., PH.D., F.I.BIOL.

Department of Paramedical Sciences,
North East London Polytechnic,
Romford Road,
London, E15 4LZ,
United Kingdom

1987

ELSEVIER
AMSTERDAM · NEW YORK · OXFORD

ISBN 0-444-80876-0
ISBN Series 0-7204-7400-0

Published by:

Elsevier Science Publishers B.V. (Biomedical Division)
P.O. Box 211
1000 AE Amsterdam
The Netherlands

Sole distributors for the USA and Canada:
Elsevier Science Publishing Company, Inc.
52 Vanderbilt Avenue
New York, NY 10017
USA

Library of Congress Cataloging in Publication Data:

Please refer to card number 62-27/2 for this series

Printed in The Netherlands

Preface

We have pleasure in presenting eight reviews in this volume, all of which cover important advances in the chemistry and biology of medicinal products.

In Chapter 1, the structure, synthesis and toxicity of ricin, a protein present in the endosperm cells of the seeds of the castor oil plant, are described and possible uses of this toxic agent in immunology as conjugates with antibodies are discussed. The biochemical and pharmacological evidence for the presence of functional histamine receptors in the mammalian central nervous system is set out in Chapter 2. Chapter 3 covers the chemical properties of molybdenum-containing enzymes such as aldehyde oxidase and xanthine oxidase, although their physiological roles are by no means clear.

Biologically active platinum complexes, which cross-link defined regions of DNA in cells, possess antitumour activity, and these are evaluated in Chapter 4. In Chapter 5, the present status of the medicinal uses of Cannabis is discussed, particularly as two cannabinoids are now official drugs. The latest methods of treating insulin-resistant diabetic patients with hypoglycaemic agents which do not act by increasing insulin release are described in Chapter 6.

Chapter 7 covers the mechanisms involved in the treatment and prophylaxis of angina pectoris and moderate hypertension by drugs which block the passage of calcium ions into and out of cells. Finally, in Chapter 8, the discovery of clinically useful inhibitors of aldose reductase, an enzyme involved in chronic diabetes, is described.

We thank our authors for surveying the ever-widening literature of several aspects of medicinal chemistry. We also offer our thanks to owners of copyright material who have given their permission for it to be reproduced in this volume and to the staff of our publishers for their help and encouragement.

G.P. Ellis
G.B. West

November 1986

Contents

Progress in Medicinal Chemistry – Vol. 24, edited by G.P. Ellis and G.B. West

1 Ricin: Cytotoxicity, Biosynthesis and Use in Immunoconjugates

J. MICHAEL LORD, Ph.D., JANE GOULD, Ph.D., DAVID GRIFFITHS, Ph.D., MARY O'HARE, Ph.D., BERNADETTE PRIOR, Ph.D., PETER T. RICHARDSON, Ph.D., and LYNNE M. ROBERTS, Ph.D.

Department of Biological Sciences, University of Warwick, Coventry, CV4 7AL, United Kingdom

INTRODUCTION

Attempts to develop effective anticancer agents are restricted by a single major constraint, the similarity between cancer cells and normal cells. Cancer cells are normal cells which have been transformed and are behaving aberrantly. The most prominent aspect of the aberrant behaviour is rapid, uncontrolled growth. As a result, cancer cells quickly invade adjacent tissues and can metastasize to distant tissues. Many of the reagents used in cancer chemotherapy have been selected on the basis of their ability to do more harm to rapidly dividing cells than to slowly or non-dividing cells. The success of these drugs has at best been limited; not only are rapidly dividing normal cells a target, but non-dividing tumour cells in silent metastases remain undisturbed. Reagents that are much more selective would have obvious advantages and ideally would be extremely toxic to malignant cells and completely harmless to normal cells. Such reagents should have the specificity of an antibody and the potency of the most toxic compounds known.

Growth rate is not the only difference between malignant and normal cells, and a further difference forms the basis of a selective therapy. Many types of tumour cell express high levels of certain surface antigens, while these molecules are expressed at a much lower level or not at all on the surface of normal cells. Monoclonal antibodies raised against these antigens can differentiate between malignant and normal cells, and bind strongly to the former but only weakly or not at all to the latter. Specificity of binding to a subset of cells can be followed by the elimination of these cells if the antibody molecule has been coupled to potent drugs or toxins. Many conjugates of this type are currently being generated and evaluated, the most popular consisting of a tumour-cell-specific monoclonal antibody linked to a potent protein toxin (immunotoxin) [1]. Several cytotoxic proteins from plants or bacteria have been utilized for immunotoxin construction. The most widely used of these toxins are ricin and abrin from the seeds of the plants *Ricinus communis* and *Abrus precatorius*, respectively, and diphtheria toxin, the exotoxin secreted by *Corynebacterium diphtheriae* lysogenic for the DNA phage β tox^{+}. In this review one of these toxins, ricin, will be discussed in some detail. The structure, synthesis and toxicity of ricin will be covered, in addition to its use as a component of immunotoxins. Constraints affecting the chemotherapeutic potential of ricin will also be considered together with current approaches aimed at structurally modifying ricin in order to remove these constraints.

OCCURRENCE AND STRUCTURE OF RICIN

Ricin is exclusively present in the endosperm cells of the seeds of the castor oil (*Ricinus communis*) plant [2,3]. Mature seeds contain maximum ricin concentration, the toxin being actively synthesized during later stages of seed maturation, and rapidly degraded during seed germination [4]. Although the toxicity of *Ricinus* seeds had been recognized since antiquity, the first attempts to identify their toxic component were made in 1887 [5]. Stillmark mixed a crude seed extract with blood and observed a rapid aggregation of the red cells. He further demonstrated that the agent responsible for the haemagglutination was a protein which he named ricin. At the time, Stillmark understandably concluded that *Ricinus* seeds were toxic because they contained the haemagglutinin, ricin. More recently, the application of increasingly sophisticated protein fractionation techniques has shown this conclusion to be an oversimplification. Stillmark's active component is a mixture of two closely related proteins in equal proportions [6–9]. One of these proteins, still termed ricin, is a potent cytotoxin but is a weak haemagglutinin, while the other, *Ricinus communis* agglutinin (RCA) is relatively non-toxic to intact cells but is a strong agglutinin [10].

Ricin is a heterodimer consisting of two distinct polypeptides held together by a single disulphide bond [7,8,11–13] (*Figure 1.1*). One of these polypeptides

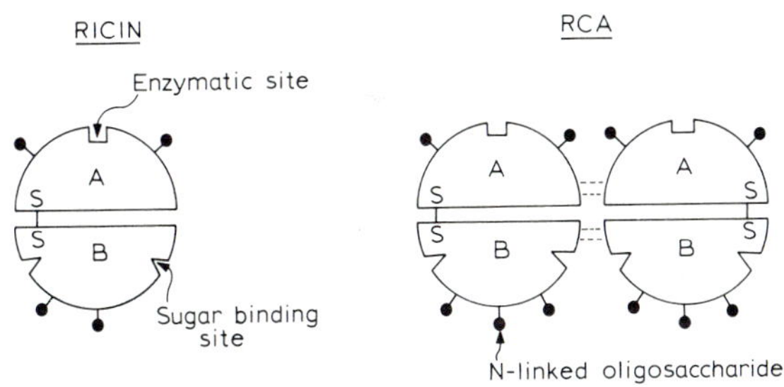

Figure 1.1. Schematic structure of ricin and RCA.

(M_r 32,000, designated the A chain) is a potent toxin, while the other (M_r 34,000, the B chain) is a galactose- or *N*-acetylgalactosamine-binding lectin [14]. RCA is a tetramer consisting of two ricin-like heterodimers, each of which contains an A chain (M_r 32,000) and a galactose-binding B chain (M_r 36,000) [10] (*Figure 1.1*). Although RCA is relatively non-toxic to intact cells, its isolated A chain is of comparable toxicity to the ricin A chain when added to a cell-free protein synthesis system [15] (see below).

Ricin and RCA are closely related proteins. It has been established for some time that antisera raised against individual ricin A or B chains cross-react with the corresponding RCA chains and *vice versa* [16–18]. Peptide fingerprinting

A chain

```
IlePheProLysGlnTyrProIleIleAsnPheThrThrAlaGlyAlaThrValGlnSer
                                           *           *
IlePheProLysGlnTyrProIleIleAsnPheThrThrAlaAspAlaThrValGluSer

TyrThrAsnPheIleArgAlaValArgGlyArgLeuThrThrGlyAlaAspValArgHis
                            *  *
TyrThrAsnPheIleArgAlaValArgSerHisLeuThrThrGlyAlaAspValArgHis

AspIleProValLeuProAsnArgValGlyLeuProIleAsnGlnArgPheIleLeuVal
 *                                      *
GluIleProValLeuProAsnArgValGlyLeuProIleSerGlnArgPheIleLeuVal

GluLeuSerAsnHisAlaGluLeuSerValThrLeuAlaLeuAspValThrAsnAlaTyr

GluLeuSerAsnHisAlaGluLeuSerValThrLeuAlaLeuAspValThrAsnAlaTyr

ValValGlyTyrArgAlaGlyAsnSerAlaTyrPhePheHisProAspAsnGlnGluAsp
          *
ValValGlyCysArgAlaGlyAsnSerAlaTyrPhePheHisProAspAsnGlnGluAsp

AlaGluAlaIleThrHisLeuPheThrAspValGlnAsnArgTyrThrPheAlaPheGly
                                        *  *
AlaGluAlaIleThrHisLeuPheThrAspValGlnAsnSerPheThrPheAlaPheGly

GlyAsnTyrAspArgLeuGluGlnLeuAlaGlyAsnLeuArgGluAsnIleGluLeuGly
                                  *
GlyAsnTyrAspArgLeuGluGlnLeu( )GlyGlyLeuArgGluAsnIleGluLeuGly

AsnGlyProLeuGluGluAlaIleSerAlaLeuTyrTyrTyrSerThrGlyGlyThrGln
 *              *                                *
ThrGlyProLeuGluAspAlaIleSerAlaLeuTyrTyrTyrSerThrCysGlyThrGln
```

B chain

```
AlaAspValCysMetAspProGluProIleValArgIleValGlyArgAsnGlyLeuCys

AlaAspValCysMetAspProGluProIleValArgIleValGlyArgAsnGlyLeuCys

ValAspValArgAspGlyArgPheHisAsnGlyAsnAlaIleGlnLeuTrpProCysLys
          *  *  *  *     *  *        *
ValAspValThrGlyGluGluPhePheAspGlyAsnProIleGlnLeuTrpProCysLys

SerAsnThrAspAlaAsnGlnLeuTrpThrLeuLysArgAspAsnThrIleArgSerAsn
             *                    *  *     *
SerAsnThrAspTrpAsnGlnLeuTrpThrLeuArgLysAspSerThrIleArgSerAsn

GlyLysCysLeuThrThrTyrGlyTyrSerProGlyValTyrValMetIleTyrAspCys
                *  *  *  *        *  *  *     *        *
GlyLysCysLeuThrIleSerLysSerSerProArgGlnGlnValValIleTyrAsnCys

AsnThrAlaAlaThrAspAlaThrArgTrpGlnIleTrpAspAsnGlyThrIleIleAsn
 *        *  *  *                             *
SerThrAlaThrValGlyAlaThrArgTrpGlnIleTrpAspAsnArgThrIleIleAsn

ProArgSerSerLeuValLeuAlaAlaThrSerGlyAsnSerGlyThrThrLeuThrVal
          *                                      *
ProArgSerGlyLeuValLeuAlaAlaThrSerGlyAsnSerGlyThrLysLeuThrVal

GlnThrAsnIleTyrAlaValSerGlnGlyTrpLeuProThrAsnAsnThrGlnProPhe

GlnThrAsnIleTyrAlaValSerGlnGlyTrpLeuProThrAsnAsnThrGlnProPhe

ValThrThrIleValGlyLeuTyrGlyLeuCysLeuGlnAlaAsnSerGlyGlnValTrp
                            *                       *
ValThrThrIleValGlyLeuTyrGlyMetCysLeuGlnAlaAsnSerGlyLysValTrp
```

```
*                 * *
IleProThrLeuAlaArgSerPheMetValCysIleGlnMetIleSerGluAlaAlaArg

PheGlnTyrIleGluGlyGluMetArgThrArgIleArgTyrAsnArgArgSerAlaPro

PheGlnTyrIleGluGlyGluMetArgThrArgIleArgTyrAsnArgArgSerAlaPro

AspProSerValIleThrLeuGluAsnSerTrpGlyArgLeuSerThrAlaIleGlnGlu

AspProSerValIleThrLeuGluAsnSerTrpGlyArgLeuSerThrAlaIleGlnGlu

SerAsnGlnGlyAlaPheAlaSerProIleGlnLeuGlnArgArgAsnGlySerLysPhe

SerAsnGlnGlyAlaPheAlaSerProIleGlnLeuGlnArgArgAsnGlySerLysPhe

SerValTyrAspValSerIleLeuIleProIleIleAlaLeuMetValTyrArgCysAla
*
AsnValTyrAspValSerIleLeuIleProIleIleAlaLeuMetValTyrArgCysAla

ProProProSerSerGlnPhe

ProProProSerSerGlnPhe
```

```
*      *
LeuGluAspCysThrSerGluLysAlaGluGlnGlnTrpAlaLeuTyrAlaAspGlySer

IleArgProGlnGlnAsnArgAspAsnCysLeuThrSerAspSerAsnIleArgGluThr
                                    *     *        *  *
IleArgProGlnGlnAsnArgAspAsnCysLeuThrThrAspAlaAsnIleLysGlyThr

ValValLysIleLeuSerCysGlyProAlaSerSerGlyGlnArgTrpMetPheLysAsn

ValValLysIleLeuSerCysGlyProAlaSerSerGlyGlnArgTrpMetPheLysAsn

AspGlyThrIleLeuAsnLeuTyrSerGlyLeuValLeuAspValArgAlaSerAspPro
                        *
AspGlyThrIleLeuAsnLeuTyrAsnGlyLeuValLeuAspValArgAlaSerAspPro

SerLeuLysGlnIleIleLeuTyrProLeuHisGlyAspProAsnGlnIleTrpLeuPro
                     *  *     *        *  *
SerLeuLysGlnIleIleValHisProPheHisGlyAsnLeuAsnGlnIleTrpLeuPro

LeuPhe

LeuPhe
```

```
             A    B                       A     B
Matches = 249  221        Mismatches = 17     41
Length  = 267  262    Matches/length = 93.3   84.4 percent
```

Figure 1.2. *Primary sequence of ricin and RCA A and B chains. The upper sequence is ricin, the lower RCA. Amino-acid differences are indicated by an asterisk.*

also suggested a close primary structure relationship between these two proteins [14] and this has been confirmed now that the complete primary sequences of ricin and RCA A and B chains are known (*Figure 1.2*) [19–22]. The RCA A chain is one residue shorter than that of ricin, corresponding to the omission of the alanine residue at position 130 in the ricin A chain [22]. In total, the A chains differ in 18 out of 267 residues and are thus 93% homologous at the amino-acid level, while the B chains differ in 41 out of 262 residues, giving 84% homology [22].

The complete three-dimensional structure of ricin has not yet been described, although preliminary X-ray crystallographic studies have been performed [23–25].

The structural characterization of ricin has been simplified by the ease with which the protein can be purified in large amounts. Since both ricin and RCA are galactose-binding lectins, they can be separated from all other proteins present in crude extracts of *Ricinus communis* seeds by affinity chromatography on Sepharose 4B (the Sepharose 4B matrix contains β-galactose). Ricin and RCA can be conveniently separated during this procedure, since ricin can be eluted from a Sepharose 4B column with *N*-acetylgalactosamine, whereas galactose is required for the elution of RCA [14]. Separation of the A and B chains of ricin is also straightforward. The interchain disulphide bond is reduced with β-mercaptoethanol in the presence of hapten sugar (which prevents the precipitation of the chains after reduction). The individual subunits can be separated by ion-exchange chromatography [15, 16]. Alternatively, since the p*I* (isoelectric point) of the ricin A chain is 7.2 and that of the B chain 4.6 [19, 20], they can be separated by chromatofocusing (Clements, G. unpublished data).

One problem encountered with ricin subunits separated by ion-exchange chromatography is that significant cross-contamination usually occurs. Recently an improved purification scheme has been described which eliminates this problem [26]. Intact ricin, non-covalently attached to Sepharose 4B, was split into its A and B chains on the column by treatment with β-mercapto-ethanol and any contaminating B chains were subsequently removed by affinity chromatography on asialofetuin-Sepharose and monoclonal anti-B-chain-Sepharose. The B chain was eluted from the Sepharose 4B column in buffer containing galactose and was further purified by ion-exchange chromatography. Contaminating A chains in this fraction were then removed by affinity chromatography on monoclonal anti-A-chain-Sepharose.

RICIN, A GLYCOPROTEIN

Both the A and B chains of ricin (and their RCA counterparts) are *N*-glycosylated [27–30]. The A and B chains each contain two (Asn-x-Ser/Thr) *N*-glycosylation sites [31]. Purified ricin A chain contains two components, of molecular weight 32,000 and 34,000. The lighter component has been designated 'A_1-chain' and the heavier component 'A_2-chain' [32]. These two components apparently represent glycosylation variants of a single A-chain polypeptide. The A_1 chain contains a single oligosaccharide side-chain with the composition (*N*-acetylglucosamine)$_2$(xylose)$_1$(fucose)$_1$(mannose)$_{3-4}$ which, presumably because of the presence of fucose [35], is not susceptible to endo-*N*-acetylglucosaminidase H, an enzyme which cleaves between the two *N*-acetylglucosamine residues [33, 34]. This oligosaccharide is similar in monosaccharide composition to that described for pineapple stem bromelain [36], which likewise is insensitive to endo-*N*-acetylglucosaminidase H digestion. The oligosaccharide side-chain of stem bromelain is, however, endo-*N*-acetylglucosaminidase F sensitive [37], whereas that of ricin A chain is not, suggesting that the two carbohydrate structures are not identical. The greater molecular weight of the A_2 chain compared with that of the A_1 chain appears to be attributable solely to higher carbohydrate content. No other differences between the two types of A chain were found during analysis by N-terminal amino-acid sequencing, amino-acid composition, immunodiffusion or inhibition of protein synthesis in a cell-free assay [35].

The B chain contains two N-linked oligosaccharide side-chains which contain only *N*-acetylglucosamine and mannose [27, 35]. One of these carbohydrate side-chains can be removed from the B chain by endoglycosidase H or F, whereas both can be removed after denaturation of the polypeptide by SDS [35, 38]. This indicates that the *N,N'*-diacetylchitobiose core of the second carbohydrate chain is normally protected by the polypeptide conformation.

TOXICITY OF RICIN

When administered parenterally to animals, ricin is one of the most toxic compounds known. There is species variation in sensitivity to the toxin and details of its toxicity and the symptoms this induces can be found elsewhere [13]. Ricin is a member of a group of toxic polypeptides consisting of two functionally distinct moieties. It is firmly established that the toxic effects of ricin are entirely attributable to the biological activity of the A chain [39–41].

Likewise, the A chains of other plant [42–44] and bacterial [45–47] heterodimeric toxins are responsible for toxicity. These toxins contain a single A chain moiety which, in each case, has catalytic activity and efficiently inactivates its intracellular target [44]. The A chain is only toxic to intact cells when combined with B chain. The function of the B chain is to bind the toxin to cell-surface receptors, in the case of ricin to appropriate surface glycoproteins or glycolipids. This is the essential first step in the transfer of ricin A chain into the cytosol, where ribosome inactivation occurs [48]. Additionally, the B chain is believed to have a second function during the intoxication process in which it facilitates the transfer of the A chain across a membrane into the cytoplasm [49]. Separated A and B chains are essentially non-toxic, the toxic A chain lacking the ability to bind to and enter cells in the absence of the B chain. The toxicity of ricin therefore results from three sequential steps: (1) binding of the whole molecule to the cell surface via the B chain; (2) penetration of at least the A chain into the cytosol, and (3) inhibition of protein synthesis caused by the interaction of the A chain with the 60 S ribosomal subunit.

BINDING TO THE CELL SURFACE

Ricin and RCA react opportunistically with cell-surface structures by binding stereospecifically and reversibly to any suitably exposed galactose residue. The galactose residues may be present on a variety of glycoproteins and glycolipids. As expected, galactose and lactose are effective inhibitors of toxicity when ricin is presented to intact cells because of their competition with the surface oligosaccharides for the toxin-sugar binding sites. The structural features of the surface oligosaccharides required for ricin binding have been studied in detail [50]. Oligosaccharides terminating in galactose–*N*-acetylglucosamine–mannose bind strongly to the toxin. The galactose binding sites are present on the ricin B chain and they are distinct from, and do not appear to depend upon, the N-linked oligosaccharide side-chains attached to the ricin B chain itself. This conclusion is based on findings made for RCA where blocking the oligosaccharide side-chains by binding to concanavalin A did not affect the extent or strength of its interaction with lactose when compared with the untreated RCA molecule [51]. Each ricin B chain molecule has two galactose-binding sites. Low-resolution X-ray crystallographic studies of ricin have shown that the B chain is a bilobal structure and that each domain binds a galactose molecule, although one of the binding sites was found to be more highly occupied than the other [24].

Eukaryotic cells may have many ricin-binding structures on their surface; binding studies have shown that HeLa cells contain 3×10^7 binding sites for ricin [52]. Internalization of surface-bound toxin is apparently very inefficient and only a very small proportion of ricin molecules which bind to cell surfaces ultimately appear to be transferred into the cell cytoplasm. While it has been estimated that the transfer of a single toxin molecule into the cytoplasm can result in cell death [53], several thousand molecules must apparently bind to the cell surface in order to ensure the internalization of this single toxin molecule [54]. The number of ricin molecules a cell can bind can normally be significantly increased by treating the cell with neuraminidase [55]. The enzymic removal of sialic acid residues from the terminal position on the oligosaccharide chain of membrane glycoproteins exposes galactose residues. Increased ricin binding to such cells is accompanied by an increased sensitivity to the toxin [56].

A number of ricin-resistant cell lines have been characterized and described. Resistance does not result from cells possessing ribosomes which cannot be inactivated by the toxin, but rather from defects in the transport mechanism which normally delivers ricin to the ribosome. These defects fall into two main categories: cells having reduced numbers of exposed toxin binding sites on their surface or cells which, although possessing a normal number of binding sites, have an impaired ability to transfer bound toxin into the cytoplasm. Most cell lines having a reduced number of surface ricin-binding sites are glycosylation mutants which either effectively sialate exposed galactose residues or else fail to add terminal galactose to their oligosaccharide side-chains. The first class of mutants have increased membrane sialic acid content, while the galactose, *N*-acetylglucosamine and mannose content is normal [57]. Treating such mutants with neuraminidase exposes surface galactose residues and restores the ricin-sensitive phenotype [52, 57–59].

A second class of ricin-resistant mutants are defective in surface oligosaccharide synthesis. For example, cells lacking *N*-acetylglucosaminyl transferase, which adds *N*-acetylglucosamine to oligosaccharide side-chains containing terminal non-reducing mannose residues, are subsequently unable to add terminal galactose residues [57, 60, 61]. All mutants of this type were found to accumulate oligomannosidic glycans in surface glycoproteins rather than complex oligosaccharide chains [62].

A ricin-resistant mouse lymphoma cell line has been isolated which was shown to possess a normal number of ricin-binding sites [63–66]. These cells internalized reduced amounts of ferritin-ricin complexes at low toxin concentrations compared with the parent cells. Resistance in this cell line is therefore due to a defect in transferring ricin from the cell surface to the cytoplasm, since the resistant cells were completely sensitive to ricin which was incorporated into liposomes and delivered directly into the cytoplasm after membrane fusion.

INTERNALIZATION OF RICIN

In order to exert its toxic effect, ricin A chain must encounter its intracellular ribosomal target [67]. For this to occur, the binding of ricin to the cell surface must be followed by translocation of the toxin, or its A chain, across a membrane. There is no evidence that this translocation step occurs at the cell surface in the form of direct transfer of the toxin across the plasma membrane and into the cytoplasm. Rather, both biochemical [13, 68–72] and ultra-structural [63, 73–79] studies have established that ricin, like other toxins [69, 70, 80–84], enters cells by endocytosis. Thus at 4 °C ricin-ferritin complexes bind to and remain at the cell surface, where they are randomly distributed and can be displaced by galactose [85]. When the temperature was increased to 37 °C, an increasing proportion of the surface-bound toxin was internalized with time and was no longer released by galactose. The fate of internalized ricin has been studied by electron microscopy using toxin conjugated to horseradish peroxidase, ferritin or colloidal gold [73, 79]. There are several potential problems with this approach. For example, the internalized conjugate may dissociate upon encountering low pH, proteolytic enzymes or some other dissociating intracellular environment. The released label may follow an intracellular route different from that of the toxin itself. Even if the conjugate remains intact, it may follow a route different from that of free toxin. In spite of a clear demonstration that this latter possibility does occur [78], an overall picture of how ricin enters cells is beginning to emerge. Both native ricin, as demonstrated by immunoperoxidase cytochemistry, and ricin conjugates are internalized by the classical receptor-mediated endocytosis pathway [86–92] *via* coated pits and coated vesicles to reach vacuolar and tubulo-vesicular portions of the endosomal system. In addition, native ricin and a purified monovalent fraction of ricin conjugated to horseradish peroxidase reached distinct Golgi cisternae, whereas ricin conjugated to colloidal gold or polyvalent ricin-horseradish peroxidase conjugates did not [78]. When endocytosis is inhibited by depleting cells of ATP, ricin entry does not occur [72].

After ricin binds to the cell surface, there is always a significant lag period before a decrease in the rate of cellular protein synthesis is observed [93]. This lag period represents the time taken for A chains to move from the cell surface to the cytosol. During the early stages of this lag period, both anti-A-chain and anti-B-chain antibodies are able to protect the cells against the toxin [68]. For the greater part of the lag period, however, the antibodies do not have any protective effect. This period, during which protein synthesis is not inhibited, represents the time the endocytosed toxin molecules are contained in intracellular compartments before being released into the cytoplasm. A lag period is not observed when cells are treated with liposomes containing ricin under conditions which allow membrane fusion and direct discharge of ricin into the cytoplasm [94–96].

Although some endocytosed ricin appears to be recycled back to the cell surface [52], a significant proportion of the remainder enters the cytoplasm and eventually kills the cell. At present, the identity of the intracellular compartment from which the toxin enters the cytoplasm is not known. In the case of diphtheria toxin, it is clear that the toxic A fragment enters the cytoplasm from an endocytotic vesicle with a low pH, probably an endosome [97, 98]. Low pH induces the exposure of a hydrophobic domain on the diphtheria toxin B fragment which is able to insert itself into the membrane [99]. This exposure may be due to a conformational change occurring in response to *cis-trans* isomerization of proline, induced by the low pH [100], and can be demonstrated by the ability of the toxin to bind [^{3}H]Triton X-100 [101]. Insertion of the B chain hydrophobic domain into the membrane forms an ion-permeable channel through which the A fragment can pass [102, 103]. As expected, the toxic effects of diphtheria toxin are completely nullified by prior treatment of cells with compounds which increase the pH of acidic intracellular vesicles [104–107], although binding to the cell surface is unaffected.

In contrast to the situation with diphtheria toxin, there is no evidence that low pH is necessary for the transfer of ricin A chain from an intracellular compartment into the cytoplasm. Indeed, ricin requires neutral or slightly alkaline pH in the medium to intoxicate cells and is not toxic when the pH is 6.5 or lower [70]. Treatments which lead to an increase in pH inside the cell increase the sensitivity of such cells to ricin [108–110]. It is likely that ricin enters the cytoplasm from neutral vesicles but the identity of such vesicles remains obscure. Equally obscure is the mechanism by which ricin A chain is translocated across the membrane. Although mechanistically different from the transport of diphtheria toxin A fragment, it seems that the ricin B chain also plays a key rôle in facilitating A-chain translocation [111]. The ricin B chain domain responsible for facilitating A-chain translocation is physically and functionally distinct from its sugar-binding domain, since galactose binding can be completely inhibited by chemical modification of B chain without affecting its A-chain translocation capacity [112].

RICIN A-CHAIN INHIBITION OF PROTEIN SYNTHESIS

Studies using intact ricin and its individual A and B chains with intact 80 S ribosomes and their 60 S and 40 S subunits have firmly established that the A chain is the toxic moiety and that it inhibits protein synthesis by inactivating a function of the 60 S ribosomal subunit [113, 114]. Prokaryotic 70 S ribosomes, on the other hand, are insensitive to the toxin [115]. The A chain inactivates 60 S subunits catalytically [13, 15] and as a result elongation factor 2 (EF-2) is unable to bind to the subunit and protein synthesis does not occur.

Kinetic experiments have shown that pure ricin A chain inactivates salt-washed ribosomes at a rate of 1500 ribosomes per min per A chain molecule. The K_m with respect to ribosomes is $(1–2) \times 10^{-7}$ M, which indicates that the A chain acts in the cytosol at close to its V_{max}. Penetration of a single A-chain molecule into the cytoplasm is believed to be sufficient to kill a cell [53]. Ricin inhibits protein synthesis by both animal and plant ribosomes, including those from *Ricinus communis* itself [116], but, in general, plant ribosomes seem to be significantly less sensitive.

Ricin A chain appears to act by inhibiting EF-2-dependent GTPase activity of the ribosome [117]. The rate of loss of translational activity seems to correlate with the loss of GTPase activity [118]. High concentrations of EF-2 and prebound aminoacyl-tRNA reduce inactivation of ribosomes by ricin A chain [41]. In turn, ricin A chain specifically binds to a single site on rat liver ribosomes, indicating that the toxin acts on the ribosome at a site identical with, or overlapping, the binding site of EF-2 [119].

In spite of considerable research effort, the nature of the enzymic activity of the A chain remains uncertain. Toxin-dependent changes in the RNA or protein moieties of 60 S ribosomal subunits have not been described. Since A chain inactivates ribosomes in the absence of cofactors, it may act hydrolytically, possibly by removing some minor functional group [120], or may introduce a 'lethal' conformational change in a ribosomal component.

The inhibitory effect of ricin on protein synthesis in cell-free systems is greatly increased in the presence of β-mercaptoethanol. This is because free A chain is the enzymatically active toxin, whereas A chain linked to B chain in whole ricin is not active [121]. In the absence of reducing agents, even high concentrations of intact toxin do not inactivate ribosomes. Presumably the catalytic site on the A chain is formed or exposed only when the A chain is released from the B chain. Free A chain, however, is non-toxic to intact cells since, in the absence of B chain, it lacks the ability to bind to and enter cells. Ricin A-chain preparations frequently show some level of toxicity, however, because the complete removal of contaminating B chain can be difficult to achieve [122].

Although RCA is several orders of magnitude less toxic than ricin when injected intraperitoneally into mice [16], it is clear that the agglutinin A chain is also a potent protein synthesis inhibitor [15] but is internalized much less efficiently than is ricin A chain.

BIOSYNTHESIS OF RICIN

Ricin and RCA are simultaneously synthesized in equivalent amounts in the endosperm cells of ripening *Ricinus communis* seeds [123]. In common with the other major storage protein components of the protein bodies, synthesis occurs during and after testa formation when these storage organelles are being rapidly formed [4]. The two lectins are homologous in both structure and function and, as would be expected, share a common biosynthetic mechanism. Although both lectins contain two distinct polypeptide subunits, these polypeptides are not the products of two distinct transcripts. Rather, both A- and B-chain sequences are derived from a precursor polypeptide which is encoded by a single mRNA species [38]. A series of co-translational and post-translational modification steps during the synthesis of the ricin precursor and its processing to yield the native heterodimer have been defined [124] and are described briefly below.

CO-TRANSLATIONAL MODIFICATIONS

The first indication that ricin A and B chains were synthesized together in a precursor polypeptide (proricin) came from cell-free synthesis studies [38, 125]. Antibodies were raised in rabbits against the individually purified A and B chains of ricin. Although anti-A-chain antibodies did not cross-react with the B chain and *vice versa*, each of the antisera recognized a single polypeptide species of molecular weight 59,000 when ripening castor bean endosperm RNA was translated *in vitro* in a rabbit reticulocyte lysate cell-free protein synthesis system [38]. When dog pancreas microsomal membranes were included in the translational system, the functional equivalent of rough endosplasmic reticulum, ribosome-studded microsomal vesicles, were generated *in vitro* and the ricin precursor was co-translationally translocated into the lumen of the microsomal vesicles [125]. In this case the membrane-translocated precursor, precipitated by anti-A-chain or anti-B-chain antibodies, appeared as a group of polypeptides in the 64,000–68,000 molecular weight range. A detailed analysis of these early biosynthetic events established that the synthesis of proricin conforms to the mechanism predicted and defined by the signal hypothesis originally elucidated for mammalian secretory proteins [126, 127].

Thus the mRNA encoding proricin also encodes a cleavable N-terminal signal peptide (preproricin). Translation of this mRNA *in vivo* begins on free cytoplasmic ribosomes and continues until the N-terminal sequence containing the signal peptide begins to emerge from the ribosome. The emerging signal peptide is recognized by a cytoplasmic ribonucleoprotein particle (signal recognition particle, SRP) [128] which binds to it and transiently arrests translation of the mRNA [129]. The SRP-ribosome complex then binds to the cytoplasmic surface of endoplasmic reticulum (ER) membrane *via* an SRP receptor (the docking protein, [130]) and a tight junction is formed between the ribosome and the rough ER membrane. The SRP-mediated translational arrest is relieved and now, as translation proceeds, the elongating polypeptide is vectorially transferred across the ER membrane, folding into its proricin conformation in the ER lumen. Translocation of the nascent ricin precursor is accompanied by the proteolytic cleavage of the N-terminal signal peptide by an ER signal peptidase [131].

The proricin precursor segregated *in vitro* into the lumen of the dog pancreatic microsomes appears as a heterogeneous group of polypeptides of molecular weight 64,000–68,000 because of a further co-translational modification, *N*-glycosylation [125]. *N*-Glycosylation of asparagine residues is catalyzed by an ER glycosyltransferase which transfers *en bloc* an oligosaccharide moiety from a lipid carrier to nascent polypeptides [132, 133]. The appearance of a group of translocated ricin precursors was due to heterogeneity in glycosylation, since treatment of this group with endo-*N*-acetylglucosaminidase H removed the oligosaccharide side-chains converting the precursors back to a single polypeptide of molecular weight 57,500 [125]. The difference in size between the non-segregated precursor synthesized *in vitro* in the absence of microsomes (M_r 59,000) and the deglycosylated segregated precursor (M_r 57,500 Da) represents the N-terminal signal peptide which is cleaved off by signal peptidase during the membrane-segregation step.

Another modification to the ricin precursor which occurs during or immediately after synthesis is disulphide bond formation [124]. In mature ricin, the B chain contains four intrachain disulphide bonds and it is joined to the A chain by a single interchain bond. These disulphide bonds are formed enzymically by ER protein disulphide isomerase [134, 135] which introduces five intrachain disulphide bonds into nascent proricin, the bond destined to become the interchain disulphide bond joining the A and B chains of mature ricin being formed between cysteine residues present in the A- and B-chain sequences in the proricin.

The identity of the glycosylated polypeptides as precursor forms of the ricin subunits was further confirmed by *in vivo* radiolabelling studies [136].

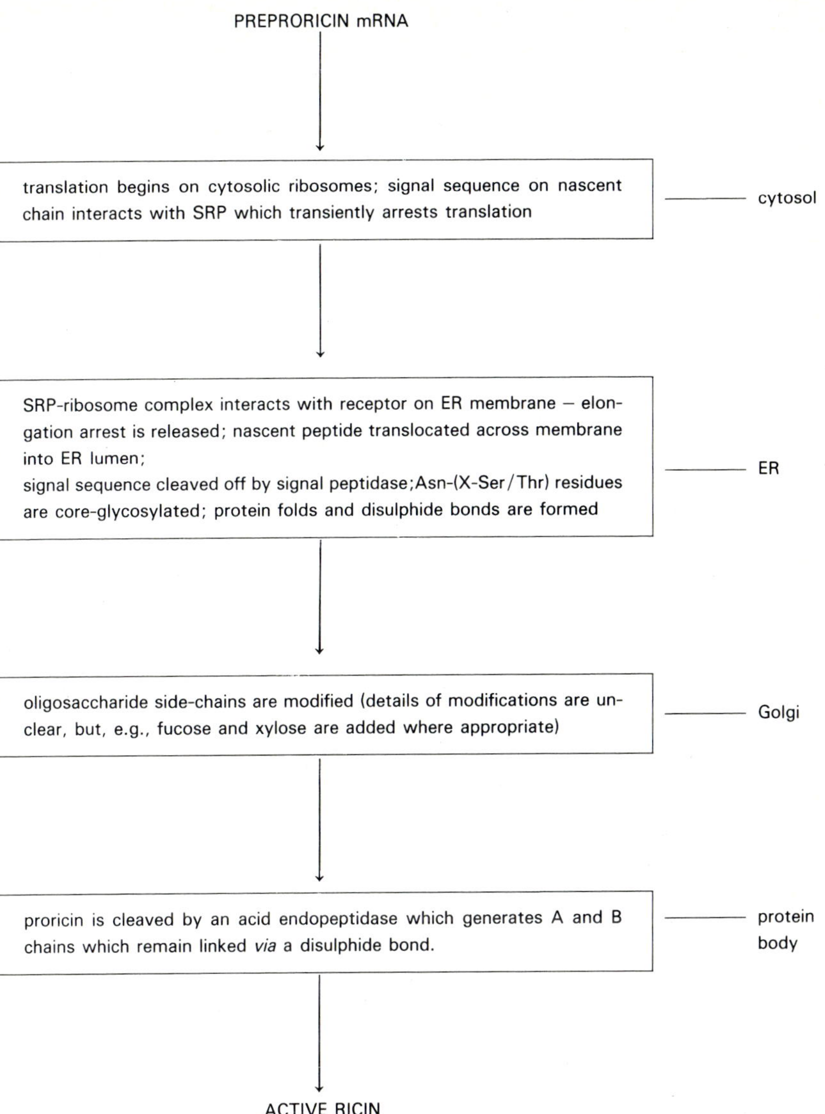

Figure 1.3. Biosynthesis and intracellular transport of ricin.

POST-TRANSLATIONAL MODIFICATIONS

The synthetic and co-translational modification steps described above result in the transient deposition of core-glycosylated proricin in the lumen of the ER. Mature ricin is stored in single membrane delimited organelles called protein bodies within *Ricinus* seed endosperm cells. The newly synthesized proricin has to be both transported from the ER to the protein bodies and processed to liberate the individual A and B chains. It is now clear that proricin initially moves from the ER to the Golgi apparatus and from there, contained within a population of Golgi-derived transporting vesicles, to the protein bodies [137]. Intracellular transport is also accompanied by structural modifications to proricin. In the Golgi apparatus, enzymic modification of the oligosaccharide side-chains occurs. Although details of the sugar modifications are unclear, they probably involve typical Golgi oligosaccharide trimming and monosaccharide additions to the side-chains [138–141], the latter including the addition of fucose and xylose to A-chain oligosaccharides [136, 142]. These sugar modifications confer partial endoglycosidase H resistance to the proricin oligosaccharide side-chains. Proricin is then transported from the Golgi apparatus in vesicles which presumably pinch off from Golgi cisternae and move to protein bodies where, after membrane fusion, they discharge their contents into the protein body matrix. Within the protein bodies proricin is processed by an acid endopeptidase which liberates free A and B chains, still covalently linked by a disulphide bond, since cleavage occurs within a disulphide loop between the A- and B-chain sequences [143].

The synthesis of ricin, illustrated schematically in *Figure.1.3*, ensures that the castor bean ribosomes are not inactivated by the toxic A chain. Because the proricin is segregated into the ER lumen as it is synthesized, a membrane barrier always exists between proricin or mature ricin and the ribosomes. The overall scheme for synthesis, transport and processing has now been established for a variety of plant protein body components, including lectins [144–147] and storage proteins [148, 149].

MOLECULAR CLONING OF PREPRORICIN

A preproricin gene from *Ricinus communis* has been characterized by both cDNA [31] and genomic [150] cloning. Southern blot analysis has indicated that the *Ricinus communis* genome includes at least six lectin genes [150]. In our laboratory we constructed a cDNA library using ripening castor bean endosperm poly(A)$^+$ RNA which had been enriched for lectin precursor mRNA by size fractionation on sucrose density gradients [151]. Eighty positive clones were isolated from the library of recombinants by hybridization using, as a probe, a mixture of synthetic oligonucleotides representing all possible sequences encoding a peptide of the ricin B chain [31]. Every positive recombinant plasmid we tested selected, by hybridization, a single mRNA species whose translational product was identified as preprolectin by immunoprecipitation. Restriction analysis of these clones showed that two classes were present representing sequences complementary to two distinct but closely related preprolectin species [22]. Representative clones from each class were sequenced. The preprolectin species were identified as either preproricin or preproRCA, respectively, by comparing the deduced amino-acid sequence with the published ricin sequence [19, 20] and the N-terminal sequence of purified RCA B chain [22]. Preproricin contains 576 amino-acid residues consisting of a 35-amino-acid N-terminal sequence which includes the cleavable signal sequence, the A-chain sequence (267 residues) which is joined to the B-chain sequence (262 residues) by a 12-amino-acid linker peptide [124, 150] (*Figure 1.4*).

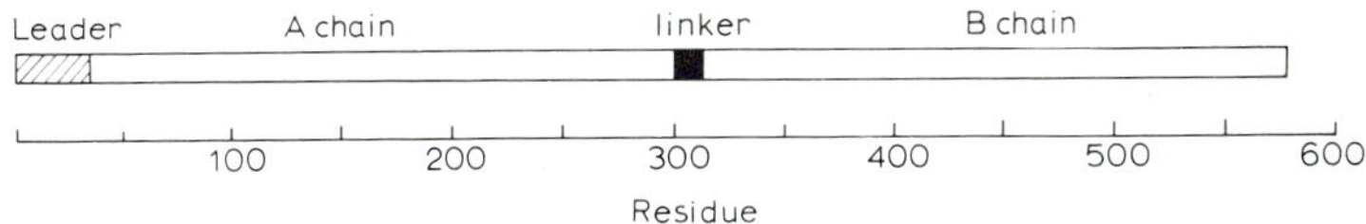

Figure 1.4. Schematic structure of preproricin.

We initially suggested that preproricin contained a 24-amino-acid signal peptide. The N-terminal methionine in this sequence represented the first in frame initiation codon 5′ to the start of mature A chain [31]. The signal peptide thus encoded was assumed to be cleaved immediately before the mature A-chain N-terminus. This putative signal peptide was not a typical eukaryotic sequence because the proposed signal peptide cleavage site differed markedly from the consensus cleavage site [152]. Furthermore, the proposed initiation codon was not in a good translational initiation environment [153]. Subsequently, a more extensive ricin 5′ gene sequence was published and it contained an additional in-frame initiation codon 33 bases upstream from the one we suggested [150]. This upstream initiation codon is in an ideal translation initiation environment and presumably represents the true translation start site. It seems unlikely that the 34-amino-acid sequence which separates this N-terminal methionine from the mature A-chain N-terminus represents the true signal peptide. A 35-residue peptide is considerably longer than most prokaryotic and eukaryotic signal peptides and from consensus sequences [152], cleavage on the carboxyl side of asparagine, the residue at −1 to the A-chain N-terminus, is unlikely. More typical cleavage sites lie within the 35-residue sequence and if co-translational signal cleavage occurs at one of these alternative sites, further processing would be required to reveal the mature A-chain N-terminus. Such post-translational N-terminal processing has been implicated during the biosynthesis of other plant protein body proteins [154, 155].

The A and B polypeptides in proricin are joined by a twelve-residue linking peptide. This linker sequence is excised in the protein bodies to release the A and B chains and it is noteworthy that the carboxy-terminal residue of the linker peptide is asparagine. Endoproteolytic cleavage after an asparagine residue is a feature of the processing of plant proproteins [156]. In the case of the ricin precursor, the endoproteinase which removes the linker peptide, or a related protein body enzyme, may also be responsible for the proposed post-translational N-terminal processing.

RICIN-BASED IMMUNOTOXINS

Monoclonal antibodies directed against tumour-associated antigens have considerable potential in cancer therapy [157, 158]. The hybridoma technology developed by Kohler and Milstein [159, 160] has led to the availability of unlimited supplies of a wide range of exquisitely selective antibodies. The selectivity of tumour-specific antibodies has revolutionized the detection and diagnosis of cancer [161–164]. Indeed, it was hoped that specific interaction between the antibodies and tumour cells *in vivo* would lead to the selective elimination of these cells. In practice, however, the performance of monoclonal antibodies in therapy has been disappointing, due primarily to their lack of toxicity [165, 166]. This, in turn, has led to attempts to increase the potency of monoclonal antibodies by linking them chemically to various cytotoxic agents. This approach is a tribute to the foresight of Paul Ehrlich, who, at the turn of the century, speculated on the potential use of antibodies as carriers of pharmacological agents [167]. The antibody therefore functions as a homing device for a conjugate whose potency is enhanced by the inclusion of conventional drugs, radioisotopes, biological response modifier or toxins. Conjugates between antibodies and protein toxins (immunotoxins) are the most potent and widely used at present, with ricin the most frequently included toxin in such hybrid molecules. The formation, properties and promise of immunotoxins in general, and of ricin-based immunotoxins in particular, have been discussed in detail in a plethora of recent reviews [1, 168–182]. Two types of ricin-based immunotoxin have been used in these studies, the first involve conjugating purified ricin A chain to the antibody, the second conjugating whole ricin.

CONJUGATE PREPARATION

Ricin or its A chain is chemically purified from *R. communis* seeds as described earlier. In the case of A chain, it is essential to remove contaminating B chain as completely as possible, and this normally entails passing purified A chain preparations down asialofetuin or anti-B-chain antibody affinity columns. Coupling of the toxin and antibody can be achieved by several chemical methods [173] and it is conventional to link them by a disulphide bond. The most common strategy for preparing A-chain immunotoxins (A-chain IT) is to use a heterobifunctional cross-linking agent carrying a disulphide bond such as *N*-succinimidyl 3-(2-pyridyldithio)propionate (SPDP) which reacts with free amino groups on the antibody (*Figure 1.5*). The derivatized antibody is then

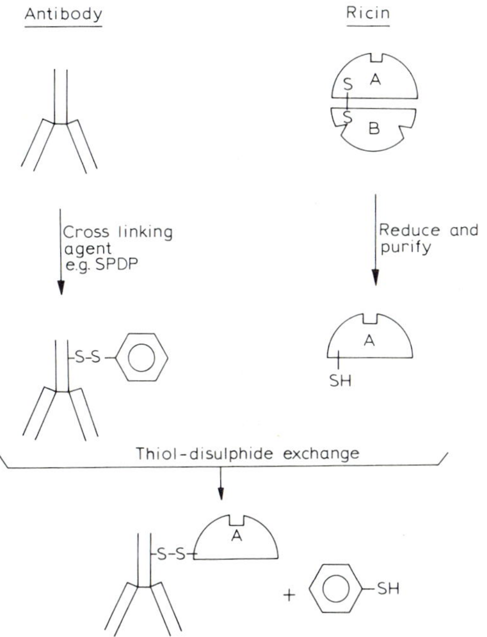

Figure 1.5. Conjugation of ricin A chain to a monoclonal antibody using the cross-linking reagent, SPDP.

mixed at neutral pH with pure A chains to allow conjugation by disulphide exchange. The resulting conjugates are separated from free A chains and uncoupled antibody by gel filtration and affinity chromatography using immobilized anti-A-chain antibodies or immobilized antigen [169–171].

For preparing whole ricin IT, the intact toxin is normally derivatized with a bifunctional cross-linking agent before being mixed with antibody prepared for cross-linking by partial reduction of disulphide bonds [183].

At present the scientific literature contains a large and rapidly expanding number of accounts of the conjugation of cell-reactive antibodies to ricin or its A chain and their cytotoxic effects on cultured cells or whole animal models. In the present account, we will not attempt to catalogue these reports but will briefly and simply outline emerging conclusions and present constraints.

RICIN A-CHAIN IMMUNOTOXINS

The simplest rationale for ricin-based IT construction is to completely purify the A chain away from the B chain, which makes whole ricin non-specifically toxic because of its ability to interact with galactose on a wide range of cells, and to conjugate the pure A chain to a monoclonal antibody directed against a tumour cell-surface antigen. In this way one hopes to produce a hybrid molecule which retains the toxicity effected by ricin A chain and which specifically binds to the antigen-bearing tumour cells. In practice, such A-chain ITs are extremely selective in their toxic effect and only eliminate cultured cells bearing the appropriate antigen. This advantage is countered by a major disadvantage: the potency of A-chain IT is variable and unpredictable, being normally weak and occasionally non-existent [179]. A number of factors may determine the effectiveness of an A-chain IT. These factors include the density of the surface antigen, the affinity of the antibody used to prepare the IT, the route of entry of the antigen-IT complex into the cell, and the nature of the target antigen itself. The major limitation of A-chain ITs which affects their toxicity, however, is the absence of B chain. Thus while the antibody can replace the cell-binding function of the B chain, and can replace it in a highly selective way, it cannot substitute for the second function of the B chain, facilitating the translocation of the A chain across an intracellular membrane.

Two ways of enhancing the toxicity of A chain ITs have been described. Firstly, the ITs frequently show a moderate to dramatic increase in potency in the presence of monensin [184, 185], ammonium chloride [186, 187] or chloroquine [188]. It seems likely that these reagents reduce the rate of proteolytic digestion of the ITs by elevating endosomal and lysosomal pH. Secondly, the addition of free B chain gives a markedly enhanced cytotoxic effect [189]. Similarly the toxicity of A-chain ITs is enhanced if B chain is provided in the form of a B-chain IT [190–192]. The synergy afforded by B-chain ITs is obtained even when the B chain has been chemically modified to attenuate its capacity to bind to cells (galactose) [112].

INTACT RICIN IMMUNOTOXINS

Intact ricin ITs are invariably outstandingly toxic to cells carrying appropriate antigens, often surpassing the native toxin in potency. While this potency is attributable to the B chain, the presence of B chain in these conjugates has an inevitable disadvantage – the specificity conferred by the antibody is overriden by the B chain which also directs the conjugate to non-target cells. A simple and very effective way of blocking the nonspecific binding property of intact ricin conjugates *in vitro* is by competition with high concentrations of free galactose or lactose [183]. For *in vivo* use, however, this protective approach cannot be used and a permanent blockade of the B-chain galactose-binding sites is required. One such method of blockade was recently described in which intact ricin ITs were prepared in which the B-chain galactose-binding sites were sterically hindered by the antibody moiety itself. Monoclonal anti-Thy 1.1 antibody linked in this way to ricin was approximately 10^4-times more toxic to Thy 1.1-expressing AKR-A lymphoma cells than it was to EL4 lymphoma cells which express the alternative Thy 1.2 allele [193]. These blocked ricin ITs still retain considerable cytotoxicity and it is clear that a more vigorous elimination of the ricin B-chain galactose binding capacity is required.

GENETICALLY ENGINEERED IMMUNOTOXINS

With cloned ricin genes available, the most definitive way of altering ricin B chain to make it suitable for inclusion in intact ricin ITs will be by genetic manipulation. In this way the B chain will be structurally modified so that it loses completely its ability to bind galactose but retains its ability to translocate A chain across a membrane. An important indication of the feasibility of this approach was the recent demonstration that these two functions of the B chain reside on separate domains in the molecule. Thus while chloramine-T-treated ricin B chains showed a dramatic reduction in their ability to bind galactose, their ability to potentiate the killing of cells treated with an A-chain IT was virtually unimpaired [112].

Several groups are currently attempting to identify the amino acids in the B chain that are responsible for cell-binding activity and certain important residues have been defined [194]. A high-resolution X-ray structure of ricin crystals is approaching completion and a full structural definition of the galactose-binding sites will soon be available. This knowledge will enable B-chain nucleotide sequences which specify galactose-binding residues to be deleted or altered by site-directed mutagenesis [195]. The modified gene sequences, incorporated into appropriate expression vectors, will be introduced into prokaryotic or eukaryotic cells and the modified recombinant B chain produced in large quantity.

Other structural modifications may be created by DNA technology to enhance the efficacy of ricin-based ITs. For example, intact ricin or ricin A chain ITs administered *in vivo* are rapidly cleared from the circulation because the high-mannose oligosaccharide side-chains of ricin A chain are recognized by mannose receptors on Kupffer cells [196, 197]. Chemical or enzymatic deglycosylation of ricin A chain without protein denaturation has proved to be difficult [35] but could be accomplished by either expressing cloned A chain cDNA in a non-glycosylating prokaryotic system or by mutation to remove *N*-glycosylation sites prior to expression in eukaryotic cells.

Another advantage in producing ricin by expressing the cloned gene in a heterologous system is that the purity of the product can be guaranteed. As noted earlier, conventially purified ricin A chain contains trace but significant B chain contamination, a problem readily overcome if the A-chain cDNA is expressed in a prokaryotic system to generate biologically active product.

Ultimately, both the antibody and toxin components of ITs may be expressed together. This will be achieved by fusing the immunoglobulin heavy chain and toxin genes and subsequently transfecting a myeloma cell line which synthesizes immunoglobulin light chain and is thus able to constitute the relevant antigen-binding site with the recombinant heavy chain [198]. The feasibility of this approach has been demonstrated by the expression of an antibody-enzyme hybrid [199]. These chimaeric antibodies will be further modified by the engineering of a suitable cleavage site to permit A-chain release, and the inclusion of B-chain sequences required for membrane translocation.

These genetic manipulations will attempt to generate the ideal IT – a conjugate which kills target cells *in vivo* with great potency while causing little or no harm to normal cells. If this goal is achieved, while cancer chemotherapy will doubtless remain a difficult and unpredictable area, ITs may add an exciting and effective alternative to the available reagents.

REFERENCES

1 E.S. Vitetta and J.W. Uhr, Cell, 41 (1985) 653.
2 R.E. Tulley and H. Beevers, Plant Physiol., 58 (1976) 710.
3 R.J. Youle and A.H.C. Huang, Plant Physiol., 58 (1976) 703.
4 L.M. Roberts and J.M. Lord, Planta, 152 (1981) 420.
5 H. Stillmark (1887) Uber Ricin, ein gifiges Ferment aus den Samen von *Ricinus communis* L. und anderen Euphorbiaceen, Inaug.-Dissert., University of Dorpat, Estonia.
6 G.L. Nicolson and J. Blaustein, Biochim. Biophys. Acta, 266 (1972) 543.
7 S. Olsnes and A. Pihl, Eur. J. Biochem., 35 (1973) 179.
8 S. Olsnes and A Pihl, Biochemistry, 12 (1973) 3121.
9 T.-S. Lin and S. S-L. Li, Eur. J. Biochem., 105 (1980) 453.
10 S. Olsnes, E. Saltvedt and A. Pihl, J. Biol. Chem., 249 (1974) 803
11 S. Olsnes and A Pihl, Nature (London), 238 (1972) 459.
12 M. Funatsu in Proteins, Structure and Function, eds. M. Funatsu, K. Hiromi, K. Imahori, T. Murachi and K. Nrita (John Wiley, New York, 1972) p. 103.
13 S. Olsnes and A. Pihl in Molecular Action of Toxins and Viruses, eds. P. Cohen and S. van Heyningen (Elsevier Biomedical Press, Amsterdam, 1982) p. 51.
14 G.L. Nicolson, J. Blaustein and M. Etzler, Biochemistry, 13 (1974) 196.
15 D.B. Cawley, M.L. Hedblom and L.L. Houston, Arch. Biochem. Biophys., 190 (1978) 744.
16 E. Saltvedt, Biochim. Biophys. Acta, 451 (1976) 536.
17 A.M. Pappenheimer, S. Olsnes and A.A. Harper, J. Immunol., 113 (1974) 835.
18 S. Olsnes and E. Saltvedt, J. Immunol., 114 (1975) 1743.
19 G. Funatsu, S. Yoshitake and M. Funatsu, Agric. Biol. Chem., 42 (1978) 501
20 G. Funatsu, M. Kimura and M. Funatsu, Agric. Biol. Chem., 43 (1979) 2221.
21 H. Bull, S. S.-L. Li, E. Fowler and T. T.-S. Lin, Int. J. Peptide Protein Res., 16 (1980) 208.
22 L.M. Roberts, F.I. Lamb, D.J.C. Pappin and J.M. Lord, J. Biol. Chem., 260 (1985) 15682.
23 J.E. Villafranca and J.D. Robertus, J. Mol. Biol., 116 (1977) 331.
24 J. E. Villafranca and J.D. Robertus, J. Biol. Chem., 256 (1981) 554.
25 C.H. Wei and C. Koh, J. Biol. Chem., 253 (1978) 2061.
26 R.J. Fulton, D.C. Blakey, P.P. Knowles, J.W. Uhr, P.E. Thorpe and E.S. Vitetta, J. Biol. Chem., 261 (1986) 5314.
27 S. Nanno, M. Ishiguro, G. Funatsu and M. Funatsu, Jap. J. Med. Sci. Biol., 24 (1971) 39.
28 M. Funatsu, G. Funatsu, M. Ishiguro, S. Nanno and K. Hara, Proc. Jap. Acad., 47 (1971) 718.
29 S. Olsnes, K. Refsnes, T.B. Christensen and A. Pihl, Biochim. Biophys. Acta, 405 (1975) 1.
30 G. Funatsu, T. Mise, H. Matsuda and M. Funatsu, Agric. Biol. Chem., 42 (1978) 851.
31 F.I. Lamb, L.M. Roberts and J.M. Lord, Eur. J. Biochem., 148 (1985) 265.
32 S. Ramakrishnan, M.R. Eagle and L.L. Houston, Biochim. Biophys. Acta, 719 (1982) 341.
33 A.L. Tarentino, T.H. Plummer and F. Maley, J. Biol. Chem., 247 (1972) 2629.
34 A.L. Tarentino and F. Maley, J. Biol. Chem., 249 (1974) 811.
35 B.M.J. Foxwell, T.A. Donovan, P.E. Thorpe and G. Wilson, Biochim. Biophys. Acta, 840 (1985) 193.
36 H. Ishihara, N. Takahasi, S. Oguri and S. Tejima, J. Biol. Chem., 254 (1979) 10715.
37 J.H. Elder and S. Alexander, Proc. Natl. Acad. Sci. USA, 79 (1983) 4540.
38 A.G. Butterworth and J.M. Lord, Eur. J. Biochem., 137 (1983) 57.
39 L. Montanero, S. Sperti, A. Mattioli, G. Testoni and F. Stirpe, Biochem. J., 146 (1975) 127.

40 R.D. Noland, H. Grasmuk and J. Drews, Eur. J. Biochem., 64 (1976) 69.
41 G. Fernandex-Puentes, S. Benson, S. Olsnes and A. Pihl, Eur. J. Biochem., 64 (1976) 437.
42 S. Olsnes and A.K. Abraham. Eur. J. Biochem., 93 (1979) 447.
43 S. Olsnes, F. Stirpe, K. Sandvig and A. Pihl, J. Biol. Chem., 257 (1982) 13263.
44 S. Olsnes, K. Sandvig, A. Sundan, K. Eiklid and A. Pihl, in Receptor Mediated Targeting of Drugs, eds. (G. Gregoriadis, G. Poste, J. Senior and A. Trouet (Plenum, New York, 1985) p. 87
45 J.R. Collier, Bacteriol. Rev., 39 (1975) 54.
46 A.M. Pappenheimer, Annu. Rev. Biochem., 46 (1977) 69.
47 S.H. Leppla, O.C. Martin and L.A. Muehl, Biochem. Biophys. Res. Commun., 81 (1978) 532.
48 G.L. Nicholson, Nature (London), 251 (1974) 628.
49 R.J. Youle, G.J. Murray and D.M. Neville, Cell, 23 (1981) 551.
50 J.U. Baenziger and D. Fiete, J. Biol. Chem., 254 (1979) 9795.
51 S.K. Podder, A. Surobia and B.K. Bachhawat, Eur. J. Biochem., 44 (1974) 151.
52 K. Sandvig, S. Olsnes and A. Pihl, Eur. J. Biochem., 82 (1978) 13.
53 K. Eiklid, S. Olsnes and A. Pihl, Exp. Cell Res., 126 (1980) 321.
54 S. Olsnes, K. Sandvig and A. Pihl, in Protides of the Biological Fluids, ed. H. Peters (Pergamon Press, New York, 1979) p. 395.
55 G.L. Nicolson, M. Lacorbiere and W. Eckhart, Biochemistry, 14 (1975) 172.
56 S.W. Rosen and R.C. Hughes, Biochemistry, 16 (1977) 4908.
57 C. Gottlieb and S. Kornfeld, J. Biol. Chem., 251 (1976) 7761.
58 S. Olsnes and K. Refsnes, Eur. J. Biochem., 88 (1978) 7.
59 S. Olsnes, K. Sandvig, K. Eiklid and A Pihl, J. Supramol. Struct., 9 (1978) 15.
60 P. Stanley, S. Narasimhan, L. Siminovitch and H. Schachter, Proc. Natl. Acad. Sci. USA, 72 (1975) 3323.
61 A. Meager, A. Ungkitchannkit and R.C. Hughes. Biochem. J., 154 (1976) 113.
62 R.C. Hughes and G. Mills, Biochem. J., 211 (1983) 575.
63 G.L. Nicolson, J.R. Smith and R. Hyman, J. Cell Biol., 78 (1978) 565.
64 J.C. Robbins, R. Hyman, V. Stallings and G.L. Nicolson, J. Natl. Cancer Inst., 58 (1977) 1027.
65 R. Hyman, M. Lacorbiere, S. Stavarde and G.L. Nicholson, J. Natl. Cancer Inst., 52 (1974) 963.
66 G.L. Nicholson, J.C. Robbins and R. Hyman, J. Supramol. Struct., 4 (1976) 15.
67 S. Olsnes, Naturwissenschaften, 59 (1972) 497.
68 K. Sandvig and S. Olsnes, Exp. Cell Res., 121 (1979) 15.
69 K. Sandvig and S. Olsnes, J. Biol. Chem., 257 (1982) 7495.
70 K. Sandvig and S. Olsnes, J. Biol. Chem., 257 (1982) 7504.
71 S. Olsnes and K. Sandvig, in Receptor Mediated Endocytosis: Receptors and Recognition, eds. P. Cuatrecasas and T.F. Roth, Vol. 15 (Chapman & Hall, London, 1983) p. 188.
72 S. Olsnes and K. Sandvig, in Endocytosis, eds. I. Pastan and M.C. Willingham (Plenum New York, 1985) p. 195.
73 G.L. Nicolson, M. Lacorbiere and T.R. Hunter, Cancer Res., 35 (1975) 144.
74 J. Gonatas, A. Stieber, S. Olsnes and N.K. Gonatas, J. Cell Biol., 87 (1980) 579.
75 P.A. Gonnella and M.R. Neutra, J. Cell Biol., 99 (1984) 909.
76 D. Carriere, P. Casellas, G. Richer, P. Gros and F.K. Jansen, Exp. Cell Res., 156 (1985) 327.
77 B. van Deurs, L. Ryde Pedersen, A. Sundan, S. Olsnes and K. Sandvig, Exp. Cell Res., 159 (1985) 287.

78 B. van Deurs, T.I. Tonnessen, O.W. Petersen, K. Sandvig and S. Olsnes, J. Cell Biol., 102 (1986) 37.
79 N.K. Gonatas, A. Stieber, S.U. Kim, D.I. Graham and S. Avrameas, Exp. Cell Res., 94 (1975) 426.
80 J.L. Middlebrook, R.B. Dorland and S.H. Leppla, J. Biol. Chem., 253 (1978) 7325.
81 R.K. Draper and M.I. Simon, J. Cell Biol., 87 (1980) 849.
82 D. Fitzgerald, R.E. Morris and C.B. Saelinger, Cell, 21 (1980) 867.
83 R.B. Dorland, J.L. Middlebrook and S.H. Leppla, Exp. Cell Res., 134 (1981) 319.
84 M.H. Marnell, M. Stookey and R.K. Draper, J. Cell Biol., 93 (1982) 57.
85 G.L. Nicholson, Nature (London), 251 (1974) 628.
86 M.S. Brown and J.L. Goldstein, Proc. Natl. Acad. Sci. USA, 76 (1979) 3330.
87 B.M.F. Pearce and M.S. Bretscher, Annu. Rev. Biochem., 50 (1981) 85.
88 A. Dautry-Varsat, A. Ciechanover and H.F. Lodish, Proc. Natl. Acad. Sci. USA, 80 (1983) 2258.
89 R.B. Dickson, L. Beguinot, J.A. Hanover, N.D. Richert, M.C. Willingham and I. Pastan, Proc. Natl. Acad. Sci. USA, 80 (1983) 5335.
90 M. Forgac, L. Cantley, B. Wiedenmann, L. Altstiel and D. Branton, Proc. Natl. Acad. Sci. USA, 80 (1983) 1300.
91 R. Newman, C. Schneider, R. Sutherland, L. Vodinelich and M. Greaves, Trends Biomed. Sci., 7 (1982) 397.
92 J.L. Goldstein, M.S. Brown, G.W. Anderson, D.W. Russell and W.J. Schneider, Annu. Rev. Cell Biol., 1 (1985) 1.
93 K. Refsnes, S. Olsnes and A. Pihl, J. Biol. Chem., 249 (1974) 3557.
94 A. Gardas and I. MacPherson, Biochim. Biophys. Acta, 584 (1979) 538.
95 G.J. Dimitriadis and T.D. Butters, FEBS Lett., 98 (1979) 33.
96 G. Fernandex-Puentes and L. Carrasco, Cell, 20 (1980) 769.
97 A.M. Pappenheimer, Annu. Rev. Biochem., 56 (1977) 69.
98 T. Uchida in: Molecular Action of Toxins and Viruses, eds. P. Cohen and S. van Heyningen (Elsevier Biomedical Press, Amsterdam, 1982) p. 1.
99 P. Boquet and A.M. Pappenheimer, J. Biol. Chem., 251 (1976) 5770.
100 M. Deleers, N. Beugnier, P. Falmagne, V. Cabian and J.M. Ruysschaert, FEBS Lett., 160 (1983) 82.
101 P. Boquet, M.S. Silverman and A.M. Pappenheimer, Proc. Natl. Acad. Sci. USA, 79 (1976) 4449.
102 J.J. Donovan, M.I. Simon, R.K. Draper and N. Montal, Proc. Natl. Acad. Sci. USA, 78 (1981) 172.
103 B.L. Kagan, A. Finkelstein and M. Colombini, Proc. Natl. Acad. Sci. USA, 78 (1981) 4950.
104 K. Kim and N.B. Grogman, J. Bacteriol., 90 (1965) 1557.
105 J.L. Duncan and N.B. Groman, J. Bacteriol., 98 (1969) 963.
106 R.K. Draper and M.I. Simon, J. Cell Biol., 87 (1980) 849.
107 K. Sandvig and S. Olsnes, J. Biol. Chem., 256 (1981) 9068.
108 K. Sandvig, S. Olsnes and A. Pihl, Biochem. Biophys. Res. Commun., 60 (1979) 648.
109 E. Mekada, T. Uchida and J. Okada, J. Biol. Chem., 256 (1981) 1225.
110 B. Ray and H.C. Wu, Mol. Cell Biol., 2 (1982) 535.
111 R.J. Youle and D.M. Neville, J. Biol. Chem., 257 (1982) 1598.
112 E.S. Vitetta, J. Immunol., 136 (1986) 1880.
113 S. Olsnes and A. Pihl, in The Specificity and Action of Animal, Bacterial and Plant Toxins. ed. P. Cuatrecasas (Chapman and Hall, London, 1976) p. 131.

114 S. Sperti, L. Montanaro, A. Mattiolia and F. Stirpe, Biochem. J., 136 (1973) 813.
115 M. Greco, L. Montanaro, F. Novello, L. Saccone and F. Stirpe, Biochem. J., 142 (1974) 695.
116 S.M. Harley and H. Beevers, Proc. Natl. Acad. Sci. USA, 79 (1982) 5935.
117 S.C. Benson, S. Olsnes, A. Pihl, J. Skorve and A. Abraham, Eur. J. Biochem., 59 (1975) 573.
118 S.C. Benson, Biochem. Biophys. Res. Commun., 75 (1977) 845.
119 M.L. Hedblom, D.B. Cawley and L.L. Houston, Arch. Biochem. Biophys., 177 (1976) 46.
120 L.L. Houston, Biochem. Biophys. Res. Commun., 85 (1978) 131.
121 S. Olsnes, K. Sandvig, K. Refsnes and A. Pihl, J. Biol. Chem., 251 (1976) 3985.
122 S. Ramakrishnan, M.R. Eagle and L.L. Houston, Biochim. Biophys. Acta, 719 (1982) 341.
123 D.J. Gifford, J.S. Greenwood and J.D. Bewley, Plant Physiol., 69 (1982) 1471.
124 L.M. Roberts, F.I. Lamb and J.M. Lord in Membrane Mediated Cytotoxicity, UCLA Symposium Series, Vol. 45, B. Bonevida and J.R. Collier (Alan R. Liss, New York) (in press).
125 L.M. Roberts and J.M. Lord, Eur. J. Biochem., 119 (1981) 31.
126 G. Blobel and B. Dobberstein, J. Cell Biol., 67 (1975) 835.
127 P. Walter, R. Gilmore and G. Blobel Cell, 38 (1984) 5.
128 P. Walter and G. Blobel, Nature (London), 299 (1982) 691.
129 P. Walter and G. Blobel, J. Cell Biol., 91 (1981) 557.
130 D.I. Meyer, E. Krause and B. Dobberstein, Nature (London), 297 (1982) 647.
131 G. Hortin and I. Boime, Cell, 24 (1981) 453.
132 M.D. Snider and O.C. Rogers, Cell, 36 (1984) 753.
133 R. Kornfeld and S. Kornfeld, Annu. Rev. Biochem., 45 (1985) 631.
134 L.W. Bergman and W.M. Kuehl, J. Biol. Chem., 254 (1979) 5690.
135 L.T. Roden, B.J. Miflin and R.B. Freedman, FEBS Lett., 138 (1982) 121.
136 J.M. Lord, Eur. J. Biochem., 146 (1985) 411.
137 J.M. Lord, Eur. J. Biochem., 146 (1985) 403.
138 W. Dunphy and J.E. Rothman, Cell., 42 (1985) 13.
139 M.G. Farquahr, Annu. Rev. Cell Biol., 1 (1985) 447.
140 R.B. Kelly, Science (Washington DC.), (1985) 25.
141 R. Schekman, Annu. Rev. Cell Biol., 1 (1985) 115.
142 J.M. Lord and S.M. Harley, FEBS Lett., 189 (1985) 72.
143 S.M. Harley and J.M. Lord, Plant Sci., 41 (1985) 111.
144 T.J.V. Higgins, P.M. Chandler, C. Zurawski, S.C. Button and D. Spencer, J. Biol. Chem., 258 (1983) 9544.
145 T.J.V. Higgins, M.J. Chrispeels, P.M. Chandler and D. Spencer, J. Biol. Chem., 258 (1983) 9550.
146 M.J. Chrispeels, Planta, 158 (1983) 140.
147 D.J. Bowles, S.E. Marcus, D.J.C. Pappin, J.B.C. Findlay, E. Eliopoulos, P.R. Maycox and J. Burgess, J. Cell Biol., 102 (1986) 1284.
148 N.C. Nielsen, Philos. Trans. R. Soc. London, Ser. B, 304 (1984) 287.
149 D. Boulter, Philos. Trans. R. Soc. London, Ser, B, 304 (1984) 323.
150 K.C. Halling, A.C. Halling, E.M. Murray, B.F. Ladin, L.L. Houston and R.F. Weaver, Nucleic Acids Res., 13 (1985) 8019.
151 F.I. Lamb, Ph.D Thesis, University of Warwick (1984).
152 G. von Heijne, Eur. J. Biochem., 133 (1983) 17.
153 M. Kozak, Cell, 44 (1986) 283.

154 M.L. Crouch, K.M. Tenbarge, A.E. Simon and R. Ferl, J. Mol. Appl. Genet., 2 (1983) 273.
155 J.S. Graham, G. Peace, J. Merryweather, K. Titani, L. Ericssola and C. A. Ryan, J. Biol. Chem., 260 (1985) 6555.
156 J.M. Lord and C. Robinson, in Plant Proteolytic Enzymes, ed. M.J. Dalling (CRC Press, Boca Raton) (in press).
157 A. Klausner, Biotechnology, 4 (1986) 185.
158 A.I. Freeman and E. Mayhew, Cancer, 58 (1986) 573
159 G. Kohler and C. Milstein, Nature (London), 256 (1975) 495.
160 G. Kohler, S.C. Howe and C. Milstein, Eur. J. Immunol., 6 (1976) 292.
161 R. Levy and R.A. Miller, Fed. Proc., 42 (1983) 2650.
162 K.D. Lloyd, in Basic and Clinical Tumour Immunology, ed. R.B. Herberman (Martinus Nijhoff, Boston, 1983) p. 159.
163 C. Macek, J. Am. Med. Assoc., 247 (1982) 2463.
164 R.K. Oldham, J. Clin. Oncol., 1 (1983) 582.
165 R.B. Bankert, Cancer Drug Deliv., 1 (1984) 251.
166 M.J. Mastrangelo, D. Berd and H.C. Maguire, Cancer Treat. Rep., 68 (1984) 207.
167 P. Ehrlich, in Collected Studies on Immunology Vol. II (John Wiley, New York, 1906) p. 442.
168 D.C. Edwards and P.E. Thorpe. Trends Biochem. Sci., 5 (1981) 313.
169 F.K. Jansen, H.E. Blythman, D. Carriere, P. Casellas, O. Gros, P. Gros, J.C. Laurent, F. Paolucci, B. Pau, P. Poncelet, G. Richer, H. Vidal and G.A. Voisin, Immunol. Rev., 62 (1982) 185.
170 D.M. Neville and R.J. Youle, Immunol. Rev., 62 (1982) 75.
171 P.E. Thorpe and W.C.J. Ross, Immunol. Rev., 62 (1982) 119.
172 V. Raso, Immunol. Rev., 62 (1982) 93.
173 S. Olsnes and A. Pihl, Pharmacol. Ther., 156 (1982) 335.
174 E.S. Vitetta, K.A. Krolich and J.W. Uhr, Immunol. Rev., 62 (1982) 159.
175 E.S. Vitetta, K.A. Krolich, M. Miyama-Inaba, W. Cushley and J.W. Uhr, Science, (Washington, DC) 219 (1983) 644.
176 K. Sikora, H. Smedley and P.E. Thorpe, Br. Med. Bull., 40 (1983) 233.
177 R.J. Collier and D.A. Kaplan, Sci. Am., 251 (1984) 44.
178 J.W. Uhr, J. Immunol., 133 (1984) 1.
179 P.E. Thorpe, in Monoclonal Antibodies '84: Biological and Clinical Applications, eds. A. Pinchera, G. Doria, F. Dammacco and A. Bargellesi (Editrice Kurtis, Rome, 1985) p. 475.
180 J.M. Lord, L.M. RobeC.Jrts, P.E. Thorpe and E.S. Vitetta, Trends Biotech., 3 (1985) 175.
181 K.A. Krolich, ASM News, 51 (1985) 569.
182 E.S. Vitetta and J.W. Uhr, Annu. Rev. Immunol., 3 (1985) 197.
183 R.J. Youle and D.M. Neville, Proc. Natl. Acad. Sci. USA, 77 (1980) 5483.
184 P. Casellas, B.J.P. Bournie, P. Gros and F.K. Jansen, J. Biol. Chem., 259 (1984) 9359.
185 V. Raso and J. Lawrence, J. Exp. Med., 160 (1984) 1234.
186 P. Casellas, J.P. Brown, P. Gros, I. Hellstrom, F.K. Jansen, P. Poncelet, R. Roncucci, H. Vidal and K.E. Hellstrom, Int. J. Cancer, 30 (1982) 437.
187 M. Kronke, E. Schlick, T.A. Waldman, E.S. Vitetta and W.C. Greene, Cancer Res., 46 (1986) 3295.
188 S. Ramakrishnan and L.L. Houston, Science (Washington, DC), 223 (1984) 58.
189 D.P. MacIntosh, D.C. Edwards, A.J. Cumber, D. Parnell, C.J. Dean, W. C. J. Ross and J.A. Forrester, FEBS Lett. 164 (1983) 17.
190 E.S. Vitetta, W. Cushley and J.W. Uhr, Proc. Natl. Acad. Sci. USA 80 (1983) 6332.

191 E.S. Vitetta, R.J. Fulton and J.W. Uhr, J. Exp. Med. 160 (1984) 341.
192 R.J. Fulton, J.W. Uhr and E.S. Vitetta, J. Immunol. 136 (1986) 3103.
193 P.E. Thorpe, W.C.J. Ross, A.N.F. Brown, C. Meyers, A.J. Cumber, B. M. J. Foxwell and J.A. Forrester, Eur.B J. Biochem. 140 (1984) 63.
194 T. Mise, T. Shimoda and G. Funatsu, Agric. Biol. Chem. 50 (1986) 151.
195 P. Carter, Biochem. J. 237 (1986) 1.
196 N.R. Worrell, D.N. Skilliter, A.J. Cumber and R.J. Price, Biochem. Biophys. Res. Commun. 137 (1986) 892.
197 B.J.P. Bourrie, P. Casellas, H.E. Blythman and F.K. Jansen, Eur. J. Biochem. 155 (1986) 1.
198 V.T. Oi and S.L. Morrison, Biotechniques 4 (1986) 214.
199 M.S. Neuberger, G.T. Williams and R.O. Fox, Nature (London), 312 (1984) 604.

Progress in Medicinal Chemistry – Vol. 24, edited by G.P. Ellis and G.B. West

2 Histamine Receptors in the Mammalian Central Nervous System: Biochemical Studies

S.J. HILL, B.Sc., Ph.D.

Department of Physiology and Pharmacology, Medical School, Queen's Medical Centre, Clifton Boulevard, Nottingham, NG7 2UH, United Kingdom

INTRODUCTION

Histamine is widely distributed in mammalian brain [1–4] and is found in both neurones [5, 6] and non-neuronal compartments such as mast cells [7, 8] and capillary endothelial cells [9]. More recently the neuronal pathways containing histamine and its synthesizing enzyme, histidine decarboxylase, have been visualized using immunohistological techniques [10–14]. These studies have shown that histamine- and histidine-decarboxylase-immunoreactive cell bodies are restricted to the ventral part of the posterior hypothalamus and the region of the mammillary nuclei but send projections to almost all regions of the diencephalon and telencephalon. This arrangement of the histaminergic neuronal system resembles the widespread distribution of noradrenergic and serotonergic fibres, which also emanate from discrete cell body regions. This suggests that histamine, like noradrenaline and 5-hydroxytryptamine, may be involved in a large variety of physiological functions. The functional rôle or rôles that histamine may play in the CNS, however, remains largely subject to speculation.

Histamine has been implicated in a variety of vegetative functions and behaviours ranging from cardiovascular [15–20] and temperature [21–26] control to rôles in arousal [27] and the regulation of neuroendocrine mechanisms [28–30]. Most of our information concerning a possible involvement of histamine in these responses is derived from the observation that intracerebral injection of histamine or its analogues can elicit changes in particular central functions or behaviour (for example, core temperature [21–26] or locomotor activity [27]) which are sensitive to antagonism by histamine receptor antagonists. Interpretation of these findings is made difficult, however, because of the uncertainties concerning the selectivity of action of the drugs used. This problem is inherent in *in vivo* studies because of the limited selectivity of the pharmacological tools available and the inability to achieve a known concentration of the pharmacological agent at the site of action. The importance of these factors is perhaps best illustrated with reference to some recent *in vitro* studies on cat cerebral arteries [31, 32]. In this tissue, histamine can elicit a contractile response *via* a non-receptor tyramine-like mechanism. The response to histamine can be inhibited by cocaine, an inhibitor of the high-affinity uptake of noradrenaline (uptake 1), and by the H_1-receptor antagonist, diphenhydramine [31, 32]. However, in this latter case the inhibitory effect of diphenhydramine is not due to H_1-receptor antagonism, but appears to be a consequence of the uptake-inhibiting properties of this antihistamine [31, 33]. The ability to inhibit noradrenaline and 5-hydroxytryptamine uptake is shared by a number of other H_1-receptor antagonists, including mepyramine [33], chlorpheniramine and brompheniramine [34]. In all cases, uptake blockade by H_1-antihistamines occurs at higher antagonist concentrations than are required for H_1-receptor antagonism. For example, the equilibrium dissociation constant, K_B, for mepyramine inhibition of 5-hydroxytryptamine uptake is 0.09 μM [33], while that for H_1-receptor antagonism is 0.8 nM [35]. It is clear, however, that, without a careful quantitative analysis of the interaction between histamine and H_1-antagonists in cat cerebral arteries, the response to histamine might have been misinterpreted as an H_1-mediated response.

The aim of the present review is to examine the evidence for the presence of histamine receptors in the mammalian central nervous system and to review their properties. For the reasons outlined above, the scope of this review has been limited to recent *in vitro* biochemical studies where a quantitative comparison can be made between the properties of central and peripheral receptor systems. It is accepted, however, that the determination of the physiological rôle or rôles of histamine in mammalian brain will rely eventually on the availability and careful *in vivo* use of highly selective agonists and antagonists for the different classes of histamine receptor.

CLASSIFICATION OF HISTAMINE RECEPTORS AND CHEMICAL TOOLS

Histamine receptors have been classified into two major subtypes, H_1 and H_2, on the basis of quantitative studies on isolated peripheral tissues. Histamine H_1-receptors mediate the contractile actions of histamine on numerous visceral smooth muscles, most notably from the trachea, ileum and uterus of the guinea-pig [36–39]. These responses are antagonized by the classical H_1-antihistamines [36–39] such as mepyramine (1) [36] and diphenhydramine (3) [40] (see *Figure 2.1*). Histamine also stimulates the secretion of acid by stomach, increases the rate of contraction of guinea-pig isolated atria and inhibits electrically evoked contractions of rat isolated uterine horn [41]. However, these responses are not affected by H_1-receptor antagonists and have been defined as histamine H_2-receptor responses following the development of specific antagonists to these responses such as burimamide [41], cimetidine [42] and ranitidine [43]. The distribution and classification of histamine H_1- and H_2-receptors in various mammalian peripheral tissues have been reviewed elsewhere [44–46a].

Studies with H_1- and H_2-receptor agonists and antagonists in mammalian brain have indicated that the pharmacological profile of the autoreceptor controlling the release of histamine from brain slices is different from that expected for a classical H_1- or H_2-mediated response, and an H_3-receptor subtype has been proposed to account for these findings [47–49]. Thus, burimamide (an H_2-antagonist [41]), impromidine (an H_2-agonist [50]) and betahistine (an H_1-agonist [51]) have been proposed as potent antagonists of the putative H_3-receptor [47–49, 52]. However, at the present time there are no selective antagonists or agonists available for this particular receptor subtype [49]. (See note added in proof, p. 84.)

H_1-RECEPTOR ANTAGONISTS

Following the initial discovery of the first antihistamine substances in France in 1937 [53], a large number of compounds, of widely differing chemical structure, have been shown to possess H_1-receptor antagonist properties. The structural formulae of some of the more potent competitive H_1-antagonists are shown in *Figure 2.1*. Mepyramine (1) is the compound which has been principally utilized as a probe for H_1-receptor responses in both central and peripheral tissues because of its high affinity and relative selectivity. However,

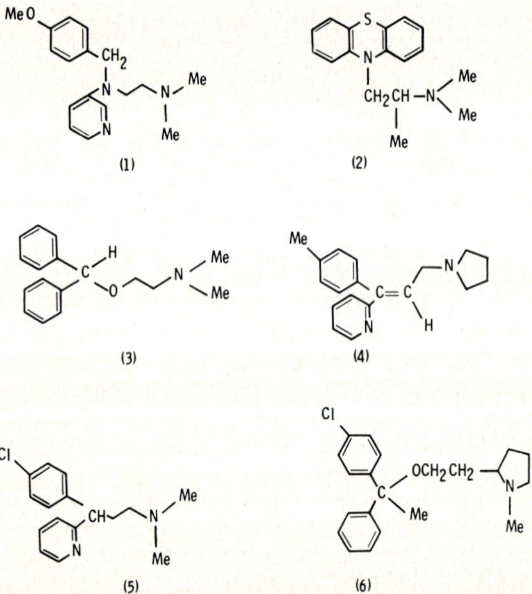

Figure 2.1. Structural formulae of some H_1-receptor antagonists: mepyramine (1), promethazine (2), diphenhydramine (3), trans-*triprolidine (4), chlorpheniramine (5) and clemastine (6).*

the selectivity of mepyramine and other compounds for the H_1-receptor is very much dependent upon the concentration used. Thus, although mepyramine can be considered to act selectively on H_1-receptors in the concentration range 1 to 100 nM (K_D = 0.8 nM [35]), at higher concentrations it will begin to antagonize muscarinic (K_D = 10 μM [54, 55]) and histamine H_2-receptors (K_D = 5 μM [56]), to inhibit histamine *N*-methyltransferase (K_D = 21 μM [57]) and indoleamine uptake (K_D = 0.09 μM [33]), and to have membrane-stabilizing properties [58]. The ability to antagonize muscarinic receptors is a property shared by a large number of H_1-antagonists and for drugs such as promethazine (2) and diphenhydramine (3) the concentration at which occupancy of muscarinic receptors becomes significant is only one order of magnitude greater than that required to produce a similar occupancy of H_1-receptors [55]. It is therefore essential to make use of a range of H_1-antagonists of different chemical structures in studies aimed at characterizing H_1-receptor responses in different tissues and to undertake a quantitative assessment of their potencies as H_1-antagonists. A comparison of the equilibrium dissociation constants obtained for several H_1-antihistamines in guinea-pig brain and ileal smooth muscle is shown in *Table 2.1*.

A number of the H_1-antagonists exist as geometric or optical isomers. *trans*-Triprolidine (4), which has pyridyl and pyrrolidino groups in a *trans* configuration about the double bond, is three orders of magnitude more potent on guinea-pig ileum than is the *cis*-isomer [62]. Chlorpheniramine (5) shows stereoselective activity on both central and peripheral H_1-receptors and is consequently a powerful tool for receptor classification. D(+)-Chlorpheniramine, which has the *S* configuration [64], is approximately 200-fold more potent than L(−)-chlorpheniramine in guinea-pig ileum [35, 63] and brain [51, 61]. However, not all chiral antihistamine drugs exhibit stereoselective antagonism on histamine H_1-receptors, and this suggests that the asymmetric centre close to the side-chain nitrogen in promethazine (2), isothipendyl [65] and clemastine (6) [66] is not important for a stereoselective interaction with H_1-receptors.

Triprolidine (4) (K_D = 0.1 nM [62]) and the tricyclic antidepressant doxepin (11-(3-dimethylaminopropylidene-6,11-dihydrodibenzoxepin) (K_D = 0.06 nM [67, 68]) are two of the most potent H_1-antagonists known and are thus potentially very valuable agents for investigating H_1-responses in mammalian brain. However, the very high affinity of these compounds for the H_1-receptor can itself lead to difficulties in biochemical studies where the incubation volume is invariably small and the tissue concentration high. Thus, the fact that only very low concentrations of triprolidine or doxepin are needed for significant occupancy of H_1-receptors means that depletion of the free concentration of

Table 2.1. EQUILIBRIUM DISSOCIATION CONSTANTS OF H_1-RECEPTOR ANTAGONISTS ON PERIPHERAL AND CENTRAL H_1-RECEPTORS

Values represent measurements made on guinea-pig ileum [35], slices of guinea-pig cerebral cortex (cyclic AMP accumulation) [51] and cerebellum (inositol phosphate accumulation) [59,60] and homogenates of guinea-pig whole brain ([^{3}H]mepyramine binding) [61].

	Equilibrium dissociation constant, K_D (nM)				
	Guinea-pig ileum		*Guinea-pig brain*		
Antagonist	*Contraction*	*[^{3}H]Mepyramine binding*	*Cyclic AMP accumulation*	*Inositol phosphate accumulation*	*[^{3}H]Mepyramine binding*
Mepyramine (1)	0.8	0.4	0.6	0.4	0.8
Promethazine (2)	1.2	0.4	0.6	3.2[a]	1.4
Methapyrilene	2.6	2.7	1.8	5.5	4.5
(+)-Chlorpheniramine (5)	0.8	1.1	0.4		0.8
(−)-Chlorpheniramine (5)	179	165	204		200

[a] Donaldson and Hill, unpublished data.

antagonist by tissue binding can occur to a large extent and lead to underestimates of antagonist potency. This has been demonstrated directly for doxepin in studies of [^{3}H]mepyramine binding in rat brain [69]. Great care is therefore required in the use of these very high affinity H_1-antagonists.

In recent years H_1-antagonists which do not readily cross the blood-brain barrier have been developed in an attempt to reduce the well-known sedative properties associated with this class of compounds. These include terfenadine (7) [70, 71], astemizole (8) [72], mequitazine (9) [73–75] and temelastine (10) [76] whose structures are given in *Figure 2.2*. It is striking that for most of these compounds, particularly astemizole and terfenadine, the onset of action is slow and the duration of action is exceptionally long [70, 72]. For example, studies

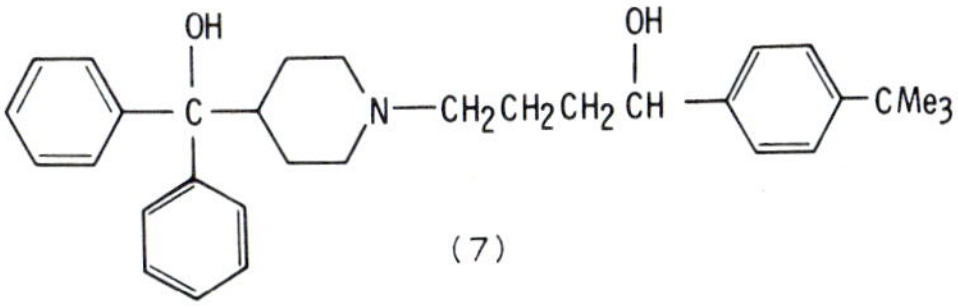

OH
C
N—CH2CH2CH2 CH
OH
CMe3
(7)

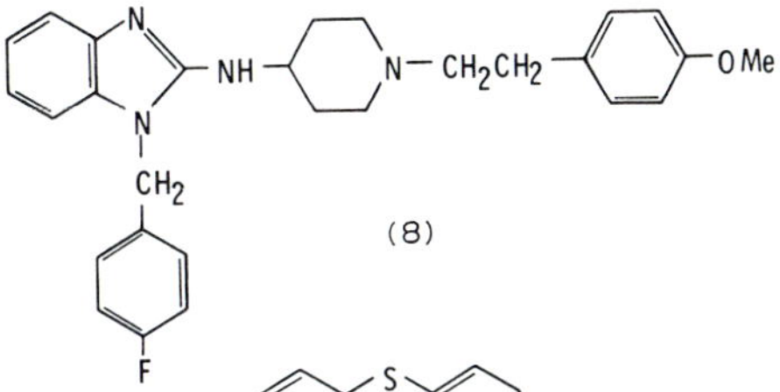

N
NH
N— CH2CH2
O Me
N
CH2
F
(8)

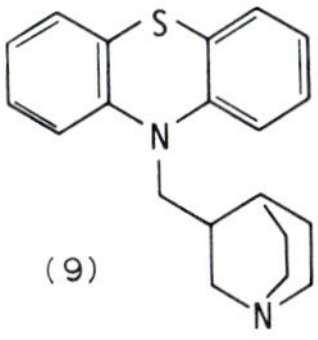

S
N
(9) 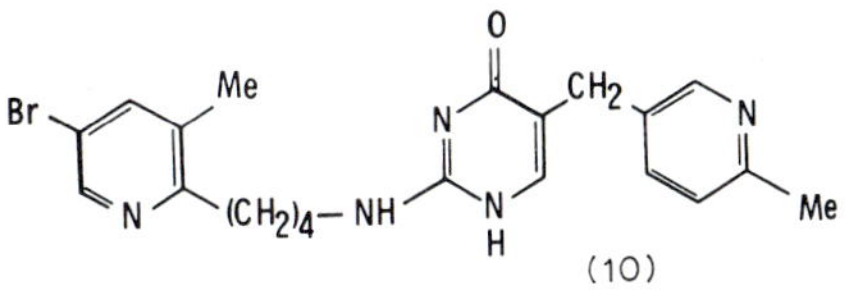
N

O
Br
Me
CH2
N
N
(CH2)4— NH
N
H
N
Me
(10)

Figure 2.2 Structural formulae of some non-sedating H_1-receptor antagonists: terfenadine (7), astemizole (8), mequitazine (9) and temelastine (10).

with astemizole have shown that it does not appreciably dissociate from H_1-receptors in guinea-pig cerebellar membranes *in vitro* following washout of the drug for more than 3 h [72]. Furthermore, the occupancy of H_1-receptors in lung declines only 4–6 days after oral administration of astemizole [72].

H_2-RECEPTOR ANTAGONISTS

The first compound to be described as an H_2-receptor antagonist was burimamide, in 1972 [41]. This compound has a relatively low affinity (K_D = 7.8 μM [41]) for the H_2-receptor when compared with the values normally expected for useful H_1-receptor antagonists (i.e., K_D values in the range 0.1 to 10 nM). However, unlike the H_1-antagonists, which usually have non-specific actions at micromolar concentrations, burimamide appears to be much more selective in this concentration range and begins to interfere with β-adrenoceptor, muscarinic and H_1-receptor responses only at concentrations above

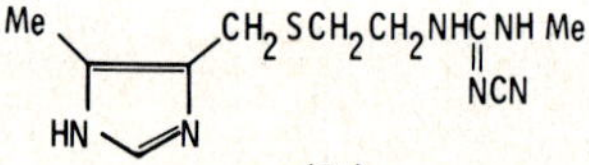

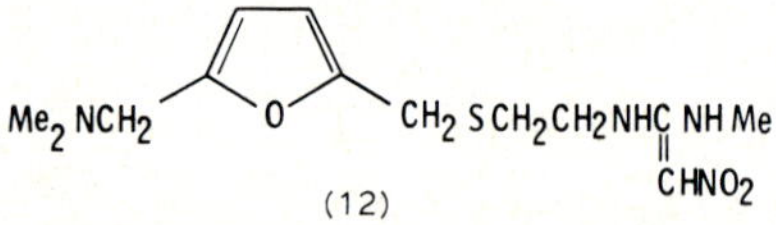

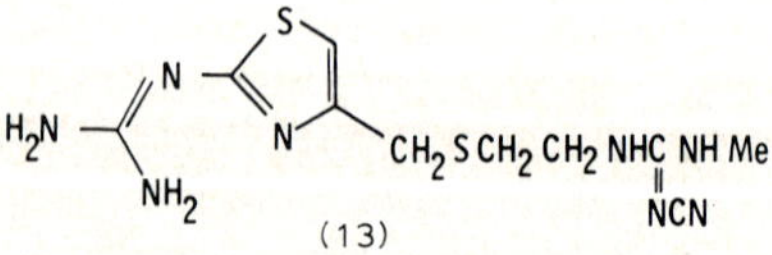

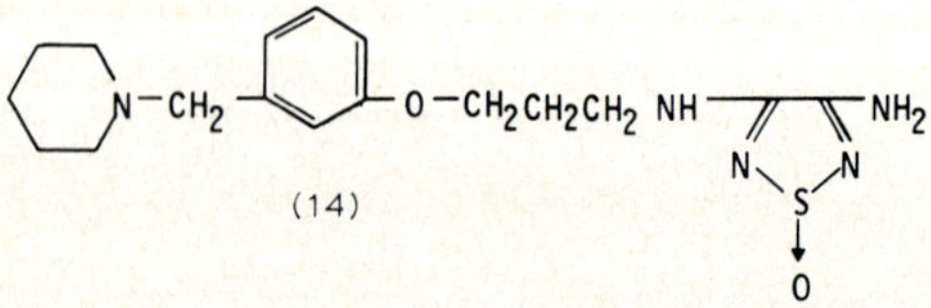

Figure 2.3. Structural formulae of some H_2-receptor antagonists; cimetidine (11), ranitidine (12), tiotidine (13) and L-643, 441 (14).

0.1 mM [41]. Other H_2-antagonists have since been developed from burimamide, which also contain the imidazole ring, and these include metiamide [77] and cimetidine (11) [42]. The two other compounds which have been employed extensively in studies of H_2-receptors in the mammalian central nervous system are ranitidine (12) (a furan derivative) [43] and tiotidine (13) (a guanidinothiazole derivative) [78]. The structures of a number of these H_2-antagonists are given in *Figure 2.3*. Also included in *Figure 2.3* is the structure of L-643,441 (14), which has been reported to be an apparently irreversible H_2-receptor antagonist [79]. This latter compound may be of utility in future studies for determining the relative efficacies and affinities of H_2-receptor agonists in functional studies of central H_2-receptors. The equilibrium dissociation constants obtained for the four major competitive H_2-antagonists on spontaneously-beating guinea-pig right atria and in three biochemical assays for H_2-antagonism in brain tissues are given in *Table 2.2*.

Table 2.2. EQUILIBRIUM DISSOCIATION CONSTANTS OF H_2-RECEPTOR ANTAGONISTS ON PERIPHERAL AND CENTRAL H_2-RECEPTORS

Values represent measurements made on guinea-pig atria [42,43,78,80], homogenates of guinea-pig hippocampus (adenylate cyclase) [81], slices of rabbit cerebral cortex (cyclic AMP accumulation) [82] and homogenates of guinea-pig cerebral cortex ([^{3}H]tiotidine binding) [83].

	Equilibrium dissociation constant, K_D *(μM)*			
		Brain		
Antagonist	*Atria*	*Adenylate cyclase*	*Cyclic AMP accumulation*	*[^{3}H]Tiotidine binding*
Tiotidine (13)	0.015	0.026	0.012	0.025
Cimetidine (11)	0.79	0.60	1.2	0.47
Ranitidine (12)	0.063		0.07	0.39
Metiamide	0.92	0.87	2.3	0.51

H_1- AND H_2-RECEPTOR AGONISTS

A range of agonists are now available for use in studies of H_1- and H_2-receptor responses. A large number of the possible methyl-substituted histamine derivatives have been synthesized and assayed quantitatively for H_1- and H_2-receptor agonist activity on guinea-pig ileum (H_1) and guinea-pig atrium (H_2) *in vitro* [65]. N^α-Methyl- and N^α,N^α-dimethylhistamine (refer to *Figure 2.4* for position

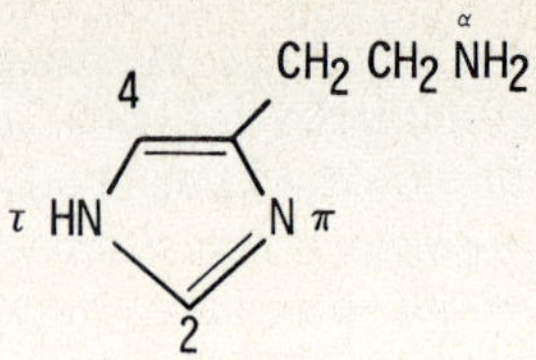

Figure 2.4. Structural formula of histamine (15).

of methyl groups in histamine (15)) are the most potent of these derivatives but do not show any marked selectivity for the H_1 or H_2 receptor subtypes. N^{τ}-Methylhistamine is one of the main natural metabolites of histamine and is produced following methylation of histamine by histamine *N*-methyltransferase, which is the major route of histamine metabolism in brain tissue, [5, 6]. Both this metabolite (N^{τ}-methylhistamine) and the N^{π}-methyl derivative of histamine are exceedingly weak agonists of both H_1- and H_2-receptors [65] and are thus useful as inactive control substances in both *in vitro* and *in vivo* studies. 2-Methylhistamine shows selectivity for the H_1-receptor, having *circa* 20% of the potency of histamine as an H_1-receptor agonist on guinea-pig ileum but only 4% of the potency of histamine as an H_2-receptor agonist on guinea-pig atrium [65]. Conversely, 4-methylhistamine shows selectivity as an agonist for the H_2-receptor. Thus, the relative potency (histamine = 100) of 4-methylhistamine for atrial H_2-receptors is 41, while that for ileal H_1-receptors is only 0.23 [41, 84]. This suggests that 4-methylhistamine is some 200-fold more potent as an H_2-receptor agonist. However, relative potencies (with respect to histamine) expressed in this way do not take into account the fact that, in some tissues, the effective concentrations of histamine necessary for activation of H_1- and H_2-receptors, respectively, may be very different. This problem has been elegantly demonstrated in guinea-pig ileum by Barker and Hough [85], who showed that the actual concentrations necessary for activation of H_2- and H_1-receptors by 4-methylhistamine differ by only a factor of 5. Thus, although 4-methylhistamine can be considered as having a selective action on H_2-receptors, its selectivity of action in a given tissue will depend upon the relative concentrations and transducing efficiency of the H_1- and H_2-receptor systems present in the tissue. Caution is therefore needed in characterizing responses mediated by H_1- and H_2-receptors with selective agonists alone.

2-Pyridylethylamine (16) and 2-thiazolylethylamine (17) are the other compounds most commonly utilized as H_1-receptor agonists. 2-Thiazolylethylamine is the most potent and selective of the H_1-agonists available, having a relative potency for ileal H_1-receptors (histamine = 100) of 26

[65, 84]. However, the selectivity of 2-thiazolylethylamine (17) for H_1-receptors is not particularly marked, and stimulation of H_2-receptors is very often observed at high concentrations [82]. N^{α}-Methyl-2-pyridylethylamine (betahistine) is also a selective agonist of H_1-receptors in guinea-pig ileum and, like 2-pyridylethylamine, acts as a weak partial agonist on atrial H_2-receptors [65]. Betahistine is used clinically as a peripheral vasodilator for the treatment of Meniere's disease [86]. Interestingly, the N^{α}, N^{α}-diethyl derivative of 2-pyridylethylamine has been reported to act as a partial agonist on H_1-receptors in the longitudinal smooth muscle of guinea-pig ileum [87].

In contrast to the lack of highly selective agonists for the H_1-receptor, two selective agents exist for the H_2-receptor, namely dimaprit (18) [88] and impromidine (19) [50]. Impromidine, in particular, is a very potent and specific H_2-agonist. On atrial H_2-receptors it has been shown to be 48-times more potent than histamine [50]. In contrast, it is 10^{-5}-times less potent than histamine on ileal H_1-receptors [50]. Dimaprit is also a very selective H_2-agonist, but is less potent than histamine on H_2-receptors in guinea-pig atrium and rat uterus [88]. Studies in guinea-pig right atrium have shown that both dimaprit and impromidine have a higher efficacy than histamine on

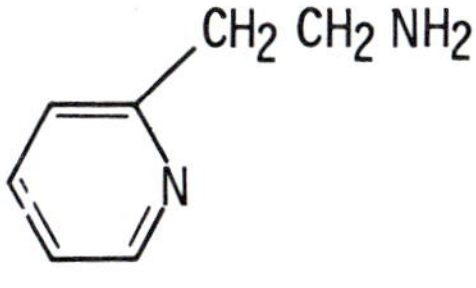

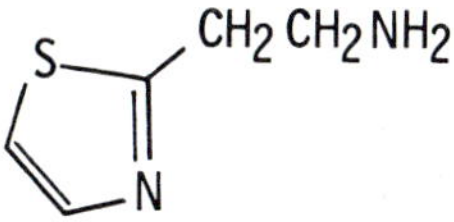

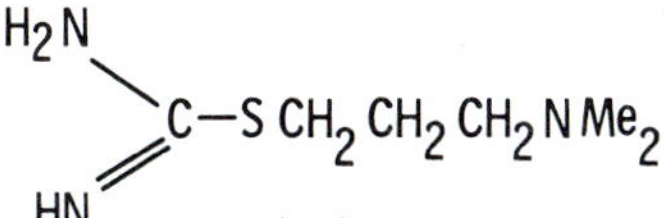

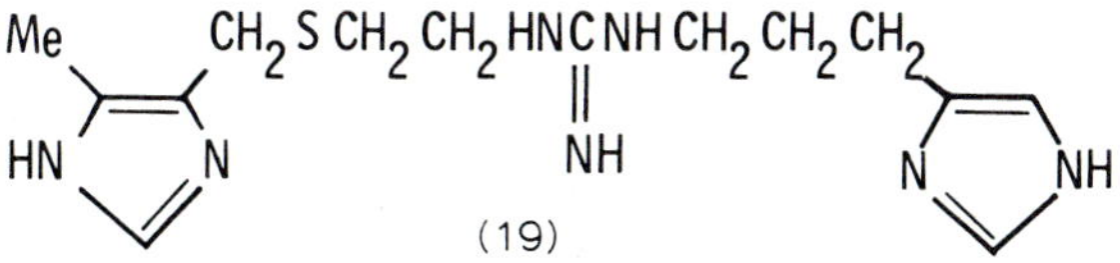

Figure 2.5. Structural formulae of two selective H_1-agonists: 2-pyridylethylamine (16) and 2-thiazolylethylamine (17), and two selective H_2-agonists: dimaprit (18) and impromidine (19).

H_2-receptors, but that the high potency of impromidine is primarily due to the high affinity of this compound for H_2-receptors ($K_D = 0.02\ \mu M$) [89].

The chemical structures of some selective H_1- and H_2-agonists are given in *Figure 2.5* and the relative potencies of a range of histamine receptor agonists of H_1-, H_2- and the putative H_3-receptor are given in *Table 2.3*.

Table 2.3. RELATIVE POTENCIES OF HISTAMINE RECEPTOR AGONISTS FOR H_1-, H_2- AND H_3-RECEPTOR SUBTYPES

Values were obtained on guinea-pig ileum (H_1) [65], atrium (H_2) [65] and rat cerebral cortical slices (H_3) [47–49, 52]. The relative agonist potency is given as the ratio: (EC_{50} value histamine/EC_{50} value agonist) × 100. Partial agonist activity is denoted by an asterisk.

	Relative agonist potency (histamine = 100)		
Agonist	*H_1*	*H_2*	*H_3*
Histamine	100	100	100
N^{τ}-Methylhistamine	0.4	<0.1	<4
N^{α}-Methylhistamine	72	74	270
N^{α},N^{α}-Dimethylhistamine	44	51	170
2-Methylhistamine	16.5	4.4	<0.08
4-Methylhistamine	0.23	43	<0.008
2-Pyridylethylamine	5.6	2.5*	
Betahistine	8.0	1.5*	antagonist
2-Thiazolylethylamine	26	2.2	<0.008
Dimaprit	$<10^{-4}$	71	<0.008
Impromidine	$<10^{-3}$	4810	antagonist

H_3-RECEPTOR AGONISTS AND ANTAGONISTS

The definition of the autoreceptor controlling the release of histamine from rat cerebral cortical slices as a putative H_3-receptor is based on quantitative studies with existing H_1- and H_2-receptor compounds, which show marked differences from the pharmacological characteristics expected for an interaction with classical H_1- or H_2-receptors. Thus, only histamine, N^{α}-methylhistamine and N^{α}, N^{α}-dimethylhistamine have any marked agonist activity on this response [47], while the selective H_1-agonist betahistine [52] and the selective H_2-agonist impromidine [47] act as antagonists in this system ($K_D = 6.9$ and $0.07\ \mu M$ for betahistine and impromidine, respectively). The weak H_2-receptor antagonists burimamide and SKF 91486 (3-[4(5)-imidazolyl]propylguanidine) also appear to be rather potent inhibitors of this response and both have dissociation constants around 0.07 μM [47]. In the case of SKF 91486 this value is 300-fold lower than the corresponding value obtained from inhibition of H_2-receptors [90]. The more potent H_2-antagonists cimetidine, ranitidine and tiotidine are, however, much less potent inhibitors of the autoreceptor controlling histamine release [47].

STUDIES WITH RADIOACTIVE LIGANDS

One approach which has been used extensively to investigate neurotransmitter receptors in the mammalian central nervous system is the study of the binding characteristics of radioactively labelled receptor ligands. Numerous radioactive ligands are now available for binding studies on central histamine receptors (particularly the H_1-receptor); however, not all of them succeed in labelling their target receptor with high specificity. Great care is therefore required in interpreting the results of binding studies with radioactive ligands and the criteria of saturability and specificity must be fulfilled before binding sites can be equated with receptor recognition sites [91, 92]. Thus, a component of binding should increase to a limiting value with increasing concentrations of the radioactive ligand. Furthermore, pharmacologically effective concentrations of drugs which act at a given receptor should displace the saturable component of binding, whilst pharmacologically effective concentrations of drugs which act on different receptors should be ineffective.

H_1-RECEPTOR ANTAGONISTS

[^{3}H]Mepyramine was introduced in 1977 as the first selective radioligand for the histamine H_1-receptor [61, 93] and there are now a large number of studies in the literature which vindicate its use for this purpose. Since that time, a number of other compounds have been introduced as ligands for the histamine H_1-receptor and these include the antidepressants [^{3}H]doxepin [94, 95] and [^{3}H]mianserin [96]. More recently, a quaternary derivative of diphenhydramine, (+)-*N*,4-[^{3}H]methyldiphenhydramine [97], and a high specific activity [^{125}I]iodobolpyramine [98] have been developed. The high selectivity of [^{3}H]mepyramine, however, has meant that it has been the ligand of choice for the majority of studies on H_1-receptors in membrane preparations.

Initial studies with [^{3}H]mepyramine were performed in homogenates of the longitudinal smooth muscle of guinea-pig small intestine [93, 35]. This is the tissue on which most of the quantitative pharmacological studies of H_1-antagonists have been performed, and a detailed characterization [35] of the binding characteristics of [^{3}H]mepyramine has confirmed that the binding of low concentrations of [^{3}H]mepyramine to guinea-pig intestinal smooth muscle is to sites with the characteristics of histamine H_1-receptors. The saturability of binding, the stereospecificity of the inhibition of binding by the isomers of chlorpheniramine and the good agreement of the dissociation constants obtained for H_1-antagonists from binding measurements with those from organ bath studies (on the inhibition of the contractile response to histamine) speak strongly for this. However, under certain assay conditions in this tissue, namely high pH coupled with high concentration of [^{3}H]mepyramine, secondary antagonist-sensitive binding sites for mepyramine become apparent [35]. In addition, analysis of antagonist inhibition curves at pH 7.5 show that for certain antagonists, most notably mepyramine, the Hill coefficient is significantly less than the value of unity expected for a simple drug-receptor equilibrium. It is still uncertain whether the low-affinity component in these curves results from a separate population of sites or from some secondary pharmacological effect (e.g., membrane stabilization). However, whatever the exact mechanism, it is clear that some 90% of the binding of low concentrations (1 nM) of [^{3}H]-mepyramine can be identified with some confidence as binding to histamine H_1-receptors [35].

In homogenates of guinea-pig brain, there is an appreciable saturable component of [^{3}H]mepyramine binding [61, 99]. In these membranes, there is a good quantitative correlation between the binding affinities of H_1-antagonists determined from inhibition of [^{3}H]mepyramine binding and those obtained from inhibition of peripheral H_1-mediated contractile responses [61]. Thus, the properties of high-affinity [^{3}H]mepyramine-binding sites in guinea-pig brain are consistent with the presence of H_1-receptors which are distributed unevenly in different brain regions, with highest levels found in cerebellum and lowest levels in brain stem and spinal cord [61, 99]. High-affinity [^{3}H]mepyramine-binding sites have now been detected in the brains of a number of other mammalian species, including man [99, 100–104]. However, there appear to be marked species differences in the regional distribution of these sites [99, 100–104]. For example, in man the frontal, parietal, temporal and cingulate cortices, together with the amygdaloid nuclei of the limbic system, show the highest density of specific [^{3}H]mepyramine-binding sites, while the cerebellum (which shows the highest density in the guinea-pig) has a very low density of sites. The regional distribution of histamine H_1-receptors in human brain, as deduced from binding studies with [^{3}H]mepyramine [99] or [^{3}H]doxepin [105], is shown in *Table 2.4*.

In addition to differences in regional localization, there are also differences

Table 2.4. REGIONAL DISTRIBUTION OF HISTAMINE H_1-RECEPTORS IN HUMAN BRAIN

The distribution of specifically bound [^{3}H]mepyramine [99] or high-affinity [^{3}H]doxepin binding [105] in human brain expressed as a percentage of the binding obtained in hypothalamus.

	% of specific hypothalamic binding obtained with:	
Region	*[^{3}H]mepyramine*	*[^{3}H]doxepin*
Cerebrum		
Cingulate cortex	422	321
Frontal cortex	478	273
Parietal cortex	444	237
Temporal cortex	377	336
Occipital cortex	200	188
Limbic system		
Hippocampus	155	94
Amygdala	311	299
Basal ganglia		
Caudate	133	76
Putamen	100	63
Globus pallidus	44	18
Thalamus	78	61
Midbrain	78	38
Pons	44	17
Cerebellum	28	20
Medulla	56	35
Hypothalamus	100	100

in the binding characteristics of the high-affinity [^{3}H]mepyramine-binding sites. Thus, the affinities of antagonists determined from inhibition of [^{3}H]-mepyramine binding in rat brain membranes do not all agree well with the values obtained in guinea-pig central and peripheral tissues [68, 99, 100], although in most respects the characteristics of [^{3}H]mepyramine binding are those expected for selective labelling of H_1-receptors [102]. Similar differences have now been reported in other species, including man [99, 105], rabbit [99], cat [103] and monkey [104]. The equilibrium dissociation constants obtained for a range of H_1-antagonists in several species are set out in *Table 2.5*. It is striking that for mepyramine, triprolidine and the isomers of chlorpheniramine the dissociation constants vary markedly between species. However, this is not the case for all antagonists and for promethazine and doxepin the values are very similar in monkey, rat, guinea-pig and man [68, 99, 100]. These observations suggest that histamine H_1-receptors in the brains of different species may

Table 2.5. SPECIES VARIATION IN THE AFFINITY OF H_1-RECEPTOR ANTAGONISTS FOR H_1-RECEPTORS IN MAMMALIAN BRAIN

Values represent equilibrium dissociation constants (nM) obtained from inhibition of [^{3}H]mepyramine- or [^{3}H]doxepin-binding obtained in membrane preparations from human [99,105], guinea-pig [61,68,98,106], rat [68,99,100], mouse [99] and rabbit [99] brain.

Antagonist	*Human*	*Guinea-pig*	*Rat*	*Mouse*	*Rabbit*
Mepyramine	1.0	0.8	9.1	3.2	3.8
Triprolidine	3.7	0.2	5.6	9.1	20
(+)-Chlorpheniramine	4.2	0.8	9.1	9.1	21
(−)-Chlorpheniramine	350	200	500	730	2100
Doxepin	0.4	0.1	0.1		
Promethazine	2.6	1.4	1.8		
Chlorpromazine	3.0	2.3	7.1		

not be structurally identical. Alternatively, it is possible that the marked species differences, for example, between guinea-pig and rat brain, are due to differences in the membrane environment of the receptor rather than due to different molecular forms of the H_1-receptor protein. However, in digitonin-solubilized preparations of the H_1-receptor [107, 108], the differences in binding properties of guinea-pig and rat brain are retained, suggesting that there are real differences in the receptor proteins [108]. In bovine and human cerebral cortex, however, radiation inactivation target size analysis indicated that the molecular size of the H_1-receptor was identical (160 kDa) [109]. Species differences in the potencies of triprolidine and (+)-chlorpheniramine have also been observed in membranes of the rat and guinea-pig brain using [^{3}H]doxepin, which has a similar affinity for H_1-receptors in these two species [108].

A range of H_1-agonists have been shown to displace [^{3}H]mepyramine binding in membranes prepared from guinea-pig cerebral cortex [51]. In general, much higher concentrations of agonists are required to displace [^{3}H]mepyramine binding than are required for displacement by H_1-antagonists. The most potent H_1-selective agonists inhibit the binding of 1 nM [^{3}H]mepyramine with IC_{50} values ranging from 33 to 100 μM, while those analogues which are almost devoid of H_1-agonist activity (e.g., 4-pyridylethylamine, N^{τ}-methylhistamine) have IC_{50} values in the 1 to 3 mM range [51]. Studies of agonist binding, however, are potentially more interesting than similar studies of antagonist binding, the pharmacological properties of which are dependent entirely on receptor occupancy, because they may give an insight into the mechanisms involved in receptor activation or desensitization.

The interaction of agonists, but not antagonists, with histamine H_1-receptors in guinea-pig brain labelled by [^{3}H]mepyramine can be regulated selectively by a range of different agents, including sodium ions, divalent cations, guanine nucleotides [110], *N*-ethylmaleimide [111] and 1,4-dithiothreitol [112]. Sodium ions and to a lesser extent lithium ions decrease the affinity of histamine and H_1-agonists for high-affinity [^{3}H]mepyramine-binding sites in guinea-pig whole brain homogenates by one order of magnitude [110]. GTP and the GTPase-resistant analogue, Gpp[NH]p, also decrease the affinity of histamine for H_1-binding sites in guinea-pig brain [110], suggesting that histamine H_1-receptors may interact with a guanine nucleotide regulatory protein. A similar observation has been made in human astrocytoma cells, where treatment with guanine nucleotides shifts the curve for agonist inhibition of [^{3}H]mepyramine binding to lower agonist concentrations [113]. This effect in astrocyte membranes is accompanied by an increase in the slope parameter (Hill coefficient) from a value less than unity (indicative of a heterogeneity of agonist binding) to a value not significantly different from the value (unity) expected for a simple interaction with a single class of binding sites [113]. Treatment of astrocytoma cells with pertussis toxin does not alter the ability of histamine to inhibit [^{3}H]mepyramine binding in the presence of guanine nucleotides [113]. This suggests that the guanine nucleotide regulatory protein linked to H_1-receptors in these cells is not equivalent to the N_i subunit of adenylate cyclase. The effect of sodium ions on H_1-agonist binding in guinea-pig brain is retained in digitonin-solubilized preparations, but the effect of GTP is lost [108].

In contrast to the results obtained with monovalent ions and guanine nucleotides, treatment of membranes with either *N*-ethylmaleimide [111] or 1,4-dithiothreitol [112] or inclusion of manganese (or magnesium) in the incubation medium [110] increases the binding affinity of H_1-agonists for [^{3}H]mepyramine-binding sites in guinea-pig brain. In the case of both *N*-ethylmaleimide and 1,4-dithiothreitol this effect is accompanied by a decrease in the slope index of histamine competition curves to a value less than unity [111, 112]. Two-site analysis of H_1-agonist inhibition of [^{3}H]mepyramine binding suggests that both the sulphydryl-alkylating agent, *N*-ethylmaleimide, and the disulphide-bond-reducing agent, 1,4-dithiothreitol, stabilize a proportion of the H_1-receptors in a high-affinity state [111, 112]. The differential effect of 1,4-dithiothreitol on the curves for inhibition of [^{3}H]mepyramine binding by histamine and mepyramine is shown in *Figure 2.6*.

Studies of the temperature dependence of the binding of [^{3}H]mepyramine to membrane fractions of guinea-pig cerebellum have indicated that, although the specific binding site capacity and binding affinity of mepyramine is little altered between 4 and 30 °C, there is a large change in the magnitude of the rate constants for association and dissociation [114]. At 4 °C the dissociation of

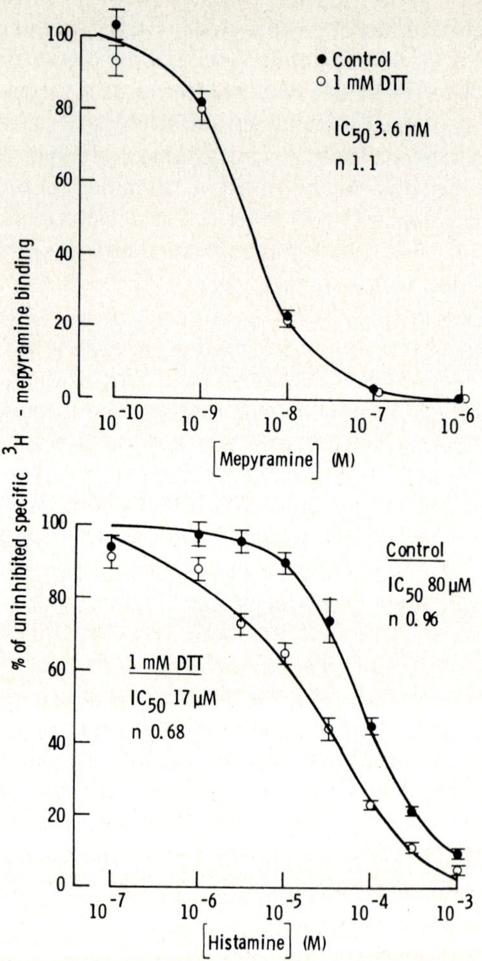

*Figure 2.6. Influence of 1 mM DTT on the inhibition of the binding of 1 nM [^{3}H]mepyramine by mepyramine and histamine in guinea-pig cerebellar membranes. Values for the IC_{50} estimate and Hill coefficient (*n*) were obtained by non-linear regression [112].*

[^{3}H]mepyramine from H_1-receptors is very slow and negligible over a 120 min period [114]. This slow dissociation at low temperatures has meant that [^{3}H]mepyramine has been of great utility for the autoradiographic localization of H_1-receptors in brain slices [115–117]. [^{125}I]Iodobolpyramine has a similar low dissociation rate at 25°C and has also been used for autoradiographic studies [98].

Several studies have attempted to localize the binding of [^{3}H]mepyramine to particular cell types. Kainate lesions of guinea-pig cerebellum have suggested that the H_1-receptor binding sites in the molecular layer are associated with neuronal cell populations [117]. However, other lesion studies in guinea-pig striatum, cerebral cortex and hippocampus suggest that a large proportion of the [^{3}H]mepyramine-binding sites may be associated with non-neuronal cell types such as glia or blood vessels [118]. In this respect it is interesting that substantial levels of histamine H_1-receptor binding sites have been located in purified preparations of microvessels from bovine cerebral cortex [119] and human astrocytoma cells [113].

A large number of psychotropic drugs have been reported to bind to histamine H_1-receptor binding sites in mammalian brain with high affinity. Of the neuroleptics studied, the phenothiazines such as chlorpromazine, fluphenazine and thioridazine, and the thioxanthines α- and β-flupenthixol were the most potent [99, 102, 106]. A large number of tricyclic antidepressants are very potent antagonists of H_1-receptor-mediated functional responses [67, 120] and bind with high affinity to [^{3}H]mepyramine-binding sites in guinea-pig, rat and human brain [99]. In particular, doxepin and amitriptyline are two of the most potent H_1-antagonists presently available [67]. The high affinity of doxepin and mianserin for histamine H_1-receptors has attracted attention to the possible use of tritiated versions of these agents as radioactive ligands for the H_1-receptor [94–96]. Unfortunately, [^{3}H]mianserin has a high affinity for 5-hydroxytryptamine 5-HT_2 receptors [96] and this limits its effective use as a probe for histamine H_1-receptors to those areas (e.g., guinea-pig cerebellum) which are virtually devoid of 5-HT_2 sites. [^{3}H]Doxepin has been used to a limited extent in both guinea-pig [94] and rat [94, 95] brain; however, as with [^{3}H]mianserin, the binding of [^{3}H]doxepin to brain membranes is complex and is complicated by the presence of a substantial secondary binding component [94, 95].

H_2-RECEPTOR ANTAGONISTS

[^{3}H]Cimetidine was introduced as a potential radioligand for the histamine H_2-receptor of mammalian brain in 1978 [121] and has been used as a probe for H_2-receptors by a number of different workers [121–125]. However, it is now clear that specific [^{3}H]cimetidine binding in guinea-pig and rat brain is to an imidazole recognition site rather than to the intended target, the histamine H_2-receptor [126–128]. It is, however, perhaps worth reviewing the evidence which led to the demise of [^{3}H]cimetidine as a selective ligand for H_2-receptors, since it illustrates the importance of fulfilling the criterion of specificity before equating binding sites with receptor recognition sites.

In both guinea-pig cerebral cortical membranes [121, 126, 128] and rat brain homogenates [123–125, 127] a saturable component of binding can be identified which binds [^{3}H]cimetidine with high affinity. Furthermore, in rat membranes there is a good quantitative correlation between the potencies of cimetidine, metiamide and burimamide as inhibitors of [^{3}H]cimetidine binding and their ability to inhibit the H_2-receptor-mediated chronotropic response in guinea-pig atrium [127]. However, it is striking that the potent, non-imidazole, H_2-antagonists tiotidine and ranitidine are virtually inactive (IC_{50} values > 1 mM) in displacing [^{3}H]cimetidine from its binding site [127, 128]. Moreover, a number of imidazole-containing compounds which are devoid of H_2-agonist or antagonist activity have been shown to be rather potent inhibitors of [^{3}H]cimetidine binding in guinea-pig cerebral cortex [128]. Cu^{2+}, Pd^{2+} and Ag^{2+} ions have been shown to increase significantly the specific [^{3}H]cimetidine-binding-site capacity in guinea-pig and rat brain and to alter the relative and absolute potencies of various H_2-receptor agonists and antagonists in displacing the tritiated ligand [124, 128, 129]. It has been suggested that under these conditions [^{3}H]cimetidine may be labelling a biologically relevant H_2-binding site [124]; however, this site still remains insensitive to non-imidazole compounds such as dimaprit, ranitidine and tiotidine in the presence of Cu^{2+} ions [128]. Taken together, the evidence clearly indicates that the high-affinity binding of [^{3}H]cimetidine to brain membranes does not represent selective labelling of histamine H_2-receptors and emphasizes the importance of using a range of antagonists, of widely differing chemical structure, in the pharmacological characterization of binding sites. [^{3}H]-Ranitidine [130] has also proved to be unsuitable as a radioligand for labelling H_2-receptors, since its binding in guinea-pig heart membranes is insensitive to inhibition by cimetidine and metiamide.

[^{3}H]Tiotidine has recently been reported to label selectively histamine H_2-receptors in guinea-pig brain [83], although an earlier attempt in rat and guinea-pig hippocampus was not successful [131]. In membranes prepared from guinea-pig cerebral cortex, [^{3}H]tiotidine has been shown to bind to a single saturable component with a dissociation constant similar to that derived from inhibition of the H_2-receptor-mediated chronotropic response in guinea-pig atrium and the stimulation of adenylate cyclase activity in guinea-pig gastric mucosa [83]. Furthermore, its binding is inhibited by a large number of H_2-receptor agonists and antagonists of diverse chemical structure [83, 132] and there is an excellent correlation between the ability of H_2-antagonists to inhibit [^{3}H]tiotidine binding and their ability to inhibit histamine-stimulated adenylate cyclase activity in gastric mucosa (*Figure 2.7*) [132]. Thus, in guinea-pig cerebral cortex, at least, there is good evidence that [^{3}H]tiotidine can label the

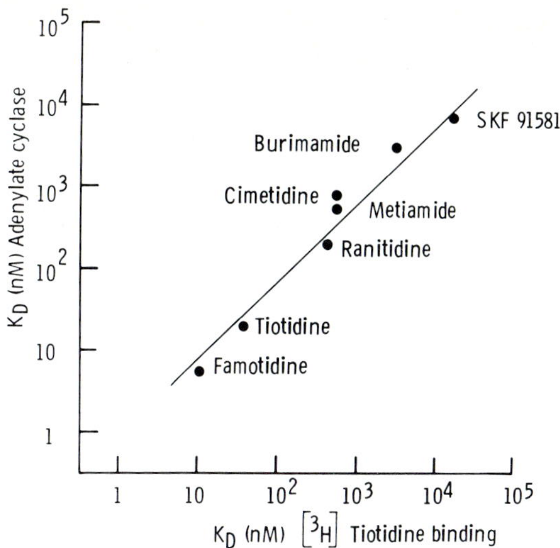

Figure 2.7. A comparison of the equilibrium dissociation constants (K_D) obtained for H_2-antagonists from inhibition of [^{3}H]tiotidine binding and from antagonism of histamine-stimulated adenylate cyclase activity [83].

H_2-receptor recognition site, although the extent of the nonspecific binding of [^{3}H]tiotidine in this tissue is very high [83]. Studies in other regions of guinea-pig brain indicate that highest levels of specific [^{3}H]tiotidine binding are observed in corpus striatum, cerebral cortex and hippocampus, with no detectable binding in pons-medullar or cerebellar membranes [133]. Labelling of peripheral tissues, which are known to possess functional H_2-receptors, has proved difficult with [^{3}H]tiotidine [132]. Thus, although H_2-binding has been demonstrated in homogenates of guinea-pig lung parenchyma [134], little success has been achieved in guinea-pig gastric mucosa or right atrium [132]. Moreover, in kidney membranes [^{3}H]tiotidine appears to label non-H_2-binding sites [132]. It therefore seems that an alternative radioligand is still required to explore the role of the H_2-receptor in the mammalian central nervous system with any degree of confidence.

[^{3}H]HISTAMINE

Several attempts have been made to label histamine receptors in mammalian brain using [^{3}H]histamine [123, 135–138]. These studies have shown that homogenates of rat brain possess high-affinity binding sites for [^{3}H]histamine, with an equilibrium dissociation constant for histamine in the nanomolar range. Early studies with H_1- and H_2-receptor antagonists concluded that the pharmacological specificity of the [^{3}H]histamine sites in rat brain was neither H_1 nor H_2 [123, 136]. A similar conclusion has also been drawn in studies of [^{3}H]histamine binding in guinea-pig gastric mucosal cells [139]. Furthermore, it is very striking that the high affinity of histamine (K_D = 7 nM) deduced in binding studies with [^{3}H]histamine in brain membranes is several orders of magnitude different from the low affinities of histamine for the H_1- and H_2-receptor deduced directly from binding studies with [^{3}H]mepyramine [51, 61] and [^{3}H]tiotidine [83] or indirectly from functional studies with irreversible H_1- and H_2-receptor antagonists [89, 140]. However, more recent work with a broader selection of H_2-receptor antagonists in homogenates of rat cerebral cortex has suggested that low concentrations of [^{3}H]histamine might label a high-affinity agonist state of the H_2-receptor [138, 141–143]. Thus, these latter authors, using a much lower concentration of [^{3}H]histamine (1.4 nM) than in previous studies, have reported an excellent correlation between the ability of a wide range of H_2-receptor antagonists to inhibit the specific binding of [^{3}H]histamine in rat cerebral cortical membranes and their ability to inhibit the H_2-mediated chronotropic response in guinea-pig right atrium [138, 142]. H_2-Agonists also inhibit [^{3}H]histamine binding in cerebral cortical membranes, but do so in a characteristic and biphasic manner which is not observed with compounds that lack H_2-agonist activity [138, 143].

Solubilization of the [^{3}H]histamine-binding sites in rat cerebral cortex [137] with digitonin appears to markedly change the binding characteristics of H_2-receptor agonists and antagonists [138]. Thus, although the binding parameters obtained for [^{3}H]histamine, itself, are similar in particulate and soluble preparations the correlation between [^{3}H]histamine binding and H_2-receptor antagonism is lost for H_2-antagonists following solubilization [138]. In solubilized preparations, H_2-agonist binding is also modified and reverts to a simpler situation which is consistent with labelling of a single class of binding sites [138]. It has been proposed that H_2-agonists and antagonists might bind at different sites in the membrane such that the inhibition of [^{3}H]histamine binding by H_2-antagonists reflects an allosteric interaction which is uncoupled following solubilization [138, 142, 143]. This suggestion would explain the marked change in potency of H_2-antagonists as inhibitors of [^{3}H]histamine binding following dissolution of the binding sites in digitonin, but is difficult to reconcile with the competitive H_2-receptor antagonism observed with these compounds in studies of adenylate cyclase activity [81, 144].

The relevance of the high-affinity [^{3}H]histamine-binding sites in rat cerebral cortex for H_2-receptor–effector coupling or other conformational changes (e.g., desensitization) remains to be established. Furthermore, it may be premature to rule out the possible involvement of other sites with high affinity for histamine, e.g., the putative H_3-receptor [49], in the overall binding profile of [^{3}H]histamine, particularly in the soluble preparations. However, whatever the exact rôle of these sites, it is interesting that the binding of [^{3}H]histamine can be regulated by guanine nucleotides in both particulate and soluble preparations [136, 137]; this finding is indicative of an interaction with a guanine-nucleotide regulatory protein such as the N_s protein linked to adenylate cyclase [145].

The studies with radioactive ligands, particularly [^{3}H]mepyramine and [^{3}H]tiotidine, outlined above have shown that there are binding sites in mammalian brain with the characteristics of H_1- and H_2-receptors, respectively. However, binding studies alone do not provide any information concerning the coupling of the detected recognition sites to functional responses. The observed regulation of H_1-agonist binding by guanine nucleotides may provide an indication of the coupling of this particular recognition site to a regulatory protein, but the demonstration of the operational integrity of the receptor-effector system in a given brain region requires the measurement of a functional response. In the remainder of this review, the properties of a number of histamine-stimulated biochemical responses will be examined. Particular attention will be focused on the effect of histamine-receptor stimulation on the production of the intracellular second messengers, adenosine 3′, 5′-(cyclic) monophosphate (cyclic AMP) [146], inositol trisphosphate [147] and diacylglycerol [148].

HISTAMINE AND CYCLIC AMP ACCUMULATION

A number of neurotransmitter receptors are closely linked to ion channels and produce fast (millisecond) changes in postsynaptic membrane potential. Others, however, can induce slower changes in membrane potential or other neuronal properties by processes involving the production of an intracellular chemical messenger. The intracellular messenger used by many neurotransmitters and neuromodulators is cyclic AMP, whose intracellular concentration is determined by the relative activities of the enzymes adenylate cyclase and phosphodiesterase, which are respectively responsible for synthesizing and degrading this cyclic nucleotide. The activity of adenylate cyclase can be regulated by stimulation of neurotransmitter receptors coupled to regulatory proteins which either stimulate or inhibit catalytic activity [145, 149–151]. An increase in cyclic AMP levels initiates changes in cell function by activating specific cyclic AMP-dependent protein kinases [152–154]. These phosphorylate key proteins (e.g., synapsin I) which may modify a wide range of processes including axoplasmic transport, membrane ion transport and neurotransmitter release [152–154].

Studies in brain slices have shown that histamine is one of the most powerful agents in stimulating cyclic AMP accumulation in the mammalian central nervous system [146]. Early studies in the rabbit showed that histamine elicited very large increases in the accumulation of cyclic AMP in cerebral cortex (12–16-fold increase above basal levels), hypothalamus (20–30-fold), brain stem (35-fold) and cerebellum (3–10-fold) [155, 156]. The effect in cerebral cortex has subsequently been studied by a number of other groups and increases in cyclic AMP accumulation as high as 74-fold above basal levels have been reported [82, 157, 158]. Furthermore, the response to histamine in rabbit cerebral cortical slices appears to depend on the age of the animal and increases markedly during the first 8 days postpartum before declining to adult levels [157].

Histamine-stimulated cyclic AMP accumulation has been extensively investigated in guinea-pig brain slices. In guinea-pig cerebral cortex and hippocampus, histamine produces large increases in the accumulation of cyclic AMP, of similar magnitude to that observed in rabbit cerebral cortex [51, 159–166]. Early studies indicated that the response to histamine in cortical slices could be effectively blocked by H_1-receptor antagonists such as (+)-bromopheniramine, promethazine and mepyramine [161, 163, 164]; however, it is notable that these antagonists were invariably applied at relatively high concentrations, at which nonspecific actions are possible. Other studies indicate that both H_1- and H_2-receptors may be involved in the cyclic AMP response to histamine in guinea-pig brain [162], although, as will be discussed later, the involvement of H_1-receptors is complex [51, 165, 166].

Synergistic interactions have been observed between histamine and noradrenaline on cyclic AMP accumulation in guinea-pig cerebral cortical slices [161, 167–169]. A similar interaction between histamine and noradrenaline has been observed in slices of guinea-pig hippocampus [61]. The interaction between histamine and noradrenaline in cerebral cortex is sensitive to inhibition by the adenosine receptor antagonist, theophylline [146, 170, 171]. This suggests that adenosine may be involved in this interaction. Interestingly, there is a report which suggests that histamine can increase the release of adenosine from guinea-pig cerebral cortical slices [172]. Furthermore, it is well established that adenosine-amine combinations have a much greater than additive effect on cyclic AMP accumulation in guinea-pig brain slices [146].

In rat and mouse cerebral cortical slices, histamine produces only marginal effects on cyclic AMP accumulation [146]. However, in rat cerebral cortex significant changes have been observed in response to histamine in the presence of a phosphodiesterase inhibitor such as isobutylmethylxanthine [146]. In neocortex this response to histamine is sensitive to inhibition by the H_2-receptor antagonist, metiamide [146]. The only other species studied so far, apart from the rabbit and guinea-pig, to show a marked sensitivity of the cyclic-AMP-generating system to histamine appears to be the chick [173, 174]. In chick cerebral cortical slices, histamine produces a large stimulation of cyclic AMP accumulation which is sensitive to inhibition by H_2-receptor antagonists [173, 174]. However, the properties of the receptor involved in cyclic AMP generation in the chick are somewhat different from those of the classical H_2-receptor-mediating peripheral responses in guinea-pig atrium [174] or adenylate cyclase activity in guinea-pig brain homogenates.

H_2-RECEPTOR-LINKED ADENYLATE CYCLASE

In mammalian cells, adenylate cyclase is a particulate enzyme which is localized in the plasma membrane [175]. This has meant that it is possible to investigate the effect of neurotransmitter receptor stimulation of adenylate cyclase activity in broken cell preparations provided that the necessary substrates and cofactors are supplied in the incubation medium. It is generally accepted that MgATP is the physiological substrate for the enzyme, although some other divalent cation-ATP complexes (e.g., Mn^{2+}) may be effective substrates in some cells [175]. The other essential ingredient for agonist stimulation or inhibition of adenylate cyclase activity is the guanine nucleotide, GTP [149–151, 176].

It is now generally agreed that agonist-induced regulation of adenylate cyclase activity involves three distinct functional units: the receptor; a transducer, which transmits information from the receptor to the active site of the enzyme; and the catalytic unit in which the active site of the enzyme resides [149–151, 176]. It is well established that the function of the transducer molecules is dependent upon the presence of the guanine nucleotide GTP [149, 150, 176]. Two guanine nucleotide regulatory components of adenylate cyclase (transducer molecules) have now been identified. One is responsible for stimulating (N_s) adenylate cyclase and can be activated by cholera toxin. It has been shown that adenylate cyclase is active while GTP is bound to the N_s-GTP binding component but reverts to an inactive state when GTP is hydrolysed to GDP [176, 177]. Cholera toxin inhibits the GTPase action of the N_s subunit and consequently stabilizes adenylate cyclase in an active conformation [176, 177]. The other GTP-regulatory protein (N_i) mediates receptor-induced inhibition of adenylate cyclase activity and its action can be prevented by

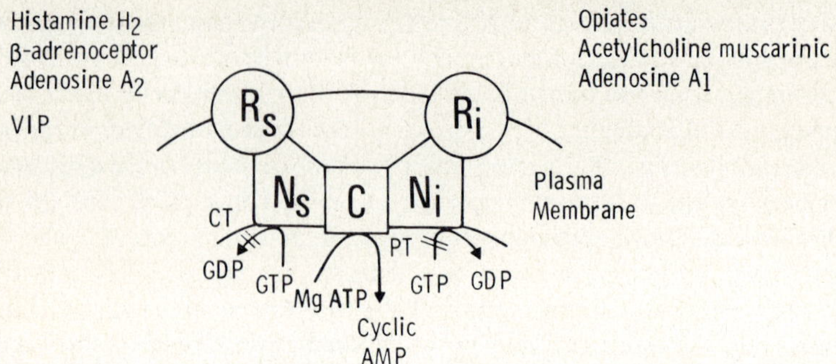

Figure 2.8. Model of bimodal regulation of adenylate cyclase by receptors (R) acting via the guanine regulatory proteins N_s and N_i to either stimulate or inhibit cyclic AMP production by the catalytic unit (C). The proposed sites of action of cholera toxin (CT) and pertussis toxin (PT) are indicated. VIP = vasoactive intestinal polypeptide. Modified from [176].

pertussis toxin from *Bordetella pertussis* [151, 176, 178]. Thus, the N_s and N_i GTP-binding proteins act as interfaces between neurotransmitter receptors and the catalytic unit of adenylate cyclase, and the ultimate action of a given receptor system on cyclic AMP production depends on which of the two distinct N proteins they regulate. The proposed model of receptor-adenylate cyclase signal transduction is illustrated in *Figure 2.8*. It seems a general rule that a given receptor only associates with one particular GTP-regulatory protein. At the present time, only histamine H_2-receptors have been shown to interact directly with adenylate cyclase via the N_s regulatory component [81, 175] and no inhibitory effect of histamine on adenylate cyclase activity has been reported.

A histamine-sensitive adenylate cyclase preparation was first described in homogenates of guinea-pig brain [179–183]. These studies showed that enzyme activity could be stimulated by low concentrations of histamine [179]. This response appears to be mediated by a typical histamine H_2-receptor, since it can be competitively antagonized by metiamide [179], cimetidine and a range of H_2-receptor antagonists [182, 183]. The potencies of these compounds as inhibitors of histamine-stimulated adenylate cyclase agree very well with their potencies as inhibitors of H_2-receptor-mediated responses in peripheral tissues [81, 175, 179, 182, 183]. Furthermore, the potent H_1-antagonist, mepyramine, inhibits the response to histamine only at high concentrations (K_D = 6.6 μM) which are known to antagonize peripheral H_2-receptors in guinea-pig atrium [182, 183]. Moreover, the relative potencies of histamine (100), dimaprit (172), N^{α}, N^{α}-dimethylhistamine (50), 4-methylhistamine (50) and 2-methylhistamine

(9) on broken cell preparation of guinea-pig hippocampus are similar to their potencies on peripheral H_2-receptors [81, 179, 182].

Histamine-stimulated adenylate cyclase activity is observed in only three of the regions so far examined in guinea-pig brain, namely hippocampus, cerebral cortex and striatum, and no effect has been observed in cerebellum, hypothalamus, thalamus, midbrain and pons-medulla [179]. Interestingly, this is exactly the distribution of H_2-receptors deduced from binding studies with [^{3}H]tiotidine [133].

An extensive biochemical characterization of the histamine-stimulated adenylate cyclase response has been performed in homogenates of guinea-pig dorsal hippocampus [180]. The response in this region is almost entirely dependent upon the presence of GTP [180]. In the presence of GTP, histamine causes a maximal stimulation of adenylate cyclase activity ranging from 1.7- to 3.0-fold above basal activity [180]. This effect of histamine appears to be due to an increase in the maximal velocity of the adenylate cyclase reaction, since the affinity of the enzyme for the substrate MgATP is not altered by the presence of histamine [180]. It has been proposed that activation of the catalytic unit of adenylate cyclase by the N_s regulatory protein is dependent upon the presence of free Mg^{2+} ions acting at a low-affinity allosteric site [176]. In keeping with this hypothesis, it has been shown that in guinea-pig hippocampus, histamine increases the affinity of this allosteric site for free Mg^{2+} ions and this may be an important component of the mechanism by which histamine increases the velocity of the adenylate cyclase reaction [180].

Histamine-stimulated adenylate cyclase activity has been reported in brain regions of a number of other species, including: rat neocortex [179], hippocampus [182] and hypothalamus [184]; rabbit whole brain [185] and cerebral cortex [158, 186] and monkey hippocampus and frontal cortex [187]. Unfortunately, the receptor mediating the stimulation of adenylate cyclase by histamine has not been fully characterized in these species. Thus, in mammalian brain it is only the histamine-stimulated adenylate cyclase of guinea-pig cerebral cortex and hippocampus which has been clearly shown to be mediated by H_2-receptors. However, similar characterizations have been made in guinea-pig and human ventricle [188–191] and guinea-pig gastric mucosa [83] and these studies have confirmed the rôle of H_2-receptors in stimulating adenylate cyclase activity in broken cell preparations.

The H_2-stimulated adenylate cyclase response is sensitive to inhibition by a range of psychotropic drugs. A large number of neuroleptic drugs are effective inhibitors of histamine-stimulated cyclase activity in rabbit cerebral cortex [158, 192] and guinea-pig hippocampus and cerebral cortex [81, 193]. The most potent compounds in homogenates of guinea-pig hippocampus appear to be the phenothiazine neuroleptics, which include chlorpromazine, fluphenazine and thioridazine [193]. However, the ability to inhibit H_2-stimulated cyclase activity is shared by a number of non-phenothiazine compounds, including thiothixene, clozapine and haloperidol [193]. It has been suggested, however, that only chlorpromazine shows competitive antagonism of this response and all other neuroleptics inhibit adenylate cyclase activity in a non-competitive manner [81]. D-Lysergic acid diethylamide (D-LSD) and 2-bromo-LSD have both been shown to antagonize competitively H_2-linked adenylate cyclase activity in guinea-pig hippocampus [182]. D-LSD has an apparent dissociation constant, K_D, of 1 μM in this assay, while the 2-bromo derivative is approximately 10-times more potent (K_D = 0.07 μM) than D-LSD or indeed many of the classical H_2-antagonists, such as cimetidine and metiamide [182].

A great deal of interest has been generated by the finding that a large number of structurally diverse antidepressant drugs share the ability to inhibit histamine-sensitive adenylate cyclase in homogenates of guinea-pig cerebral cortex and hippocampus [81, 181, 193]. Thus, representative tricyclic and tetracyclic antidepressants, with very different ring structures, have been shown to be potent and competitive antagonists of the H_2-linked cyclase in broken cell preparations of mammalian brain. These include doxepin, mianserin, iprindole, amitriptyline and imipramine [181, 193].

Indeed, on the basis of the cyclase assay, amitriptyline is one of the most potent H_2-antagonists presently available (K_D 0.05 μM) [181, 193]. The dissociation constants for many of the antidepressants are sufficiently low to suggest that the activation of adenylate cyclase by histamine may be inhibited by therapeutically effective concentrations of these compounds [81, 193]. This biochemical action of the tricyclic and tetracyclic antidepressants may thus represent part of the molecular basis of clinical antidepressant activity [193]. It should also be noted that most tricyclic and tetracyclic antidepressants are much more potent inhibitors of H_1-receptor responses [67, 120] than they are of the H_2-cyclase response, and this property of these compounds may mediate the sedative actions of these drugs [71, 73, 75].

The H_2-linked adenylate cyclase of guinea-pig brain can also be inhibited by a range of H_1-receptor antagonists [175, 183]. In every case, much higher concentrations are needed than are necessary for peripheral H_1-receptor antagonism. However, certain of these compounds, most notably promethazine and cyproheptadine, are more potent inhibitors of histamine-stimulated adenylate cyclase activity than is cimetidine [182, 183]. These results are at variance with the low H_2-antagonist potency of promethazine obtained on peripheral H_2-receptors in guinea-pig atrium [175, 194]. Furthermore, histamine-stimulated adenylate cyclase activity in guinea-pig hippocampus appears also to be competitively and potently inhibited by a range of inhibitors of imidazole *N*-methyltransferase such as quinacrine, amodiaquine and metoprine [183].

The anomalously high affinities of certain H_1-antagonists, e.g., promethazine, and other non-H_2-receptor antagonists, determined in adenylate cyclase measurements in broken cell preparations, has prompted two research groups to re-evaluate the H_2-mediated cyclic AMP response in intact cellular preparations [195, 196]. Studies in guinea-pig hippocampal slices and dissociated cell preparations showed that the potencies of the classical H_2-receptor antagonists, cimetidine, ranitidine, metiamide and tiotidine, were similar in intact and broken cell preparations [195, 196]. However, the values obtained for H_1-antagonists, antidepressants, neuroleptics and imidazole *N*-methyltransferase inhibitors differed markedly between the two preparations

Table 2.6. COMPARISON OF THE POTENCY OF INHIBITORS OF H_2-RECEPTOR-MEDIATED CYCLIC AMP ACCUMULATION IN BRAIN SLICES, DISSOCIATED CELLS AND HOMOGENATES OF GUINEA-PIG HIPPOCAMPUS

Values represent measurements made on H_2-mediated adenylate cyclase activity in homogenates of guinea-pig hippocampus [182,183,193], on impromidine-stimulated cyclic AMP accumulation in hippocampal slices [195] and on H_2-mediated cyclic AMP accumulation in dissociated hippocampal tissue [196].

	Inhibition constant (K_i, μM)		
Antagonist	*Homogenates*	*Slices*	*Dissociated tissue*
H_2-antagonists			
Cimetidine	0.9	0.6	0.5
Metiamide	1.0	0.8	
Tiotidine	0.03		0.03
H_1-antagonists			
Promethazine	0.03	3.0	
Mepyramine	2.2		no effect
Cyproheptadine	0.04	5.7	
Diphenhydramine	0.6		11
Antidepressants			
Imipramine	0.2	> 10	3.3
Amitriptyline	0.05	3.5	1.9
Doxepin	0.2		1.4
Mianserin	0.07	10.0	2.8
Miscellaneous			
Chlorpromazine	0.04	5.9	3.0
Haloperidol	0.08	> 10	29
Quinacrine	0.3	> 10	

[195, 196]. For most of these latter compounds, the estimated dissociation constants in whole cell preparations were at least two orders of magnitude greater than the values obtained in brain homogenates (*Table 2.6*). In particular, the antidepressant drugs appear to be rather weak H_2-receptor antagonists in intact cell systems and this result weakens the argument that these compounds derive their clinical efficacy from blockade of H_2-receptors in mammalian brain.

The reason for the discrepancy in the discriminatory properties of H_2-receptors mediating cyclic AMP accumulation in slices or dissociated cells and homogenates is unclear. It is possible that the H_2-receptor recognition sites are damaged during membrane preparation or that the receptor adopts a modified conformation in the presence of relatively high concentrations of compounds such as ATP, Mg^{2+}, GTP and EGTA which are necessary for optimal adenylate cyclase activity [195, 196]. In this respect, it is interesting that a preliminary report has suggested that [^{3}H]tiotidine binding to H_2-receptors in guinea-pig cerebral cortex is affected by mono- and divalent cations and that the demonstration of specific [^{3}H]tiotidine in lung parenchyma is very dependent upon the experimental conditions [132]. An alternative explanation is that in homogenates, antidepressants and drugs such as quinacrine have access to parts of the receptor-effector system that are not normally available to the drugs in intact cells. Thus, in broken cell preparations these drugs may become potent inhibitors of histamine-stimulated adenylate cyclase activity by virtue of an allosteric action on regulatory units of the receptor-effector system. This latter explanation is particularly intriguing in view of the postulated allosteric inhibition of [^{3}H]histamine binding by H_2-antagonists in homogenates of rat cerebral cortex [138, 142, 143]. However, as mentioned above, this explanation is not compatible with the reported competitive nature of the inhibition of adenylate cyclase activity by most compounds [181–183, 193], although it has been noted [195] that this explanation cannot be entirely excluded in view of the complex interactions of histamine with GTP regulatory proteins and free Mg^{2+} ions leading to stimulation of adenylate cyclase activity [180].

H_1-RECEPTOR-MEDIATED POTENTIATION OF CYCLIC AMP ACCUMULATION

Studies of histamine-stimulated adenylate cyclase activity in broken cell preparations of both peripheral and central tissues suggest that the effect of histamine on cyclic-AMP-generating systems is exclusively associated with H_2-receptor stimulation [83, 179–183, 188–191]. The data obtained with selective H_2-receptor antagonists in slices of guinea-pig hippocampus [165]

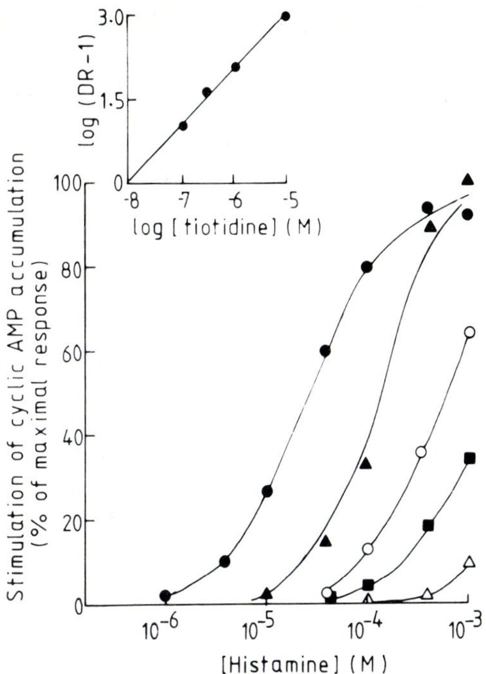

Figure 2.9. Inhibition by 3.2 × 10^{-8} M (▲), 8 × 10^{-8} M (0), 8 × 10^{-7} M (■) and 8 × 10^{-6} M (△) tiotidine of the histamine-induced accumulation of cyclic AMP in slices of rabbit cerebral cortex; (●) control curve. The inset represents a Schild plot of the same data [82].

and rabbit cerebral cortex [82] suggest that this is also true of the cyclic AMP response to histamine obtained in brain slices. Thus, all of the H_2-antagonists tested in these preparations produced parallel displacements of the concentration response curves for histamine to higher agonist concentrations, consistent with competitive antagonism of a homogeneous population of H_2-receptors [82, 165] (*Figure 2.9*). Furthermore, Schild analysis of the antagonist data gave straight lines with slope parameters not significantly different from unity [82]. This is the value expected for a competitive interaction with a single class of receptor sites. The values obtained for the dissociation constants of the H_2-antagonists from cyclic AMP measurements in brain slices also agreed exceedingly well with the values obtained on a typical H_2-mediated response in peripheral tissues [82, 165].

Studies with H_1-receptor antagonists, however, suggest that the response to histamine in guinea-pig hippocampus and rabbit cerebral cortex may be more complex and partly involve H_1-receptor stimulation [82, 165]. In hippocampal slices, the concentration response curve for histamine is modified in a complex fashion by the H_1-selective antagonist, mepyramine [165], such that the response to histamine in low concentrations remains essentially unaltered while those to the amine at concentrations above 10 μM appear to be inhibited in a competitive fashion. A similar result is observed in rabbit cerebral cortex [82]. This complex effect of mepyramine, coupled with the fact that H_2-antagonists can completely inhibit the cyclic AMP response to histamine in these brain regions, suggests that, if an H_1-receptor-mediated effect is present, it is dependent upon the prior or simultaneous activation of an H_2-receptor component of the response.

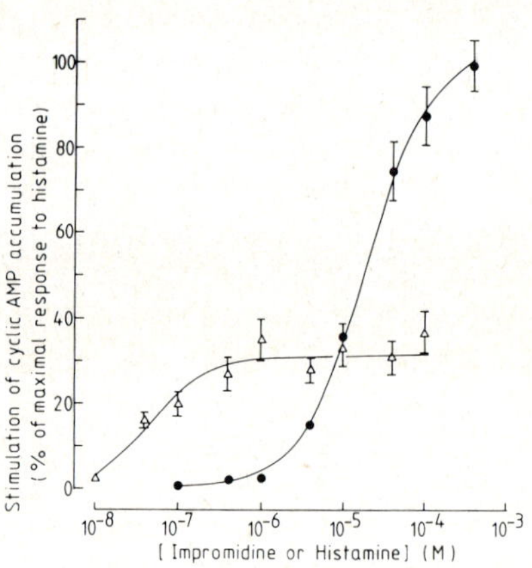

Figure 2.10. Effect of histamine (●) and the H_2-selective agonist impromidine (△) on cyclic AMP accumulation in slices of rabbit cerebral cortex. Values are expressed as a percentage of the maximum response to histamine [82].

Support for an interaction between two components in the final cyclic AMP response to histamine in guinea-pig hippocampal and rabbit cerebral cortical slices has been provided with H_1- and H_2-selective agonists. In hippocampal slices the H_2-selective agonists, impromidine and dimaprit, produced maximal cyclic AMP responses which were much less than the maximal response elicited by histamine [165, 195]. The extent of the stimulation appears to vary between experiments, ranging from 24% [165] to 62% [195] of the response to histamine. However, the extent of the cyclic AMP accumulation elicited by

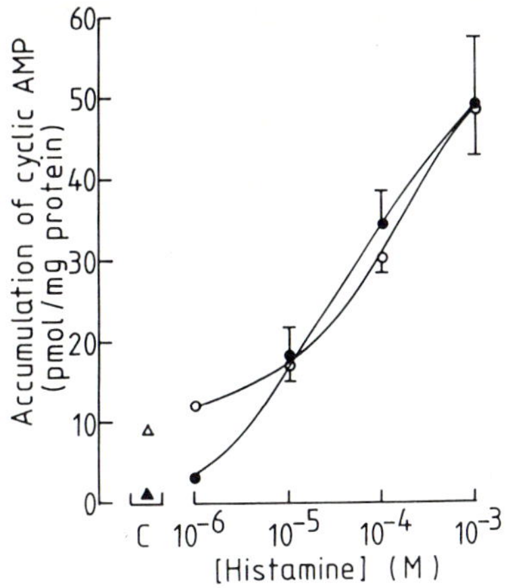

Figure 2.11. Concentration-response curves for histamine-stimulated cyclic AMP accumulation obtained in the presence (○) and absence (●) of 1 μM impromidine in slices of rabbit cerebral cortex [82]. The basal accumulation of cyclic AMP obtained in the presence (△) and absence (▲) of impromidine is indicated at C.

impromidine and dimaprit in this tissue was almost the same [195]. In rabbit cerebral cortical slices, impromidine similarly produces a maximal response which is substantially less than that obtained with histamine, being 31% of the response to 0.4 mM histamine (*Figure 2.10*) [82]. Studies with histamine and impromidine in combination suggest that the low maximal response obtained with H_2-agonists is due to a selective stimulation of the H_2-component of the cyclic AMP response to histamine, rather than a consequence of partial agonist properties [82]. *Figure 2.11* shows the data from an experiment in rabbit cerebral cortex which demonstrates that in the presence of a maximally effective concentration of impromidine, the upper portion of the concentration-response curve for histamine is not significantly different from that obtained in the absence of the selective H_2-agonist.

The H_1-agonist, 2-thiazolylethylamine, stimulates cyclic AMP accumulation in guinea-pig hippocampal slices only at relatively high concentrations [165]. The relative potency of 2-thiazolylethylamine in this preparation '7' with respect to histamine (= 100) [165] is intermediate between the values of 26 and 0.3 expected for stimulation of H_1- and H_2-receptors, respectively, and suggests that much of the response is dependent upon H_2-receptor stimulation. However, following maximal stimulation of the H_2-component with dimaprit, 2-thiazolylethylamine further elevates cyclic AMP accumulation and the relative potency obtained, 23, is similar to the value obtained for H_1-receptor stimulation in guinea-pig ileum [165]. A similar result was obtained with 2-thiazolylethylamine in rabbit cerebral cortical slices [82], where the relative potency of the H_1-agonist was substantially increased following maximal activation of H_2-receptors with impromidine. These data suggest that in both tissues visualization of the H_1-response is dependent upon prior or simultaneous activation of adenylate cyclase activity by H_2-receptor stimulation and that the primary rôle of the H_1-component of the cyclic AMP response is to amplify the effect of H_2-receptor stimulation.

The augmentation of the cyclic AMP response to H_2-agonists, such as impromidine and dimaprit, by histamine and H_1-agonists is completely sensitive to inhibition by H_1-receptor antagonists [82, 165]. Thus, potentiation of the cyclic AMP response to impromidine or dimaprit in slices of rabbit cerebral cortex and guinea-pig hippocampus represents a useful assay for H_1-receptor stimulation in mammalian brain and has been used successfully to screen for H_1-agonist and antagonist activity in these two brain regions [52, 82, 175, 195, 197].

In slices of guinea-pig cerebral cortex and hippocampus, combinations of adenosine and histamine produce much greater than additive effects on cyclic AMP accumulation [51, 146, 164, 166]. In cerebral cortical slices, there appears to be little or no direct stimulation of cyclic AMP accumulation by histamine acting on H_2-receptor [161, 163, 166]. Thus, the response to histamine alone can be completely antagonized by the H_1-antagonist mepyramine but not by the H_2-antagonist cimetidine. Indeed, the response to histamine alone appears to be largely due to augmentation of the response to endogenous adenosine, since the response can be markedly reduced by addition of the adenosine-metabolizing enzyme, adenosine deaminase [166]. It therefore seems that, in slices of guinea-pig cerebral cortex, H_1-receptor stimulation leads to an indirect potentiation of the direct effect of adenosine A_2 [198] receptor activation on adenylate cyclase activity.

Histamine and 2-thiazolylethylamine, but not dimaprit, augment the cyclic AMP response to exogenously applied adenosine. Furthermore, the response to H_1-agonists in this brain region can be competitively antagonized by a range of H_1-receptor antagonists [166]. A detailed comparison of the affinities of histamine H_1-antagonists determined from inhibition of this functional response with those obtained from [^{3}H]mepyramine binding has been reported [166]. For all of the antagonists tested there was a good quantitative agreement between the affinities obtained from inhibition of the histamine-induced augmentation of the cyclic AMP response to adenosine and those determined from binding studies with [^{3}H]mepyramine in guinea-pig brain or from inhibition of histamine-induced contractile activity in guinea-pig ileum. Thus, the properties of functional peripheral and central H_1-receptors, as mirrored by antagonist affinities, are similar.

This response in guinea-pig cerebral cortex has also been used to evaluate H_1-agonist activity in the mammalian central nervous system [51]. These studies have shown that, while histamine, 2-thiazolylethylamine and N^{α}-methylhistamine act as full H_1-agonists in this tissue, 2-methylhistamine and N^{α}, N^{α}-dimethylhistamine produce only 46% of the maximum response to histamine. Thus, 2-methylhistamine and N^{α}, N^{α}-dimethylhistamine appear to act as partial agonists in this system. 2-Pyridylethylamine and its *N*-methyl derivative, betahistine, also seem to have partial agonist activity in guinea-pig cerebral cortex [51]. A similar finding has been reported for betahistine in guinea-pig hippocampal slices for the augmentation of impromidine-stimulated cyclic AMP production [52]. The relative potencies of those compounds which induce an appreciable increase in cyclic AMP production agree reasonably well with the relative potencies obtained for peripheral H_1-receptor responses. Furthermore, there is a surprising agreement between the EC_{50} values for the H_1-elicited cyclic AMP response in cerebral cortex and the IC_{50} values obtained from binding studies with [^{3}H]mepyramine [51]. This suggests that the spare receptor reserve for the cyclic AMP response in this tissue is relatively small and probably explains why a number of compounds express partial agonist activity in guinea-pig cerebral cortex but act as full agonists in guinea-pig ileum where the receptor reserve is much larger.

In rabbit cerebral cortical and guinea-pig hippocampal slices, where there is a significant direct H_2-effect on cyclic AMP accumulation, a synergy between histamine and adenosine can also be observed [82, 164]. Furthermore, an H_1-receptor augmentation of the cyclic AMP response to adenosine can be observed following complete antagonism of the H_2-receptor response [82]. Thus, it appears that an indirect H_1-effect on cyclic AMP accumulation can be observed following stimulation of a number of different receptor systems (e.g., histamine H_2, adenosine A_2) directly coupled to the N_s unit of adenylate cyclase. In keeping with this hypothesis is the recent report that in slices of mouse cerebral cortex histamine H_1-receptor stimulation can augment the cyclic AMP response to vasoactive intestinal polypeptide (VIP) [199].

Similar, but smaller, potentiating effects of H_1-receptor stimulation on cyclic AMP responses to impromidine or adenosine have been reported in dissociated brain cells [196] and a preparation of vesicles from postsynaptic membranes of guinea-pig cerebral cortex [200–204]. This latter particulate preparation contains vesicular entities which maintain a degree of metabolic and functional integrity and have transmembrane potentials in the −58 to −78 mV range

[201, 203]. Just as in slices, the H_1-receptors have no direct effect on the vesicular cyclic-AMP-generating system, but function by augmenting the direct effect of β-adrenergic, H_2-histaminergic or adenosine A_2-receptor stimulation on adenylate cyclase activity [201].

Little is known of the intracellular events involved in the augmentation of cyclic AMP accumulation elicited by H_1-receptors in mammalian brain slices. However, it seems certain that another second messenger is involved, since the effect is not observed in membrane preparations [81, 205]. Calcium appears to be important for the response, since removal of external calcium reduces H_1-receptor-mediated cyclic AMP accumulation in guinea-pig cerebral cortical slices [206]. Inositol phospholipid breakdown or its products (inositol trisphosphate and diacylglycerol) may also be involved, since H_1-receptor stimulation is accompanied by an accumulation of inositol phosphates in slices of guinea-pig cerebral cortex [60, 207, 208]. Inositol trisphosphate may then

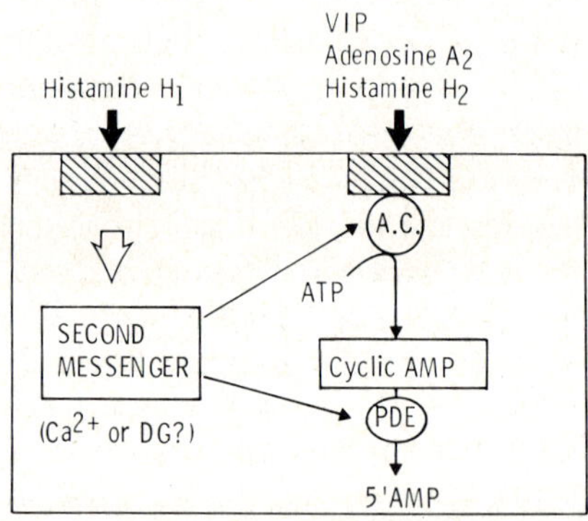

Figure 2.12. A scheme for the indirect potentiation of cyclic AMP accumulation elicited by H_1-receptor stimulation. Activation of H_1-receptors leads to the production of an intracellular second messenger (e.g., Ca^{2+} ions or diacylglycerol, DG) which subsequently acts on adenylate cyclase (AC) or phosphodiesterase (PDE) to amplify the cyclic AMP response to agonists acting on receptors directly coupled to adenylate cyclase.

mobilize intracellular calcium to produce a cytoplasmic calcium signal as in peripheral tissues [148]. Alternatively, the inositol phosphates and diacylglycerol may affect cyclic AMP metabolism directly, either to potentiate adenylate cyclase or inhibit phosphodiesterase activity. In rat pinealocytes, analogues of diacylglycerol, for example, have been shown to potentiate cyclic

AMP accumulation by activation of protein kinase C [209]. Furthermore, in vesicles prepared from guinea-pig cerebral cortex, phorbol esters which mimic the action of diacylglycerol appear to augment the cyclic AMP accumulation elicited by the adenosine analogue, 2-chloroadenosine [210]. A hypothetical scheme for the indirect effect of H_1-receptor stimulation on cyclic AMP accumulation is given in *Figure 2.12*. According to this scheme, H_1-receptor stimulation leads to the production of a second messenger which can either (1) increase the efficiency of the coupling between the directly acting receptor (e.g., histamine H_2 or adenosine A_2) and the N_s unit of adenylate cyclase, or (2) inhibit the activity of phosphodiesterase and consequently reduce the breakdown of cyclic AMP. Alternatively, H_1-receptor activation may release another neurotransmitter or neuromodulator which produces one of these two effects.

Recent electrophysiological studies in rat and guinea-pig hippocampal slices have begun to clarify the link between raised cellular cyclic AMP levels and effects on neuronal firing [211–213]. In rat hippocampal slices, stimulation of H_2-receptors with histamine or impromidine produces a small depolarization of pyramidal cells in the CA1 region of the hippocampus and a marked potentiation of neuronal excitation elicited by excitatory amino acids, intracellular current injection or synaptic stimulation [214–216]. Single action potentials or, particularly, bursts of spikes elicited by excitatory stimulation (e.g., with glutamate) in CA1 pyramidal cells are normally followed by a long-lasting after-hyperpolarization which keeps the cells away from their firing threshold for several seconds [213–216]. This effect is thought to be due to activation of a calcium-dependent potassium current which can be inhibited by H_2-receptor stimulation [213–216]. Blockade of this current by histamine or impromidine consequently leads to a profound potentiation of excitatory signals. This effect of H_2-agonists can be mimicked by intracellular application of cyclic AMP, or extracellular addition of 8-bromo-cyclic AMP [213, 217]. Furthermore, the effect of histamine on population spikes can be potentiated and prolonged by the phosphodiesterase inhibitor, RO 20-1724 [213]. In guinea-pig hippocampal slices, histamine also increases the firing rate of pyramidal cells in the CA3 region [212]. It remains to be established, however, whether a similar mechanism to that operating in the rat is involved in the guinea-pig.

If raised cyclic AMP levels are an important mechanism by which neurotransmitters and neuromodulators can regulate neuronal excitability in the higher centres of the brain, then the indirect effects of H_1-receptor stimulation on the accumulation of cyclic AMP observed in guinea-pig and rabbit brain slices provide a method for amplifying this regulatory control. Indeed, histamine, which is contained in neurones projecting diffusely over large areas of the central nervous system, could produce metabolic 'hot spots' of cyclic AMP accumulation by amplifying (*via* H_1-receptors) the actions of more localized, directly acting neuromodulators such as VIP [199]. This would be a way of controlling neuronal excitability in very localized brain regions without affecting neighbouring brain areas.

A striking feature of binding studies with [^{3}H]mepyramine in rabbit, rat and guinea-pig brain is that there appear to be marked species differences in the structure of histamine H_1-receptors in mammalian brain. A comparison of the properties of the H_1-receptor-mediated potentiation of cyclic AMP accumulation in guinea-pig and rabbit cerebral cortex also seems to support this contention [82]. Thus, the dissociation constants obtained for all H_1-antagonists tested in rabbit cerebral cortex are consistently larger than those obtained in guinea-pig ileum or cerebral cortex. In the case of triprolidine, the ratio of the antagonist potencies obtained in rabbit cerebral cortex and guinea-pig tissues is as large as 800. Similar low potencies for H_1-antagonists have been reported for histamine-stimulated glycogenolysis in mouse cerebral cortex (mepyramine K_D = 10 nM; triprolidine K_D = 5.6 nM) [218] and for histamine-stimulated inositol phosphate accumulation in rat cerebral cortex (mepyramine K_D = 2.8 nM) [219]. These data suggest that the species differences in the structure of the H_1-receptor first observed in binding studies with [^{3}H]mepyramine can now be extended to functional H_1-receptors.

HISTAMINE AND INOSITOL PHOSPHOLIPID HYDROLYSIS

Stimulation of histamine H_1-receptors can lead to a variety of biological actions, such as smooth muscle contraction, augmentation of cyclic AMP accumulation and glycogenolysis. However, the H_1-mediated glycogenolytic and cyclic AMP responses in the central nervous system appear to be restricted to certain brain regions of particular species and are not present in all of the brain regions that possess H_1-receptor-binding sites [146, 218]. For many of the cellular responses to H_1-receptor stimulation there is evidence that these are secondary to an increase in the intracellular concentration of free calcium ions [206, 218, 220]. It is therefore possible that calcium mobilization is a universal consequence of H_1-receptor activation and that the demonstration of a particular response in a given tissue depends upon the availability of the intracellular machinery (e.g., calcium-dependent protein kinases, substrate proteins) necessary to elicit that response.

An early event which appears to be associated with the activation of many

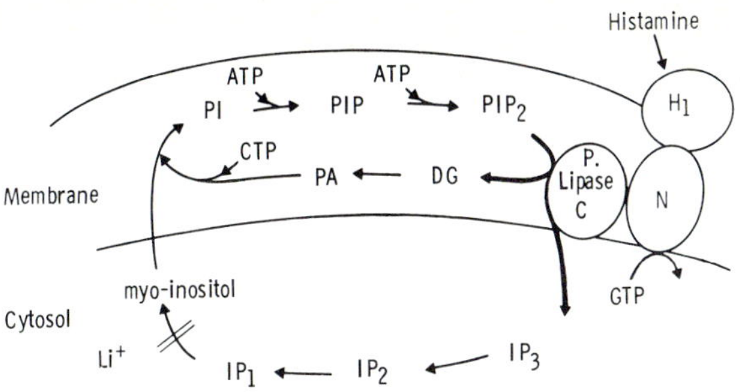

Figure 2.13. Histamine H_1-receptor-mediated inositol phospholipid hydrolysis. Stimulation of H_1-receptors leads to activation of a phospholipase C, probably via *a guanine-nucleotide regulatory protein (N), which catalyses the hydrolysis of phosphatidylinositol 4,5-bisphosphate (PIP_2) to give inositol trisphosphate (IP_3) and 1,2-diacylglycerol (DG). IP_3 is then broken down by phosphatases to eventually yield free* myo-*inositol. Lithium ions can inhibit the conversion of inositol 1-phosphate (IP_1) to* myo-*inositol. Free inositol then interacts with CDP-diacylglycerol, formed by a reaction between phosphatidic acid (PA) and CTP, to yield phosphatidylinositol (PI). Phosphorylation of PI by kinases completes the lipid cycle by reforming PIP_2. Modified from [147, 148].*

calcium-mobilizing receptors is the hydrolysis of the inositol phospholipids, phosphatidylinositol, phosphatidylinositol 4-phosphate and phosphatidylinositol 4,5-bisphosphate [147, 148, 221, 222]. Recent studies in peripheral tissues have suggested that it is the hydrolysis of phosphatidylinositol 4,5-bisphosphate which is the initial receptor-mediated event elicited by calcium-mobilizing receptors [147, 222–225]. A scheme for agonist-stimulated inositol phospholipid hydrolysis is illustrated in *Figure 2.13*. Briefly, activation of the calcium-mobilizing receptor leads to the hydrolysis of phosphatidylinositol 4,5-bisphosphate to form the lipid-soluble 1,2-diacylglycerol and the water-soluble product, inositol 1,4,5-trisphosphate. The inositol 1,4,5-trisphosphate is then subsequently broken down by phosphatases to yield inositol 1,4-bisphosphate, inositol 1-phosphate and finally free *myo*-inositol. The free inositol then interacts with CDP-diacylglycerol to reform phosphatidylinositol, which is in turn phosphorylated by kinases to produce the polyphosphoinositides and complete the cycle. The CDP-diacylglycerol is the product of a lipid cycle which channels 1,2-diacylglycerol back to phosphatidylinositol *via* phosphatidic acid. Evidence is accumulating that the catalytic activity of the enzyme (phospholipase C) which hydrolyses the inositol phospholipids is regulated by a guanine-nucleotide-binding protein [226, 227], and indeed, guanine nucleotides have been shown to stimulate the production of inositol trisphosphate in membrane fractions of mammalian brain [228].

It is now generally accepted that both of the products of phosphatidylinositol 4,5-bisphosphate hydrolysis can function as intracellular second messengers. 1,2-Diacylglycerol can affect a variety of intracellular processes by activation of protein kinase C [148, 229, 230]. Inositol 1,4,5-trisphosphate, on the other hand, has been shown to release calcium ions from non-mitochondrial stores in a number of peripheral tissues and may thus be the link between the receptor and the intracellular calcium store in many pharmacological responses [231–233]. Furthermore, it remains a possibility that inositol phospholipid hydrolysis may also have a rôle in calcium gating [221, 234]. If inositol phospholipid metabolism is closely coupled to receptor-mediated calcium mobilization, then this response may be a more general consequence of H_1-receptor stimulation than other H_1-responses.

Initial studies in rat brain showed that intracisternal injection of histamine stimulated the incorporation of $[^{33}P]P_i$ into inositol phospholipids [235, 236]. Studies with selective H_1- and H_2-agonists and antagonists further suggested that this response was mediated by H_1-receptors [236]. Increased turnover of inositol phospholipids is nevertheless a rather indirect measure of the initial receptor-mediated event, namely breakdown of phosphatidylinositol 4,5-bisphosphate. A more direct and sensitive method of monitoring inositol phospholipid breakdown is now available, however, following the discovery that Li^+ ions can cause an accumulation of inositol 1-phosphate in slices of rat cerebral cortex and parotid gland as a result of inhibition of inositol 1-phosphatase [237].

Histamine was first shown to stimulate the accumulation of $[^3H]$inositol 1-phosphate in lithium-treated slices of rat cerebral cortex preincubated with *myo*-$[^3H]$inositol [219, 237]. These studies showed that the response to histamine could be inhibited by mepyramine but not the H_2-antagonist, cimetidine. However, the response to histamine in the rat is rather small, producing only a 50–100% increase in the accumulation of $[^3H]$inositol 1-phosphate over basal levels. For this reason, most of the studies on histamine-induced inositol phosphate accumulation have been performed in slices of guinea-pig brain, where the response to histamine is somewhat larger. Two approaches have been adopted for measuring $[^3H]$inositol phosphate accumulation in guinea-pig brain slices. In the first (continuous labelling), slices have been preincubated with *myo*-$[^3H]$inositol and lithium ions, normally for 60 min, before incubation with histamine and various antagonist drugs in the continued presence of the radioactive label [60, 207]. The other approach has been to prelabel slices with *myo*-$[^3H]$inositol, and then extensively wash out the radiolabel before exposure of the slices to lithium ions and histaminergic drugs (pulse labelling) [59, 238].

In guinea-pig brain, histamine produces a marked stimulation of the accumulation of inositol 1-phosphate in slices of cerebellum, hypothalamus, cerebral cortex and hippocampus, with only a small effect in striatum [207, 239]. The response in cerebellum is perhaps the most interesting, since it is the region of guinea-pig brain which possesses the highest density of [^{3}H]mepyramine-binding sites. Indeed, there is an excellent correlation between the extent of the inositol phosphate response in a given region of guinea-pig brain and the density of H_1-receptor-binding sites deduced from ligand-binding studies [207, 239]. Thus, the largest phosphoinositide response to histamine was obtained in guinea-pig cerebellum, while the smallest was found in guinea-pig striatum, which has a low density of specific [^{3}H]mepyramine-binding sites.

The response to histamine in guinea-pig brain slices is sensitive to inhibition by mepyramine but not by cimetidine [60]. Furthermore, the H_1-selective agonist 2-thiazolylethylamine can elicit an accumulation of inositol phosphates of similar magnitude to that produced by histamine in cerebellar slices, while the selective H_2-receptor agonist has no significant effect of [^{3}H]inositol 1-phosphate levels [60, 239]. These data taken together indicate that the phosphoinositide response to histamine is mediated by H_1-receptors, a finding which has been confirmed in quantitative studies of the antagonism produced by H_1-receptor antagonists. Thus, the affinity constants obtained for mepyramine, promethazine and methapyrilene from inhibition of histamine-induced inositol 1-phosphate accumulation in guinea-pig brain are in accord with the values obtained from antagonism of other H_1-receptor-mediated responses in guinea-pig brain and periphery (*Table 2.1*) [59, 60]. In cerebellar slices, the nature of the antagonism by mepyramine has been evaluated and shown to be entirely consistent with a competitive interaction with histamine H_1-receptors. The Schild slope obtained in these experiments was 1.02, very close to the value of unity expected for a simple competitive interaction with an homogeneous population of H_1-receptor sites [59].

Early studies of histamine-induced inositol phospholipid breakdown in guinea-pig brain slices detected large accumulations of inositol 1-phosphate but very little change in the levels of the more polar inositol phosphates, inositol 1,4-bisphosphate and inositol 1,4,5-trisphosphate [59, 60]. More recent studies, however, have shown that acid conditions are necessary for the extraction of these latter inositol phosphates [238]. With this modification to the experimental protocol it is possible to demonstrate a rapid accumulation of inositol trisphosphate in slices of guinea-pig cerebellum in response to histamine [238]. The rate of accumulation of inositol trisphosphate in this brain region is more rapid than that of the less polar inositol phosphates and the highest rate of accumulation is normally achieved during the first 5 min of

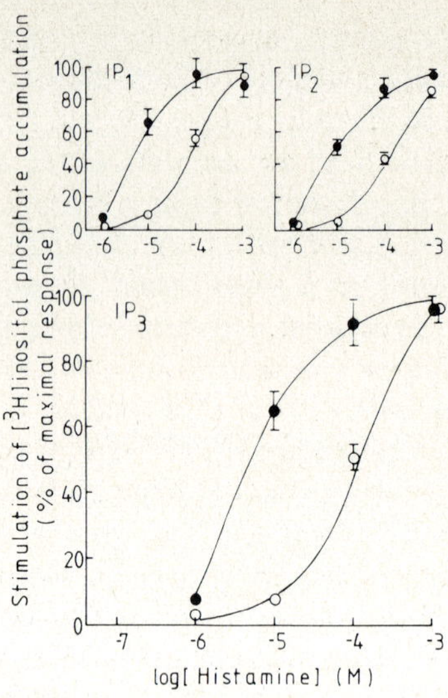

Figure 2.14. Histamine-induced accumulation of [^{3}H]inositol 1-phosphate (IP_1) [^{3}H]inositol bisphosphate (IP_2) and [^{3}H]inositol trisphosphate (IP_3) in slices of guinea-pig cerebellum. Concentration-response curves were obtained in the presence (○) and absence (●) of 3×10^{-8} M mepyramine. Data taken from [238].

stimulation with histamine. The effect of the H_1-antagonist, mepyramine, on the dose-response curves obtained for histamine-stimulated accumulation of inositol mono-, bis- and trisphosphate is illustrated in *Figure 2.14*. Mepyramine is a competitive antagonist of the effect of histamine on all three inositol phosphates and the data are consistent with these responses being mediated by typical histamine H_1-receptors. Since 1,2-diacylglycerol is the other product of inositol phospholipid breakdown, studies of inositol phosphate accumulation also provide indirect evidence for an H_1-mediated production of this lipid-soluble second messenger.

Studies in slices of rat cerebral cortex have shown that the accumulation of [^{3}H]inositol 1-phosphate elicited by histamine can be abolished by omission of calcium from the incubation medium [240]. This is not the result that would be expected if inositol phosphate were generated solely by the pathway outlined in *Figure 2.13* and does not support the hypothesis that phospholipase-C-mediated inositol phosphate production precedes, rather than results from, Ca^{2+} mobilization. However, interpretation of this finding is complicated by the effect of calcium ion removal on the incorporation of [^{3}H]inositol into the inositol phospholipids. Omission of calcium or preincubation with EGTA greatly enhances the incorporation of [^{3}H]inositol into the phospholipid layer [240]. A comparison of the effect of calcium on phosphoinositide responses to histamine and a number of other neurotransmitters in rat cerebral cortex does suggest that the response to histamine is more sensitive to calcium removal than are most responses. Thus, whereas the response to histamine is abolished in the absence of calcium ions, the response to the muscarinic agonist, carbachol, is unaffected by changing to calcium-free buffers [240]. In guinea-pig cerebral cortical slices, omission of calcium ions from the incubation medium reduces the response to histamine, but does not abolish it [207]. The different calcium requirements for H_1- and muscarinic phosphoinositide responses in rat cerebral cortex suggest that there may be differences in the way in which these two receptor systems are coupled to inositol phospholipid hydrolysis. This contention is supported by studies of the temperature dependence of the two responses in guinea-pig cerebral cortex [241]. This study shows that, whereas the H_1-receptor mediated response is markedly reduced at low temperatures, the response to carbachol is relatively unaltered by changes in temperature.

Detailed analysis of the dose-response curves for histamine in different brain regions suggests that the characteristics of the curve may differ between guinea-pig cerebral cortex and cerebellum. Thus, while the EC_{50} value for histamine-induced inositol phospholipid hydrolysis in cerebral cortex is similar to the value derived from binding studies with [^{3}H]mepyramine, the value in cerebellum is lower than the binding constant [207]. This suggests that there is no spare receptor reserve for this response in cerebral cortex, as would be expected for a response closely coupled to receptor occupation. However, in cerebellum, the low EC_{50} value suggests a more complex relationship between H_1-receptor occupancy and inositol phospholipid hydrolysis.

A similar conclusion can be drawn from the data obtained with a number of H_1-agonists. In guinea-pig cerebral cortex, 2-methylhistamine, N^{α}, N^{α}-dimethylhistamine and 2-pyridylethylamine appear to act as partial agonists of the H_1-mediated inositol phosphate response [207, 242], just as they are on the H_1-agonist potentiation of adenosine-stimulated cyclic AMP accumulation [51]. In cerebellar slices these compounds also appear to act as partial agonists. However, the maximal response they elicit (relative to histamine) in this tissue is very much larger than that obtained in cerebral cortex [207, 242]. The data obtained with these histamine analogues therefore confirm that the efficiency of H_1-receptor-effector coupling is different in guinea-pig cerebral cortex and cerebellum.

In peripheral tissues, the responsiveness of contractile responses to H_1-receptor stimulation can be selectively potentiated by the disulphide bond reducing agent, 1,4-dithiothreitol (DTT) [243, 244]. A similar effect can be observed for the H_1-elicited cyclic AMP and phosphoinositide responses in guinea-pig brain [242]. Thus, in both guinea-pig cerebral cortex and cerebellum, the dose-response curves for histamine are shifted in a parallel fashion to lower agonist concentrations by inclusion of 1,4-dithiothreitol in the incubation medium. The effect of DTT on phosphoinositide responses elicited by H_1-receptor stimulation in brain slices appears to be dependent upon the

Table 2.7. EFFECT OF 1,4-DITHIOTHREITOL (DTT) ON H_1-AGONIST-INDUCED INOSITOL PHOSPHATE RESPONSES IN GUINEA-PIG BRAIN

Values represent mean ± S.E.M. of the maximal stimulations (expressed as a percentage of the response to 1 mM histamine) and the ratio of the EC_{50} values obtained in the absence and presence of 1 mM DTT [242].

	Maximum stimulation (% of 1 mM histamine)		$\frac{EC_{50}\,(-DTT)}{EC_{50}\,(+DTT)}$
Agonist	*Control*	*DTT*	
Cerebellum			
Histamine	100.6 ± 5.6	103.3 ± 5.9	6.1 ± 1.4
2-Thiazolylethylamine	99.8 ± 3.0	107.6 ± 2.4	3.9 ± 0.4
2-Methylhistamine	83.8 ± 7.0	94.1 ± 5.2	6.8 ± 2.0
2-Pyridylethylamine	77.4 ± 5.4	93.0 ± 5.3	3.2 ± 0.7
Cortex			
Histamine	106.5 ± 3.9	109.8 ± 3.5	8.0 ± 0.8
2-Methylhistamine	33.6 ± 3.7	81.9 ± 10.9	1.8 ± 1.4
2-Pyridylethylamine	35.4 ± 7.7	80.8 ± 16.1	0.5 ± 0.6

efficacy of the agonist under study. For example, in guinea-pig cerebral cortex, DTT increases the maximal responses to the partial agonists 2-methylhistamine and 2-pyridylethylamine without affecting the position of the dose-response curves [242]. In guinea-pig cerebellum, however, where these agonists are more potent, DTT produces smaller increase in the maximum responses to 2-pyridylethylamine and 2-methylhistamine which is accompanied by a significant decrease in the EC_{50} values. The differing effects of DTT on inositol phospholipid responses to H_1-agonists in these two brain regions are summarised in *Table 2.7*. These studies suggest that DTT can increase the efficacy of H_1 agonist-receptor interactions in guinea-pig brain in addition to the alterations in agonist binding affinity observed in binding studies with [^{3}H]mepyramine in broken cell preparations [112]. DTT does not produce a similar potentiation of muscarinic-receptor-mediated phosphoinositide responses [242]; thus, the effect of DTT on H_1-agonist efficacy must be produced at a site prior to the stage at which the receptor-effector pathways are shared by the two receptor systems.

OTHER BIOCHEMICAL RESPONSES

GLYCOGENOLYSIS

Studies in the mouse and chick have shown that histamine can elicit a marked glycogenolytic response in brain tissues. For example, injection of histamine into the neonatal chick results in an increase in cyclic AMP, which in turn leads to a conversion of phosphorylase *b* to phosphorylase *a* and a fall in cerebral glycogen levels [245]. A simple technique is now available for measurement of this glycogenolytic response *in vitro* [246]. This assay enables the action of agonists on [^{3}H]glycogen levels to be evaluated in mouse brain slices previously incubated in the presence of [^{3}H]glucose. Using this technique it has been possible to show that, like noradrenaline, histamine can elicit a marked concentration-dependent hydrolysis of [^{3}H]glycogen [246].

Pharmacological analysis of the response to histamine in mouse cerebral cortical slices indicates that glycogen hydrolysis is mediated by histamine H_1-receptors [218]. Thus, whereas the H_2-receptor antagonist metiamide does not affect the histamine-induced hydrolysis of [^{3}H]glycogen, increasing concentrations of mepyramine progressively shift the dose-response curve for histamine to higher agonist concentrations. The large difference in potency of the stereoisomers of the H_1-antagonist chlorpheniramine also provides strong evidence for the involvement of the H_1-receptor subtype [218]. In keeping with this hypothesis, it is striking that the selective H_2-agonist dimaprit does not produce a glycogenolytic response in mouse cerebral cortical slices. Other histamine agonists, however, do elicit hydrolysis of [^{3}H]glycogen with relative potencies consistent with stimulation of H_1-receptors [218].

A comparison of the EC_{50} value for the glycogenolytic response to histamine with the dissociation constant obtained for histamine from binding studies with [^{3}H]mepyramine suggests that there is a substantial spare receptor reserve for this response in this tissue [218]. The finding that 2-methylhistamine, which has partial agonist activity on other systems [51, 207, 242], behaves as a full agonist on the glycogenolytic response [218] is also consistent with this argument. It is notable, however, that although betahistine behaves as a partial agonist on both the H_1-mediated cyclic AMP response of guinea-pig cerebral cortex and the H_1-elicited glycogenolytic response of mouse cerebral cortex, it has a higher intrinsic activity for the response in the mouse [52]. This is remarkable in view of the difference in the extent of the spare receptor reserves for these two responses. In guinea-pig cerebral cortex there appears to be no spare receptor reserve for the H_1-mediated cyclic AMP and inositol phospholipid responses [51, 207], while in mouse cerebral cortex there appears to be a substantial spare receptor reserve [218]. However, one explanation may be that betahistine induces part of its glycogenolytic effect through activation of non-H_1-receptors. Support for this hypothesis is provided by the finding that the response to betahistine is much less sensitive to inhibition by H_1-antagonists than the glycogenolytic response to histamine [52]. Thus, the apparent dissociation constant of 28 nM for mepyramine, obtained from inhibition of the response to betahistine, is approximately 6-times higher than the value determined using histamine as agonist [52].

The glycogenolytic response to histamine can be significantly potentiated in the presence of the phosphodiesterase inhibitor isobutylmethylxanthine [218]. In addition, dibutyryl cyclic AMP, but not dibutyryl cyclic GMP, can promote glycogen hydrolysis in slices of mouse cerebral cortex [246]. These observations suggest that cyclic AMP accumulation may be involved in the response to histamine. The H_1-nature of this response, however, indicates that, if cyclic AMP is involved, it is generated by a synergistic interaction between histamine (*via* H_1-receptors) and another endogenous neurotransmitter, e.g., adenosine or VIP [199]. The maximal glycogenolytic response to histamine can be markedly reduced by reducing the concentration of Ca^{2+} ions in the medium [218]. This finding is consistent with the postulate that the indirect effect of H_1-receptor stimulation on adenylate cyclase activity is mediated by another second messenger such as calcium ions. Alternatively, an H_1-induced rise in intracellular calcium may affect glycogen hydrolysis independently of cyclic AMP. For example, there is evidence that phosphorylase *b* kinase in brain tissues can be activated by calcium ions [247]. Furthermore, there is strong evidence that phosphorylase *b* kinase can be activated by calcium ions *via* calmodulin [248, 249] in peripheral tissues.

Desensitization of the glycogenolytic response to H_1-receptor stimulation has been observed following prolonged incubation of mouse cerebral cortical slices with histamine [250]. This is seen as a parallel shift of the concentration-response curve for histamine to higher agonist concentrations, with no significant change in the maximum response. The development of the desensitization induced by 10 μM histamine occurs with a half-time of 20 min and the responsiveness can be fully restored within 1 h of removing the desensitizing dose of histamine [250]. Desensitization of the H_1-glycogenolytic response appears to be a specific process, and the responsiveness of other glycogenolytic agents, such as noradrenaline and adenosine, is not modified. The loss in responsiveness following desensitization is accompanied by a small decrease (*circa* 20%) in the H_1-binding site capacity; however, the functional significance of this small decrease in [^{3}H]mepyramine binding sites is uncertain [250]. A reduction in the number of H_1-receptors on prolonged agonist-stimulation could explain the desensitization of the glycogenolytic response to histamine by reducing the spare receptor reserve. However, the rapid recovery from desensitization and the fact that this is not impaired by the presence of a protein synthesis inhibitor, such as cycloheximide, suggest that the desensitization process might involve changes at the level of the effector system.

CYCLIC GMP ACCUMULATION

Studies in murine neuroblastoma cells (clone NIE-115) have shown that histamine can produce a rapid and marked (up to 50-fold above basal levels) increase in the formation of cyclic [^{3}H] GMP following preloading of cells with [^{3}H]guanine [251]. The peak response to histamine is normally observed 30 s after application of histamine and then rapidly declines towards basal levels over the next 2 or 3 min [251, 252]. When neuroblastoma cells are pre-incubated with the phosphodiesterase inhibitor 3-isobutyl-1-methylxanthine, the rate of decay of cyclic [^{3}H]GMP is significantly reduced [252].

The pharmacological characteristics of the cyclic GMP response to histamine in NIE-115 neuroblastoma cells are consistent with the involvement of a typical H_1-receptor [251]. Studies with the irreversible antagonist, dibenamine, have suggested that there is no spare receptor reserve in the cyclic GMP response to histamine in this cell line [251]. The response to H_1-receptor stimulation, like that following muscarinic receptor activation, has an absolute dependence on extracellular calcium ion concentration [253]. This observation suggests that it is the influx of calcium ions which leads to a stimulation of guanylate cyclase activity and hence cyclic GMP formation. However, no change in the intracellular free concentration of calcium has been observed in NIE-115 cells previously loaded with the bioluminescent protein aequorin, following stimulation with either H_1- or muscarinic agonists [254]. Instead, it has been shown that muscarinic or H_1-receptor stimulation elicits a release of arachidonic acid which is sensitive to inhibition by quinacrine [254]. Furthermore, inhibition of this arachidonate release or interference with the metabolism of arachidonic acid *via* the lipoxygenase pathway prevents receptor-mediated cyclic GMP formation [254].

Although the cyclic GMP response to histamine in murine neuroblastoma cells is well established, very little work has been done on histamine-mediated cyclic GMP formation in mammalian brain slices. Histamine has been shown to increase cyclic GMP levels in rabbit [255] and guinea-pig [146] cerebral cortical slices, although the receptor involved has not been identified. Indeed, the only quantitative study of histamine-stimulation cyclic GMP accumulation in the mammalian peripheral and central nervous systems has been performed in blocks of bovine superior cervical ganglion [256]. In this tissue, histamine stimulates the accumulation of both cyclic AMP and cyclic GMP. The evidence suggests that these two effects are mediated by stimulation of different receptors, activation of H_1-receptors leading to cyclic GMP accumulation and stimulation of H_2-receptors leading to cyclic AMP formation [256]. The H_1-mediated increase in cyclic GMP levels appears to be mediated *via* calcium ions, since the effect on cyclic GMP accumulation requires the presence of calcium in the extracellular medium. Furthermore, the calcium ionophore A23187 causes a calcium-dependent increase in cyclic GMP formation [256].

CONCLUDING REMARKS

For the reader who has reached this far it will be clear that there is strong biochemical and pharmacological evidence for the presence of functional histamine H_1-, H_2- and, to some extent, H_3-receptors in the mammalian central nervous system. This information, together with the recent ability to visualize histamine- and histidine-decarboxylase-containing cell bodies and fibres, suggests that histamine may have an important rôle to play in the brain. However, the fact that histaminergic fibres project diffusely over large areas of the central nervous system from highly localized cell body regions in the hypothalamus and mammilary body suggests that this rôle may be modulatory rather than as a 'front line' neurotransmitter. A modulatory rôle is indicated by the synergy observed between histamine and other neurotransmitters on cyclic AMP accumulation, where H_1-receptor stimulation leads to an amplification of the response elicited by other receptor systems. Furthermore, the ability of histamine to alter the responsiveness of excitatory neurotransmitters in certain brain regions, by reducing (*via* H_2-receptors) the after hyperpolarization normally associated with depolarizing stimuli, is also consistent with this rôle.

It is well established that the classical H_1-antagonists can produce a marked sedation, decreased attention and 'mental clouding'. On the basis of the biochemical and electrophysiological studies in guinea-pig and rat hippocampus, it might be predicted that H_2-antagonists would have a similar sedating effect if they readily crossed the blood-brain barrier. Furthermore, inhibition of the facilitatory effect of histamine on excitatory electrical stimuli may provide a rôle for H_2-antagonists as anticonvulsant agents. One might also predict that mimicking the actions of histamine with selective agonists or facilitating histaminergic neurotransmission (e.g., via H_3-receptor antagonism) would lead to arousal or an improvement in mental function, a property which might have therapeutic potential in Alzheimer's disease and related disorders. Selective stimulation of histamine receptors in the brain is difficult at present, however, because of the lack of specificity of the compounds available. Thus, even the use of impromidine, which discriminates exceedingly well between H_2- and H_1-receptors, is not simple because it now appears to inhibit the putative H_3-autoreceptor. The development of suitably selective agonist compounds together with antagonists of the H_3-receptor is therefore needed before the functional rôle of histamine in the CNS can be properly assessed. (See note added in proof, p. 84.)

ACKNOWLEDGEMENTS

The work performed in the author's laboratory was supported by the Medical Research Council, The Wellcome Trust and the Smith Kline (1982) Foundation.

REFERENCES

1 H.M. Adams and H.K.A. Hye, Br. J. Pharmacol. Chemother., 28 (1976) 137.
2 K.M. Taylor and S.H. Snyder, J. Pharmacol. Exp. Ther., 179 (1971) 619.
3 K.M. Taylor, E. Gfeller and S.H. Snyder, Brain Res., 41 (1972) 171.
4 J.F. Lipinski, H.H. Schaumburg and R.J. Baldessarini, Brain Res., 52 (1973) 403.
5 J.-C. Schwartz, Annu. Rev. Pharmacol. Toxicol., 17 (1977) 325.
6 J.-C. Schwartz, H. Pollard and T.T. Quach, J. Neurochem., 35 (1980) 26.
7 L. Edvinsson, J. Cervos-Navarro, C. Owman, L.I. Larsson and A.L. Ronnberg, Neurology, 27 (1977) 878.
8 L.B. Hough, R.C. Goldschmidt, S.D. Glick and J. Padawer, in Frontiers in Histamine Research, Advances in the Biosciences, eds. C.R. Ganellin and J.-C. Schwartz (Pergamon, New York) Vol. 51 (1985) pp. 131–140.
9 I.L. Karnushina, J.M. Palacios, G. Barbin, E. Dux, F. Joo and J.-C. Schwartz, J. Neurochem., 34 (1980) 1201.

10 T. Watenabe, Y. Taguchi, S. Shiosaka, J. Tanaka, H. Kubota, Y. Terano, M. Tohyama and H. Wada, Brain Res., 295 (1984) 13.
11 P. Panula, H.-Y.T. Yang and E. Costa, Proc. Natl. Acad. Sci. U.S.A., 81 (1984) 2572.
12 N. Takeda, S. Inagaki, S. Shiosaka, Y. Taguchi, W.H. Oertel, M. Tohyama, T. Watenabe and H. Wada, Proc. Natl. Acad. Sci. U.S.A., 81 (1984) 7647.
13 H. Pollard, I. Pachot and J.-C. Schwartz, Neurosci. Lett., 54 (1985) 53.
14 H.W.M. Steinbusch and A.H. Mulder, in Ref. 8, pp. 119–130.
15 T. White, Experientia, 21 (1965) 132.
16 H.E. Brezenoff and D.J. Jenden, Int. J. Neuropharmacol., 8 (1969) 593.
17 J.N. Sinha, M.L. Gupta and K.P. Bhargava, Eur. J. Pharmacol., 5 (1969) 235.
18 L. Finch and P.E. Hicks, Naunyn-Schmiedebergs Arch. Pharmacol., 293 (1976) 151.
19 L. Finch and P.E. Hicks, Neuropharmacology, 16 (1977) 211.
20 M.C. Klein and S.B. Gertner, J. Pharmacol. Exp. Ther., 216 (1981) 315.
21 G.G. Shaw, Br. J. Pharmacol., 42 (1971) 205.
22 H.E. Brezenoff and P. Lomax, Experientia, 26 (1970) 51.
23 P. Sweatman and R.M. Jell, Brain Res., 127 (1977) 173.
24 B. Cox, M.D. Green and P. Lomax, Experientia, 32 (1976) 498.
25 J.Z. Nowak, B. Bielkiewicz and U. Lebrecht, Neuropharmacology, 18 (1979) 783.
26 A. Pilc and J.Z. Nowak, Neuropharmacology, 19 (1980) 773.
27 P.W. Kalivas, J. Pharmacol. Exp. Ther., 222 (1982) 37.
28 R.I. Weiner and W.F. Ganong, Physiol. Rev., 58 (1978) 905.
29 J.-C. Schwartz, G. Barbin, A.-M. Duchemin, M. Garbarg, C. Llorens, H. Pollard, T.T. Quach and C. Rose, in Pharmacology of Histamine Receptors, eds. C.R. Ganellin and M.E. Parsons (J. Wright, Bristol, 1982) pp. 351–391.
30 A.O. Donoso and E.O. Alvarez, Trends Pharmacol. Sci., 5 (1984) 98.
31 E.J. Marco, G. Balfagon, J. Marin, B. Gomez and S. Lluch, Naunyn Schmiedebergs Arch. Pharmacol., 312 (1980) 239.
32 G. Balfagon, R. Galvan and E.J. Marco, J. Pharm. Pharmacol., 36 (1984) 248.
33 P.A. Brown and J. Vernikos, Eur. J. Pharmacol., 65 (1980) 89.
34 P. Lidbrink, G. Jonsson and K. Fuxe, Neuropharmacology, 10 (1971) 521.
35 S.J. Hill and J.M. Young, Mol. Pharmacol., 19 (1981) 379.
36 P.B. Dews and J.D.P. Graham, Br. J. Pharmacol. Chemother., 1 (1946) 278.
37 O. Arunlakshana and H.O. Schild, Br. J. Pharmacol. Chemother., 14 (1959) 48.
38 A.S.F. Ash and H.O. Schild, Br. J. Pharmacol. Chemother., 27 (1966) 427.
39 P. Eyre and N. Chand, in Ref. 29, pp. 351–391.
40 P.B. Marshall, Br. J. Pharmacol. Chemother., 10 (1955) 270.
41 J.W. Black, W.A.M. Duncan, C.J. Durant, C.R. Ganellin and M.E. Parsons, Nature (London), 236 (1972) 385.
42 R.W. Brimblecombe, W.A.M. Duncan, G.J. Durant, J.C. Emmett, C.R. Ganellin and M.E. Parsons, J. Int. Med. Res., 3 (1975) 86.
43 J. Bradshaw, R.T. Brittain, J.W. Clitherow, M.J. Daly, D. Jack, B.J. Price and R. Stables, Br. J. Pharmacol., 66 (1979) 464P.
44 N. Chand and P. Eyre, Agents Actions, 5 (1975) 277.
45 T.O. Yellin (ed.), Histamine Receptors (Spectrum, New York, 1979).
46 C.R. Ganellin and M.E. Parsons (eds.), Pharmacology of Histamine Receptors (J. Wright, Bristol, 1982).
46a M.J. Daly and B.J. Price, Prog. Med. Chem., 20 (1983) 337.
47 J.M. Arrang. M. Garbarg and J.-C. Schwartz, Nature (London), 302 (1983) 832.

48 J.M. Arrang, M. Garbarg and J.-C. Schwartz, Neuroscience, 15 (1985) 553.
49 J.-C. Schwartz, J.M. Arrang and M. Garbarg, Trends Pharmacol. Sci., 7 (1986) 24.
50 G.J. Durant, W.A.M. Duncan, C.R. Ganellin, M.E. Parsons, R.C. Blakemore and A.C. Rasmussen, Nature (London), 276 (1978) 403.
51 P.R. Daum, S.J. Hill and J.M. Young, Br. J. Pharmacol., 77 (1982) 347.
52 J.M. Arrang, M. Garbarg, T.T. Quach, M.D.T. Tuong, E. Yeramian and J.-C. Schwartz, Eur. J. Pharmacol., 111 (1985) 73.
53 D. Bovet, Ann. N.Y. Acad. Sci., 50 (1950) 1089.
54 J.J. Reuse, Br. J. Pharmacol. Chemother., 3 (1948) 174.
55 A.S.V. Burgen and C.J. Harbird, Br. J. Pharmacol., 80 (1983) 600.
56 U. Trendelenburg, J. Pharmacol. Exp. Ther., 130 (1960) 450.
57 K.M. Taylor, Biochem. Pharmacol., 22 (1973) 2775.
58 P. Seeman and J. Weinstein, Biochem. Pharmacol., 15 (1966) 1737.
59 J. Donaldson and S.J. Hill, Br. J. Pharmacol., 85 (1985) 499.
60 P.R. Daum, C.P. Downes and J.M. Young, J. Neurochem., 43 (1984) 25.
61 S.J. Hill, P.C. Emsom and J.M. Young, J. Neurochem., 31 (1978) 997.
62 R.R. Ison, F.M. Franks and K.S. Soh, J. Pharm. Pharmacol., 25 (1973) 887.
63 F.E. Roth and W.M. Govier, J. Pharmacol. Exp. Ther., 124 (1958) 347.
64 A. Shafi'ee and G. Hite, J. Med. Chem., 12 (1969) 266.
65 C.R. Ganellin, in Ref. 29 pp. 10–102.
66 W. Th. Nauta and R.F. Rekker, in Handbook of Experimental Pharmacology, ed. M. Rocha e Silva (Springer-Verlag, Berlin) Vol. 18, part 2 (1978) pp. 215–249.
67 J. Figge, P. Leonard and E. Richelson, Eur. J. Pharmacol., 58 (1979) 479.
68 J. Aceves, S. Mariscal, K.E. Morrison and J.M. Young, Br. J. Pharmacol., 84 (1985) 417.
69 J.E. Taylor and E. Richelson, Eur. J. Pharmacol., 67 (1980) 41.
70 N.L. Wiech and J.S. Martin, Arzneim-Forsch., 32 (1982) 1167.
71 C. Rose, T.T. Quach, C. Llorens and J.-C. Schwartz, Arzneim. Forsch., 32 (1982) 1171.
72 P.M. Laduron, P.F.M. Janssen, W. Gommeren and J.E. Leysen, Mol. Pharmacol., 21 (1982) 294.
73 A. Uzan and G. Le Fur, J. Pharm. Pharmacol., 31 (1979) 701.
74 G. Le Fur, C. Malgouris and A. Uzan, Life Sci., 29 (1981) 547.
75 T.T. Quach, A.M. Duchemin, C. Rose and J.-C. Schwartz, Eur. J. Pharmacol., 60 (1979) 391.
76 E.A. Brown, R. Griffiths, C.A. Harvey and D.A. A. Owen, Br. J. Pharmacol., 87 (1986) 569.
77 J.W. Black, W.A.M. Duncan, J.C. Emmett, C.R. Ganellin, T. Hesselbo, M.E. Parsons and J.H. Wyllie, Agents Actions, 3 (1973) 133.
78 T.O. Yellin, S.H. Buck, D.J. Gilman, D.F. Jones and J.M. Wardleworth, Life Sci., 25 (1979) 2001.
79 M.L. Torchiana, R.G. Pendleton, G. Cook, C.A. Hanson and B.V. Clineschmidt, J. Pharmacol. Exp. Ther., 224 (1983) 514.
80 C.R. Ganellin, J. Appl. Chem. Biotechnol., 28 (1978) 183.
81 J.P. Green, in Handbook of Psychopharmacology, eds. L. Iversen, S.D. Iversen and S.H. Snyder (Plenum, New York) Vol. 17 (1983) pp. 385–420.
82 M. Al-Gadi and S.J. Hill, Br. J. Pharmacol., 85 (1985) 877.
83 G.A. Gajkowski, D.B. Norris, T.J. Rising and T.P. Wood, Nature (London), 304 (1983) 65.
84 G.J. Durant, C.R. Ganellin and M.E. Parsons, J. Med. Chem., 18 (1975) 905.

85 L.A. Barker and L.B. Hough, Br. J. Pharmacol., 80 (1983) 65.
86 I.J.C. Frew and G.N. Menon, Postgrad. Med. J., 52 (1976) 501.
87 T.P. Kenakin and D.A. Cook, Can. J. Physiol. Pharmacol., 58 (1980) 1307.
88 M.E. Parsons, D.A.A. Owen, C.R. Ganellin and G.J. Durant, Agents Actions, 7 (1977) 31.
89 K.M.K. Bottomley, T.J. Rising and A. Steward, Br. J. Pharmacol., 84 (1985) 47P.
90 M.E. Parsons, R.C. Blakemore, G.J. Durant, C.R. Ganellin and A.C. Rasmussen, Agents Actions, 5 (1975) 464.
91 N.J.M. Birdsall and E.C. Hulme, J. Neurochem., 27 (1976) 7.
92 E.C. Hulme, N.J.M. Birdsall, A.S.V. Burgen and P. Mehta, Mol. Pharmacol., 14 (1978) 737.
93 S.J. Hill, J.M. Young and D.H. Marrian, Nature (London), 270 (1977) 361.
94 V.T. Tran, R. Lebovitz, L. Toll and S.H. Snyder, Eur. J. Pharmacol., 70 (1981) 501.
95 J.E. Taylor and E. Richelson, Eur. J. Pharmacol., 78 (1982) 279.
96 S.J. Peroutka and S.H. Snyder, J. Pharmacol. Exp. Ther., 216 (1981) 142.
97 J.M. Treherne and J.M. Young, Br. J. Pharmacol., 87 (1986) 72P.
98 M. Korner, M.-L. Bouthenet, C.R. Ganellin, M. Garbarg, L. Gros, R.J. Ife, N. Sales and J.-C. Schwartz, Eur. J. Pharmacol., 120 (1986) 151.
99 R.S.L. Chang, V.T. Tran and S.H. Snyder, J. Neurochem., 32 (1979) 1653.
100 S.J. Hill and J.M. Young, Br. J. Pharmacol., 68 (1980) 687.
101 R.S.L. Chang, V.T. Tran and S.H. Snyder, Eur. J. Pharmacol., 48 (1978) 463.
102 V.T. Tran, R.S.L. Chang and S.H. Snyder, Proc. Natl. Acad. Sci. U.S.A., 75 (1978) 6290.
103 J.E. Taylor, T.L. Yaksh and E. Richelson, Brain Res., 243 (1982) 391.
104 B. Bielkiewicz and D.A. Cook, Can J. Physiol. Pharmacol., 63 (1985) 756.
105 S. Kanba and E. Richelson, Brain Res., 304 (1984) 1.
106 S.J. Hill and J.M. Young, Eur. J. Pharmacol., 52 (1978) 397.
107 M. Gavish, R.S.L. Chang and S.H. Snyder, Life Sci., 25 (1979) 783.
108 L. Toll and S.H. Snyder, J. Biol. Chem., 257 (1982) 13593.
109 T. Kuno, N. Kubo and C. Tanaka, Biochem. Biophys. Res. Commun., 129 (1985) 639.
110 R.S.L. Chang and S.H. Snyder, J. Neurochem., 34 (1980) 916.
111 E. Yeramian, M. Garbarg and J.-C. Schwartz, Mol. Pharmacol., 28 (1985) 155.
112 J. Donaldson and S.J. Hill, Eur. J. Pharmacol., 129 (1986) 25.
113 N. Nakahata, M.W. Martin, A.R. Hughes, J.R. Hepler and T.K. Harden, Mol. Pharmacol., 29 (1986) 188.
114 R.M. Wallace and J.M. Young, Mol. Pharmacol., 23 (1983) 60.
115 J.M. Palacios, W.S. Young III and M.J. Kuhar, Eur. J. Pharmacol., 58 (1979) 295.
116 J.M. Palacios, J.M. Warmsley and M.J. Kuhar, Neuroscience, 6 (1981) 15.
117 J.M. Palacios, J.M. Warmsley and M.J. Kuhar, Brain Res., 214 (1981) 155.
118 R.S.L. Chang, V.T. Tran and S.H. Snyder, Brain Res., 190 (1980) 95.
119 S.J. Peroutka, M.A. Moskowitz, J.F. Reinhard and S.H. Snyder, Science 208 (1980) 610.
120 E. Richelson, Nature (London), 274 (1978) 176.
121 W.P. Burkard, Eur. J. Pharmacol., 50 (1978) 449.
122 P. Devoto, A.M. Marchisio, E. Carboni and P.F. Spano, Eur. J. Pharmacol., 63 (1980) 91.
123 S.I. Kandel, G.H. Steinberg, M. Kandel, J.W. Wells and A.G. Gornall, Biochem. Pharmacol., 29 (1980) 2269.
124 D.A. Kendall, J.W. Ferkany and S.J. Enna, Life Sci., 26 (1980) 1293.
125 N. Subramanian and T.A. Slotkin, Mol. Pharmacol., 20 (1981) 240.
126 T.J. Rising, D.B. Norris, S.E. Warrander and T.P. Wood, Life Sci., 27 (1980) 199.
127 I.R. Smith, M.T. Cleverley, C.R. Ganellin and K.M. Metters, Agents Actions, 10 (1980) 422.

128 S.E. Warrander, D.B. Norris, T.J. Rising and T.P. Wood, Life Sci., 33 (1983) 1119.
129 K. Ozawa, Y. Nomura and T. Segawa, Neurochem. Int., 7 (1985) 435.
130 D.R. Bristow, J.R. Hare, J.R. Hearn and L.E. Martin, Br. J. Pharmacol., 72 (1981) 547P.
131 S. Maayani, L.B. Hough, H. Weinstein and J.P. Green, in Typical and Atypical Antidepressants: Molecular Mechanisms, eds. E. Costa and G. Racagni (Raven Press, New York, 1982) pp. 133–147.
132 T.J. Rising and D.B. Norris, in Ref. 8, pp. 61–67.
133 D.B. Norris, G.A. Gajkowski and T.J. Rising, Agents Actions, 14 (1984) 543.
134 J.C. Foreman, D.B. Norris, T.J. Rising and S.E. Webber, Br. J. Pharmacol., 86 (1985) 475.
135 J.M. Palacios, J.-C. Schwartz and M. Garbarg, Eur. J. Pharmacol., 50 (1978) 443.
136 G. Barbin, J.M. Palacios, E. Rodergas, J.-C. Schwartz and M. Garbarg, Mol. Pharmacol., 18 (1980) 1.
137 D.L. Cybulsky, S.I. Kandel, J.W. Wells and A.G. Gornall, Eur. J. Pharmacol., 72 (1981) 407.
138 J.W. Wells, D.L. Cybulsky, M. Kandel, S.I. Kandel and G.H. Steinberg, in Ref. 8, pp. 69–78.
139 S. Batzri, J.W. Harmon, J. Dyer and W.F. Thompson, Mol. Pharmacol., 22 (1982) 41.
140 R.F. Furchgott, Adv. Drug Res., 3 (1966) 21.
141 G.H. Steinberg, J.G. Eppel, M. Kandel, S.I. Kandel and J.W. Wells, Biochemistry, 24 (1985) 6095.
142 G.H. Steinberg, M. Kandel, S.I. Kandel and J.W. Wells, Biochemistry, 24 (1985) 6107.
143 G.H. Steinberg, M. Kandel, S.I. Kandel and J.W. Wells, Biochemistry, 24 (1985) 6115.
144 J.W. Black, V.P. Gerskowitch, P.J. Randall and D.G. Trist, Br. J. Pharmacol., 74 (1981) 978P.
145 M. Rodbell, Nature (London), 284 (1980) 17.
146 J. Daly, Cyclic Nucleotides in the Nervous System (Plenum Press, New York, 1977).
147 M.J. Berridge and R.F. Irvine, Nature (London), 312 (1984) 315.
148 M.J. Berridge, Biochem. J., 220 (1984) 345.
149 E.M. Ross and A.G. Gilman, Annu. Rev. Biochem., 49 (1980) 533.
150 M. Rodbell, in: Cell Surface Receptors, ed. P.G. Strange (Ellis Horwood, Chichester, 1983) pp. 227–239.
151 M. Rodbell, in Advances in Cyclic Nucleotide and Protein Phosphorylation Research, eds. P. Greengard, G. Robinson, R. Paoletti and S. Nicosia (Raven Press, New York) Vol. 17 (1984) pp. 207–213.
152 C.A. Briggs and D.A. McAfee, in More about Receptors, ed. J.W. Lamble (Elsevier/North-Holland Amsterdam, 1982) pp. 54–60.
153 E.J. Nestler and P. Greengard, Nature (London), 305 (1983) 583.
154 A.C. Naire, H.C. Hemmings and P. Greengard, Annu. Rev. Biochem., 54 (1985) 931.
155 S. Kakiuchi and T.W. Rall, Mol. Pharmacol., 4 (1968) 367.
156 S. Kakiuchi and T.W. Rall, Mol. Pharmacol., 4 (1968) 379.
157 G.C. Palmer, M.J. Schmidt and G.A. Robinson, J. Neurochem., 19 (1972) 2251.
158 M.D. Spiker, G.C. Palmer and A.A. Manian, Brain Res., 104 (1976) 401.
159 H. Shimizu, C.R. Creveling and J.W. Daly, J. Neurochem., 17 (1970) 441.
160 H. Shimizu and J.W. Daly, Eur. J. Pharmacol., 17 (1972) 240.
161 M. Chasin, F. Mamrak, G. Samaniego and S.M. Hess, J. Neurochem., 21 (1973) 1415.
162 M. Baudry, M.P. Martres and J.-C. Schwartz, Nature (London) 253 (1975) 362.
163 M. Rogers, K. Dismukes and J.W. Daly, J. Neurochem., 25 (1975) 531.
164 K. Dismukes, M. Rogers and J.W. Daly, J. Neurochem., 26 (1976) 785.

165 J.M. Palacios, M. Garbarg, G.K. Barbin and J.-C. Schwartz, Mol. Pharmacol., 14 (1978) 971.
166 S.J. Hill, P. Daum and J.M. Young, J. Neurochem., 37 (1981) 1357.
167 M. Huang, H. Shimizu and J.W. Daly, Mol. Pharmacol., 7 (1971) 155.
168 H. Schultz and J.W. Daly, J. Biol. Chem., 248 (1973) 843.
169 J.W. Daly, W. Padgett, Y. Nimitkitpaisan, C.R. Creveling, D. Cantacuzene and K.L. Kirk, J. Pharmacol. Exp. Ther., 212 (1980) 382.
170 J. Schultz and J.W. Daly, J. Biol. Chem., 248 (1973) 853.
171 M. Huang and J.W. Daly, J. Neurochem., 23 (1974) 393.
172 I. Pull and H. McIlwain, J. Neurochem., 24 (1975) 695.
173 S.R. Nahorski, K.J. Rogers and B.M. Smith, Brain Res., 126 (1977) 387.
174 J.M. Brine, C.R. Calcutt, F. Nilam, J.A. Poat, I.R. Smith and G.N. Woodruff, Br. J. Pharmacol., 80 (1983) 606P.
175 C.L. Johnson, in Ref. 29, pp. 146–216.
176 J. Codina, J. Hildebrandt, T. Sunyer, R.D. Sekura, C.R. Manclark, R. Iyengar and L. Birnbaumer, in Ref. 151, pp. 111–125.
177 J. Holmgren, Nature (London), 292 (1981) 413.
178 M. Ui, T. Katada, T. Murayama, H. Kurose, M. Yajima, M. Tamura, T. Nakamura and K. Nogimori, in Ref. 151, pp. 145–151.
179 L.R. Hegstrand, P.D. Kanof and P. Greengard, Nature (London), 260 (1976) 165.
180 P.D. Kanof, L.R. Hegstrand and P. Greengard, Arch. Biochem. Biophys., 182 (1977) 321.
181 J.P. Green and S. Maayani, Nature (London), 269 (1977) 163.
182 J.P. Green, C.L. Johnson, H. Weinstein and S. Maayani, Proc. Natl. Acad. Sci. U.S.A., 74 (1977) 5697.
183 P.D. Kanof and P. Greengard, J. Pharmacol. Exp. Ther., 209 (1979) 87.
184 P. Portaleone, G. Pagnini, A. Crispino and E. Genazzani, J. Neurochem., 31 (1978) 1371.
185 L.B. Hough and J.P. Green, Brain Res., 219 (1981) 363.
186 M.H. Makman, H.S. Ahn, L.J. Thal, L.J. Sharpless, N.S. Dvorkin, S.G. Horowitz and M. Rosenfield, Brain Res., 192 (1980) 177.
187 M.V. Newton, L.B. Hough and E.C. Azmitia, Brain Res., 239 (1982) 639.
188 C.L. Johnson, H. Weinstein and J.P. Green, Mol. Pharmacol., 16 (1979) 417.
189 C.L. Johnson, H. Weinstein and J.P. Green, Biochim. Biophys. Acta, 587 (1979) 155.
190 P.D. Kanof and P. Greengard, Mol. Pharmacol., 15 (1979) 445.
191 M.R. Bristow, R. Cubicciotti, R. Ginsburg, E.B. Stinson and C. Johnson, Mol. Pharmacol., 21 (1982) 671.
192 G.C. Palmer, H.R. Wagner, S.J. Palmer and A.A. Manian, Arch. Int. Pharmacodyn., 233 (1978) 314.
193 P.D. Kanof and P. Greengard, Nature (London), 272 (1978) 329.
194 S. Verma and J. McNeill, J. Pharmacol. Exp. Ther., 200 (1977) 352.
195 M.D.T. Tuong, M. Garbarg and J.-C. Schwartz, Nature (London), 287 (1980) 548.
196 S. Kanba and E. Richelson, Eur. J. Pharmacol., 94 (1983) 313.
197 J.Z. Nowak, J.M. Arrang, J.-C. Schwartz and M. Garbarg, Neuropharmacology, 22 (1983) 259.
198 S.H. Snyder, Annu. Rev. Neurosci., 8 (1985) 103.
199 P.J. Magistretti and M. Schorderet, J. Neurosci., 5 (1985) 362.
200 M. Chasin, F. Mamrak and S.G. Samaniego, J. Neurochem., 22 (1974) 1031.
201 J.W. Daly, E. McNeal, C. Partington, M. Neuwirth and C.R. Creveling, J. Neurochem., 35 (1980) 326.

202 E.T. McNeal, C.R. Creveling and J.W. Daly, J. Neurochem., 35 (1980) 338.
203 C.R. Creveling, E.T. McNeal, D.H. McCulloh and J.W. Daly, J. Neurochem., 35 (1980) 922.
204 S. Psychoyos, Biochem. Pharmacol., 30 (1981) 2182.
205 L.B. Hough, H. Weinstein and J.P. Green, Adv. Biochem. Psychopharmacol., 21 (1980) 183.
206 U. Schwabe, Y. Ohga and J.W. Daly, Naunyn Schmiedebergs Arch. Pharmacol., 302 (1978) 141.
207 H. Carswell, P.R. Daum and J.M. Young, in Ref. 8, pp. 27–38.
208 E.B. Hollingsworth and J.W. Daly, Biochim. Biophys. Acta, 847 (1985) 207.
209 D. Sugden, J. Vanecek, D.C. Klein, T.P. Thomas and W.B. Anderson, Nature (London), 314 (1985) 359.
210 E.B. Hollingsworth, E.B. Sears and J.W. Daly, FEBS Lett., 184 (1985) 339.
211 A.L. Mueller, B.J. Hoffer and T.V. Dunwiddie, Brain Res., 214 (1981) 113.
212 M. Olianas, A.P. Oliver and N.H. Neff, Neuropharmacology, 23 (1984) 1071.
213 H.L. Haas, Agents Actions, 16 (1985) 234.
214 H.L. Haas and A. Konnerth, Nature (London), 302 (1983) 432.
215 H.L. Haas, Agents Actions, 14 (1984) 534.
216 H.L. Haas, in Ref. 8, pp. 213–223.
217 D.V. Madison and R.A. Nicoll, Nature (London), 299 (1982) 636.
218 T.T. Quach, A.M. Duchemin, C. Rose and J.-C. Schwartz, Mol. Pharmacol., 17 (1980) 301.
219 E. Brown, D.A. Kendall and S.R. Nahorski, J. Neurochem., 42 (1984) 1379.
220 T.B. Bolton, Physiol. Rev., 59 (1979) 606.
221 R.H. Michell, Trends Biochem. Sci., 4 (1979) 128.
222 R.H. Michell, C.J. Kirk, L.M. Jones, C. Downes and J.B. Creba, Phil. Trans. R. Soc. B., 296 (1981) 123.
223 S.J. Weiss, J.S. McKinney and J.W. Putney, Biochem. J., 206 (1982) 555.
224 C.P. Downes and M.M. Wusteman, Biochem. J., 216 (1983) 633.
225 M.J. Berridge, Biochem. J., 212 (1983) 849.
226 M.A. Wallace and J.N. Fain, J. Biol. Chem., 260 (1985) 9527.
227 S. Cockcroft and B.D. Gomperts, Nature (London), 314 (1985) 534.
228 R.A. Gonzales and F.T. Crews, Biochem. J., 232 (1985) 799.
229 Y. Nishizuka, Trends Biochem. Sci., 8 (1983) 13.
230 Y. Nishizuka, Nature (London), 308 (1984) 693.
231 H. Streb, R.F. Irvine, M.J. Berridge and I. Schultz, Nature (London), 306 (1983) 67.
232 G.M. Burgess, P.P. Godfrey, J.S. McKinney, M.J. Berridge, R.F. Irvine and J.W. Putney, Nature (London), 309 (1984) 63.
233 A.P. Somlyo, Nature (London), 316 (1985) 298.
234 R.H. Michell, Biochim. Biophys. Acta, 415 (1975) 81.
235 R.O. Friedel and S.M. Schanberg, J. Neurochem., 24 (1975) 819.
236 N. Subramanian, F.J. Seidler, F.J. Whitmore and T.A. Slotkin, Life Sci., 27 (1980) 1315.
237 M.J. Berridge, C.P. Downes and M.R. Hanley, Biochem. J., 206 (1982) 587.
238 J. Donaldson and S.J. Hill, Eur. J. Pharmacol., 124 (1986) 255.
239 P.R. Daum, C.P. Downes and J.M. Young, Eur. J. Pharmacol., 87 (1983) 497.
240 D.A. Kendall and S.R. Nahorski, J. Neurochem., 42 (1984) 1388.
241 H. Carswell, A. Galione and J.M. Young, Br. J. Pharmacol., 87 (1986) 71P.
242 J. Donaldson and S.J. Hill, J. Neurochem., 47 (1986) 1476.
243 H.J. Fleisch, M.C. Krzan and E. Titus, Circ. Res., 33 (1973) 284.

244 J. Donaldson and S.J. Hill, Br. J. Pharmacol., 87 (1986) 191.
245 C. Edwards, S.R. Nahorski and K.J. Rogers, J. Neurochem., 22 (1974) 565.
246 T.T. Quach, C. Rose and J.-C. Schwartz, J. Neurochem., 30 (1978) 1335.
247 E. Ozawa, J. Neurochem., 20 (1973) 1487.
248 R.D. Assimocopoulos-Jeannet, P.R. Blackmore and J.H. Exton, J. Biol. Chem., 252 (1977) 2662.
249 J.-C. Stoclet, Biochem. Pharmacol., 30 (1981) 1723.
250 T.T. Quach, A.-M. Duchemin, C. Rose and J.-C. Schwartz, Mol. Pharmacol., 20 (1981) 331.
251 E. Richelson, Science, 201 (1978) 69.
252 J.E. Taylor and E. Richelson, Mol. Pharmacol., 15 (1979) 462.
253 E. Richelson, in Ref. 8, pp. 237–247.
254 R.M. Snider, M. McKinney, C. Forray and E. Richelson, Proc. Natl. Acad. Sci. USA, 81 (1984) 3905.
255 J.F. Kuo, T.P. Lee, P.L. Reyes, K.G. Walton, T.E. Donnely and P. Greengard, J. Biol. Chem., 247 (1972) 16.
256 R.E. Study and P. Greengard, J. Pharmacol. Exp. Ther., 207 (1978) 767.
257 J.-M. Arrang, M. Garbarg, J.-M. Lecomte, H. Pollard, M. Robba, W. Schunack and J.-C. Schwartz, Nature (London), 327 (1987) 117.

NOTE ADDED IN PROOF

Arrang et al. [257] have recently described the properties of two selective and potent ligands for the histamine H_3-receptor. Thioperamide (*N*-cyclohexy-4-(imidazol-4-yl)-1-piperidinecarbothioamide) is a competitive antagonist of H_3-receptors in brain slices which has negligible activity on H_1- and H_2-receptor-mediated responses. They have also developed $(R)\alpha$-methylhistamine as a potent and selective H_3-receptor agonist. These novel agents should prove valuable in determining the rôle of histamine H_3-receptors in the mammalian CNS.

Progress in Medicinal Chemistry – Vol. 24, edited by G.P. Ellis and G.B. West

3 Molybdenum Hydroxylases: Biological Distribution and Substrate-Inhibitor Specificity

CHRISTINE BEEDHAM, B. Pharm., Ph.D., M.P.S.

Postgraduate School of Studies in Pharmaceutical Chemistry, University of Bradford, Bradford, BD7 1DP, United Kingdom

INTRODUCTION

Aldehyde oxidase (EC 1.2.3.1) and xanthine oxidase (EC 1.2.3.2) are molybdenum-containing enzymes which catalyse the oxidation of many aldehydes and nitrogen heterocycles. In addition, they have also been shown to catalyse, under *in vitro* conditions, a number of reductions. Although aldehyde oxidase is known to oxidize physiological compounds such as N^1-methylnicotinamide and pyridoxal [1], there does not yet appear to be a specific endogenous rôle for this enzyme. It has a wide distribution throughout the animal kingdom, and Krenitsky [2] has suggested that the primary metabolic function of the enzyme is of a fundamental nature rather than a highly specialized one. The name 'aldehyde oxidase', if not a misnomer, is at the least misleading, and several authors have put forward alternative names; these include quinine oxidase [3], pterine oxidase [4], quinine dehydrogenase [5], methotrexate oxidase [6] and N^1-methylnicotinamide oxidase [7]. However, each of these proposed alternatives implies a narrow substrate specificity for the enzyme and a more accurate description is that introduced by Bray to encompass both aldehyde oxidase and xanthine oxidase, namely "molybdenum hydroxylases' [8].

On the other hand, xanthine oxidase is involved in endogenous purine catabolism, catalysing the oxidation of hypoxanthine and xanthine to uric acid [9]. Nevertheless, this enzyme also has a wide substrate specificity and different functions have been assigned to it. One of the most plausible proposals is that both molybdenum hydroxylases may provide a biochemical barrier against ingested purines and pyrimidines, rendering them harmless [10]. This is similar to the view put forward in a recent review on the rôle of the molybdenum hydroxylases as drug-metabolizing enzymes, except that the substrate specificity is not restricted to purines and pyrimidines [11].

There are probably more publications relating to xanthine oxidase than to any other enzyme studied, certainly more than those pertaining to aldehyde oxidase. This is presumably because the former enzyme is easily accessible from cow's milk rather than from animal tissue. It is not the purpose of this review to include all the data amassed on xanthine oxidase, as this has been fully covered in recent reviews [8, 12, 13]. Furthermore, most of our own work has been concerned with aldehyde oxidase. Thus, this report compares the properties of the molybdenum hydroxylases, where possible, in terms of distribution, substrate and inhibitor specificity and mechanism of oxidation.

MECHANISM OF OXIDATION

REDUCTION AND RE-OXIDATION OF MOLYBDENUM HYDROXYLASES

The following section summarizes briefly current ideas on the mechanism of oxidation and intramolecular electron transfer of the molybdenum hydroxylases. Unless otherwise stated, the studies referred to have been carried out either with bovine milk xanthine oxidase or with rabbit liver aldehyde oxidase: the enzymes have a similar molecular structure, each being composed of two identical subunits of (M_r) 140,000–150,000 containing per subunit one atom of molybdenum (Mo), one flavin adenine dinucleotide molecule (FAD) and two iron-sulphur centres [Fe_2S_2] termed Fe/S I and Fe/S II [8, 12]. Each of these centres can function as a redox group in the intramolecular transfer of electrons *via* the enzyme from a reducing substrate such as xanthine or purine to an oxidizing substrate, e.g., oxygen or β-nicotinamide adenine dinucleotide (NAD^+) [8, 12, 14]. *Figure 3.1* illustrates the sites of substrate interaction with

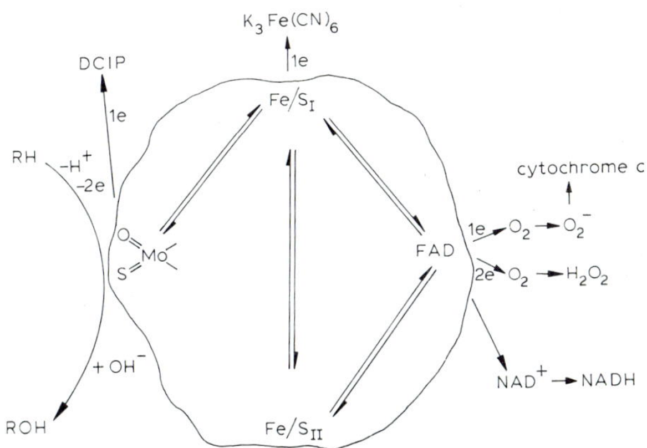

Fig. 3.1. Intermolecular and intramolecular electron transfer of the molybdenum hydroxylases.

the enzymes and also shows where intramolecular transfer between the various redox groups has been demonstrated [15–21]. The scheme is not meant to infer any particular spatial arrangement of the prosthetic groups. In fact, it has been suggested that in xanthine oxidase the Mo, Fe/S I and Fe/S II centres should be placed in a linear array, while the FAD is placed in a position to interact with both Fe-S centres but not the Mo centre [20]. In contrast, there appears to be no interaction between the two Fe-S centres in aldehyde oxidase [19].

With the exception of reduced *β*-nicotinamide adenine dinucleotide (NADH), substrates interact at the Mo centre and two electrons are transferred from the substrate to Mo(VI), reducing the metal to Mo(IV). The substrate residue reacts with an oxo ligand of Mo and a proton also reduces a terminal sulphide ligand of Mo. Hydrolysis of the Mo-substrate complex releases oxidized product, while the Mo(IV) is reoxidized by intramolecular transfer to other redox centres. The catalytic mechanism as described by Bray is depicted below [23]. Aldehyde oxidase and xanthine oxidase can each take up to six

$$\mathrm{Mo^{VI}}(=\mathrm{O})(=\mathrm{S}) + \mathrm{R{-}H} \longrightarrow \mathrm{Mo^{IV}}(\mathrm{OR})(\mathrm{SH}) \xrightarrow{H_2O} \mathrm{Mo^{IV}}(\mathrm{OH})(\mathrm{SH}) + \mathrm{ROH}$$

electrons per functional half-molecule [16, 24] and it has been assumed that the equilibration of reducing equivalents among the several redox sites is very rapid compared with the introduction or removal of electrons from xanthine oxidase and is determined solely by the relative oxidation-reduction potentials of the sites [25–27]. However, in one-electron-reduced xanthine oxidase, intramolecular transfer is not comparably faster than overall catalysis [28]. Reoxidation of the reduced enzyme can occur at any of the redox groups (see *Figure 3.1*). Artificial electron acceptors such as dichlorophenolindophenol (DCIP) or potassium ferricyanide can act as oxidizing substrates at the Mo and Fe-S centres, respectively [22, 29], but *in vivo* reoxidation of both molybdenum hydroxylases occurs via the FAD prosthetic group by reaction with either NAD^+ or oxygen [22, 30, 31]. Oxygen is assumed to be the physiological electron acceptor for aldehyde oxidase, as no interaction with NAD^+ is observed *in vitro* [32, 33]. Enzymic reduction of oxygen can proceed by a combination of one-electron and two-electron transfer to yield both superoxide anion and hydrogen peroxide as products [25, 30, 34].

Purified xanthine oxidase reacts rapidly with oxygen, but it is now widely accepted that 'native' enzyme exists mainly as a dehydrogenase, with NAD^+ as its natural acceptor [35–37]. This form, originally termed Type D by Della Corte and Stirpe [38], constitutes at least 80% of hepatic, intestinal or milk xanthine oxidase activity [37–40]. Studies on the properties of the Type D enzyme are complicated by the facile conversion of this form to at least two oxidases (Type O), depending on the method used. (i) Treatment with proteolytic enzymes causes an irreversible conversion of Type D to Type O [35]. Waud and Rajagopalan have proposed that this is due to the cleavage of an M_r 20,000 fragment, which facilitates binding of NAD^+ to the enzyme molecule [41]. A similar irreversible conversion to Type O is catalysed by a rat liver enzyme of the thiol-protein disulphide oxidoreductase type [42]. (ii) All other treatments, including sulphydryl-modifying reagents, heating and storage bring about a Type D to Type O conversion which can be reversed by thiols [38]. Consequently, this reversible conversion is thought to be due to modification of one or more vicinal thiol groups [31, 43]. In addition, intermediate dehydrogenase-oxidase (Type D/O) and dehydrogenase-associated oxidase forms have also been isolated, but conflicting results are obtained as to whether NAD^+ and oxygen react at the same FAD site [39, 44].

It was originally suggested by Stirpe and Della Corte that the Type O enzyme has no physiological significance and is merely an artifact of isolation [35]. The possibility cannot be ruled out, however, that the D–O conversion may also occur *in vivo* and could even play a rôle in the regulation of enzyme activity, linked *via* NADH production to the generation of adenosine 5′-triphosphate (ATP) [37]. Indeed, it has been shown that intestinal xanthine oxidase undergoes a rapid interconversion from Type D to Type O within 10 s of complete ischemia [45] and many of the proposed rôles of the enzyme depend on the postulated *in vivo* production of either superoxide anion or hydrogen peroxide [46–48] (see section on tissue distribution).

OXIDATION OF SUBSTRATES

Although aldehyde oxidase and xanthine oxidase both catalyse the oxidation of aldehydes to the corresponding carboxylic acid, there is very little information relating to mechanistic studies. Presumably, this is because simple aliphatic aldehydes are relatively poor substrates for these enzymes, but this is perhaps an area which warrants more detailed study as longer chain and aromatic aldehydes are rapidly oxidized by these enzymes (see [11]).

The hydroxylation of *N*-heteroaromatic substrates is a complex process involving nucleophilic attack at an electropositive carbon, which is generally adjacent to a ring nitrogen atom [10, 49, 50]. The products do not remain in the enol form but revert to a lactam, e.g., the hydroxylation of phthalazine (1) to 1-phthalazinone (2) as shown below [50].

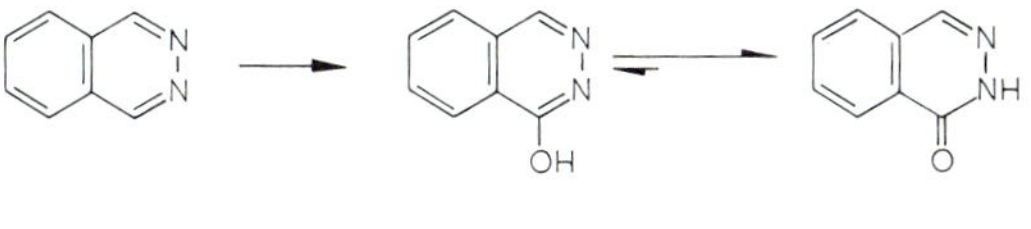

(1) (2)

Because the molecular properties of aldehyde oxidase and xanthine oxidase do not differ significantly and they contain basically similar substrate-binding sites, it is often assumed that any mechanistic conclusions regarding one enzyme, usually the latter, can be also applied to the other [14, 51]. This may not be a valid assumption to make; subtle alterations in the Mo centre may explain varying substrate specificities, but more fundamental modifications of the enzyme molecule may need to be considered to account for the differences in the position of substrate oxidation and, in particular, the rate-determining step. The oxidation of xanthine (3) or lumazine (4) catalysed by xanthine oxidase can be formulated as:

$$\mathrm{E}(i) + \text{substrate} \underset{K_{-1}}{\overset{K_1}{\rightleftharpoons}} \mathrm{E}(i)\ \text{substrate} \xrightarrow{K_2} \mathrm{E}(i+2)\ \text{product} \xrightarrow{K_3} \mathrm{E}(i+2) + \text{product}$$

where i is 0, 2 or 4 and E (i) and E $(i + 2)$ are enzyme species containing i and $i + 2$ electrons [25, 34]. Under most conditions, the breakdown of the enzyme-product complex is the rate-determining step [25, 34, 52]. Thus, there is little or no difference between the maximal oxidation rates (V_{max} values) of deuterium- or protium-substituted substrates (53, 54]. However, an isotope effect on the Michaelis-Menten constant K_m (K_m = [S] at $\frac{1}{2}$ V_{max} value) has been observed for the replacement of hydrogen by deuterium in xanthine and it is proposed that at low substrate concentrations the electron transfer process becomes slower [52, 53]. Results obtained using pteridines with either electron-withdrawing or electron-donating substituents also indicate that electron transfer can be rate-limiting [54].

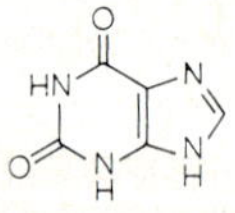

(3) 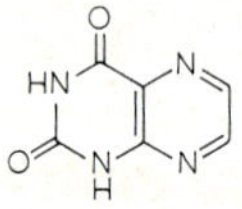 (4)

In contrast to xanthine-oxidase-catalysed oxidation, which is not very sensitive to substituent effects but appears to be facilitated by a high electron density at the reaction site on the substrate molecule [54], the reaction catalysed by aldehyde oxidase is very sensitive to substituent effects and the rate-limiting step is facilitated by a low electron density at the reaction site [51]. A similar conclusion has been reached by Ruenitz and Thomas, who demonstrated a significant isotope effect on the aldehyde-oxidase-catalysed oxidation of 6-ethyl-5*H*-dibenz[*c*,*e*]azepine (5) [55]. They suggest that hydrogen transfer from the methine carbon in the seven-membered ring is the rate-limiting step [55]. Furthermore, studies by the same group show a negligible solvent hydrogen isotope effect in the turnover number for both (5) and N^1-methylnicotinamide (7a) [56]. Unlike the reaction catalysed by xanthine oxidase, therefore, it appears that hydrolysis of the enzyme-product complex is not a rate-determining step.

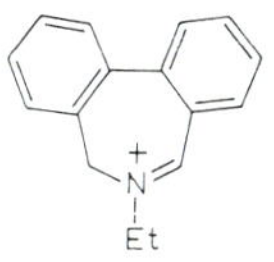

(5)

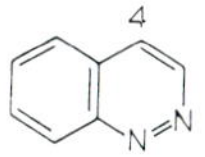

(6)

The electron density at the carbon undergoing oxidation does appear to have a greater influence on aldehyde-oxidase-catalysed attack than on that of xanthine oxidase. Thus cinnoline (6) is oxidized at carbon 4, the most electropositive carbon, by aldehyde oxidase but is refractory to oxidation by xanthine oxidase [50]. Similarly, quaternary compounds often undergo more facile oxidation by aldehyde oxidase than the corresponding non-quaternized bases [3, 5, 57], whereas the former group of compounds are poor substrates for xanthine oxidase except at high pH values [51, 58]. Quaternization of a ring nitrogen in a six-membered aromatic ring activates both the α- and the γ-carbons to nucleophilic attack and aldehyde oxidase can catalyse simultaneous oxidation at either of these positions [57, 59]. However, the product ratio is governed not only by electronic effects but also by steric factors. This is illustrated in *Table 3.1*, which shows the major products arising from rabbit liver aldehyde-oxidase-catalysed oxidation of quaternary pyridinium and quinolinium compounds [51, 57, 60, 62]. N-substituted pyridinium chlorides are oxidized predominantly at carbon 6, except when the position is sterically hindered [51, 60]. The quinolinium compounds are more susceptible to steric effects, presumably because of their larger size, and in most cases the 4-quinolone is the major product [57, 61, 62]. A further complication in considering the mechanism of aldehyde-oxidase-catalysed oxidation is that the product ratio also varies with species [57, 59], e.g., when *N*-phenylquinolinium perchlorate is incubated with guinea-pig liver enzyme, 1-phenyl-4-quinolone is the predominant product, whereas 1-phenyl-2-quinolone is the major oxidation product with rabbit liver aldehyde oxidase [57].

Unlike the iminium ions mentioned above, many of the intermediate

Table 3.1. OXIDATION PRODUCT RATIOS WITH RABBIT LIVER ALDEHYDE OXIDASE [51, 57, 60–62]

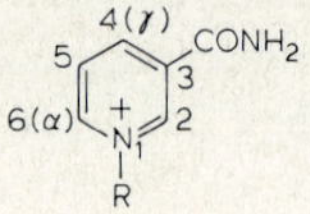

(7) 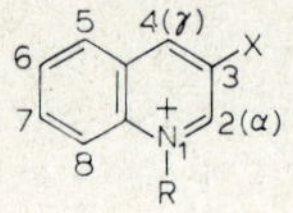 (8)

Compound	R	X	Predominant product		Product ratio
			α	γ	α : γ
7a	Me		+		
7b	Et		+		
7c	Pr		+		
7d	*i*Pr		+		3.5 : 1
7e	*t*Bu			+	
7f	Ph		+		> 16 : 1
7g	4-MeC_6H_4		+		6.5 : 1
7h	4-$MeOC_6H_4$		+		1.5 : 1
7i	4-FC_6H_4		+		> 15 : 1
7j	4-ClC_6H_4		+		> 8 : 1
7k	4-OHC_6H_4		+		6.9 : 1
7l	$PhCH_2$		+		
7m	1,3,5-$Me_3C_6H_2$			+	
7n	1,3,5-$Me_3C_6H_2CH_2$		+		
8a	Me	H	+		21 : 1
8b	Ph	H	+		4.5 : 1
8c	$PhCH_2$	H		+	1 : 3.4
8d	2,6-$Cl_2C_6H_4CH_2$	H		+	
8e	Me	$CONH_2$		+	
8f	$PhCH_2$	$CONH_2$		+	

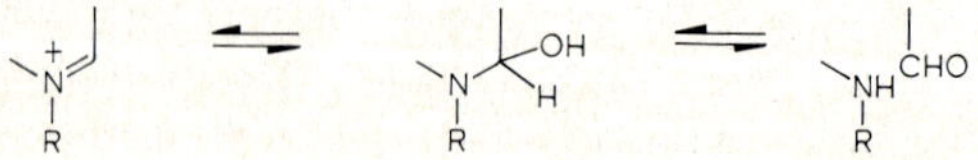

(9) (10) (11)

metabolites of drugs are unstable and it is often difficult to prove categorically whether it is the iminium ion intermediate (9), the hydrated carbinolamine (10) or a ring-opened aldehyde (11) which is the substrate for aldehyde oxidase

[64–67]. In the case of azapetine (12) and nicotine (13), the iminium ion intermediate rather than the carbinolamine appears to be the form of substrate preferred by aldehyde oxidase [63–66]. On the other hand, the ring-opened aldehyde intermediate (aldophosphamide) (14) of cyclophosphamide is oxidized by the enzyme [67], although the isomeric 4-hydroxycyclophosphamide (15) also undergoes oxidation by a soluble fraction enzyme [11]. Oxidation of the carbinolamine form to a cyclic lactam would not seem to involve nucleophilic enzyme attack, and yet we have shown that the stable pseudobase of 3-methylquinazolin-2-one (16), is an efficient substrate of aldehyde oxidase and competitively inhibits the oxidation of both quaternary and non-quaternized substrates [62].

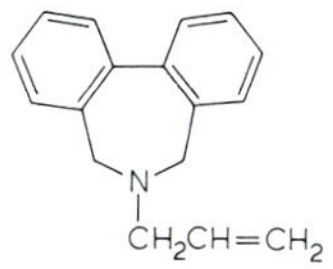
N
$CH_2CH=CH_2$

(12)

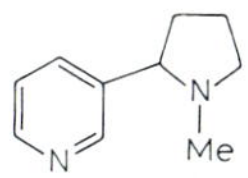
N
N
Me

(13)

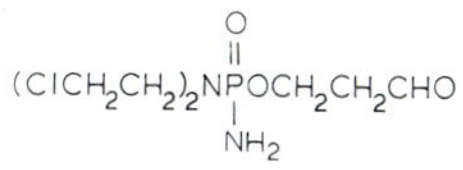
O
$(ClCH_2CH_2)_2NPOCH_2CH_2CHO$
NH_2

(14)

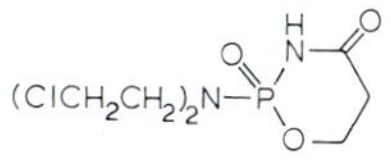
H
N
O
O
$(ClCH_2CH_2)_2N$—P
O

(15)

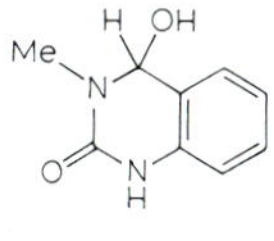
H OH
Me
N
O
N
H

(16)

BIOLOGICAL DISTRIBUTION

CELLULAR LOCALIZATION

A major part of the molybdenum hydroxylase activity is found in the cytosol; thus, the aldehyde oxidase activity from sources as diverse as rabbit liver [14, 32, 68], guinea-pig polymorphonuclear leucocytes (PMN) [69] and potato tubers [70] is located primarily in the 100 000 × *g* supernatant fraction, as is hepatic xanthine oxidase [71, 72]. However, there is also recent evidence to support the existence of these enzymes, particularly aldehyde oxidase, in microsomal pellets prepared from horse, human or bovine liver [72–75]. In addition, Igo, Mackler and Duncan reported rat liver aldehyde oxidase to be primarily a mitochondrial enzyme [76], although this experiment has not been repeated. Recent investigations have also found aldehyde oxidase activity predominantly in the cytosol of rat liver [77], but in this study, as in most cases, the mitochondrial fraction was not tested and was routinely discarded. It is interesting to note that the work by Igo, Mackler and Duncan on the cellular localization was actually performed using rat liver and not hog preparations [76] as is often quoted [32]. By contrast, milk xanthine oxidase is associated mainly with the milk lipid globules or liposomes [78, 79] which are released by the secretory cells of the mammary gland [80]. Homogenization of bovine milk can cause partial disruption of the lipid globule membrane to release and activate some of the previously membrane-bound enzyme [81, 82].

On the other hand, it has been proposed that most of the xanthine oxidase in homogenized milk is entrapped in liposomes in a form which is resistant to gastric digestion [83]. Not surprisingly, there are opposing views as to whether the enzyme can (i) resist hydrolysis in the stomach and/or (ii), because of its large size (290 kDa), be absorbed from the small intestine [84–87]. In support of the theory, Schousten and De Jong have found statistically increased plasma levels of xanthine oxidase in rabbits after the administration of fortified milk preparations [87]. Nevertheless, the hypothesis first proposed by Oster, that the absorption of intact bovine milk xanthine oxidase from homogenized milk plays a rôle in the initiation of atherosclerosis, must still be regarded as equivocal in view of the controversial evidence [83, 84, 88–91].

The highest concentration of xanthine oxidase in mammals is found in the milk and lactating mammary gland [78, 92], although it is not known whether all species secrete the enzyme in milk. For obvious reasons, nearly all the information available refers to xanthine oxidase from bovine milk, but in fact both rat and guinea-pig milk have a higher enzyme content [92]. *Table 3.2*

Table 3.2. COMPARATIVE DISTRIBUTION OF XANTHINE OXIDASE AND ALDEHYDE OXIDASE ACTIVITY IN MILK AND LIVER OF MAMMALIAN SPECIES [92, 93, 95]

All activities are expressed relative to rat enzyme activity (100%).

	Enzyme activity		
	Xanthine oxidase		*Aldehyde oxidase*
Species	*Milk* [a]	*Liver* [b]	*Liver* [c]
Rat	100	100	100
Guinea-pig	80	38	957
Cow	56	45	27
Rabbit	53	32	497
Mouse	42	42	43
Man	4	36	14
Human colostrum	11	–	–
Cat	4	165	58
Dog	2	77	< 18
Pig	0	0	83

[a] Mean rat milk xanthine oxidase activity = 0.187 U/ml.
[b] Mean rat liver xanthine oxidase activity = 4.21 U/mg.
[c] Mean rat liver aldehyde oxidase activity = 2.33 U/mg.

liver xanthine oxidase and aldehyde oxidase activity of various mammalian species. There seems to be no correlation between the amount of xanthine oxidase present in milk and liver of the same species. In fact, some of the species (e.g., cat, monkey) with the highest liver content exhibit very low milk xanthine oxidase activity [92, 93]. Xanthine oxidase shows a greater variation in milk than in mammalian liver, whilst the species differences in hepatic aldehyde oxidase activity are even more pronounced. In some species (e.g., cow, buffalo, goat), there is an inverse relationship between the amount of xanthine oxidase and fat content of the milk [94]. Human milk shows low and variable xanthine oxidase activity [93], which is presumably why some workers have been unable to detect any enzyme activity [95], whereas more sensitive assays employing antibodies raised to bovine milk xanthine oxidase have shown a positive reaction to the supernatant fraction of human lipid globule membranes [78]. Higher levels of the enzyme have been found in human colostrum [93, 96], which is the fluid produced by the mammary gland during the first 2–3 days post-partum. In buffaloes, xanthine oxidase activity also gradually increases with colostrum and then decreases with the changeover to the secretion of milk

[94]. It therefore appears that newborn infants have a greater requirement for maternal xanthine oxidase in the first few days after birth and it has been proposed that the high xanthine oxidase levels in colostrum may be to effect maximum absorption of dietary iron from the underdeveloped infant gut [93, 97]. Furthermore, studies in the mouse have shown that liver xanthine oxidase is not present at birth, but is only apparent after 3 weeks, whereas the main isozyme of aldehyde oxidase can be detected in the 20-day foetus [98].

Molecular properties of the milk enzyme such as amino-acid content have been shown to vary between species [93], but it is not yet clear whether the xanthine oxidase in milk is similar to or even identical with that in other tissues within a species. Krenitsky, Tuttle, Cattau and Wang [92] observed that bovine milk xanthine oxidase is not typical of mammalian tissue xanthine oxidase, on the basis of its electron acceptor specificity. However, both the milk and liver enzymes act as NAD^+-dependent dehydrogenases *in vivo*, which can be converted into an oxidase form during the isolation procedure [35, 37, 99, 100]. The differences noted [92] may be due to variable interconversion between the dehydrogenase and oxidase forms. Other workers have found the enzymes from bovine liver, lung, mammary gland and milk to be indistinguishable on the basis of sensitive immunological methods [78], although this may not always be the case with other species. There are no reports of any aldehyde oxidase activity in milk.

TISSUE DISTRIBUTION

One of the problems in considering tissue distribution and (in the following section) species distribution of aldehyde oxidase is the number of substrates employed to monitor activity. No clear endogenous rôle has yet emerged for this enzyme, although it is known to oxidize both physiological compounds and a wide range of xenobiotics [11]. Consequently, aldehyde oxidase activity has been compared with diverse compounds such as N^1-methylnicotinamide (7a), pyridoxal (17), 6-methylpurine (18), methotrexate (19), quinine (20) and vanillin (21), all of which may have different tissue or species specificity [1, 92, 101–104]. Variations may be more pronounced between species than between different tissues. Nevertheless, of all the mammalian tissues studied, the liver contains the highest levels of aldehyde oxidase, with other organs such as the lung, kidney and small intestine containing less than 50% of the hepatic enzyme activity [92, 101, 102]. In the rabbit, using either methotrexate (19) or quinine (20), the lung is second to the liver in activity, followed by the uterus, small intestine and kidney [101–103], whereas the guinea-pig kidney has about 45% of the guinea-pig liver activity towards either 6-methylpurine (18) or

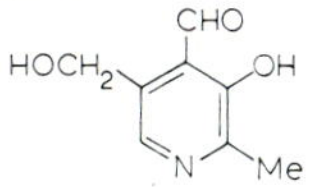

(17) 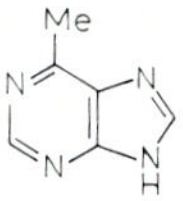 (18)

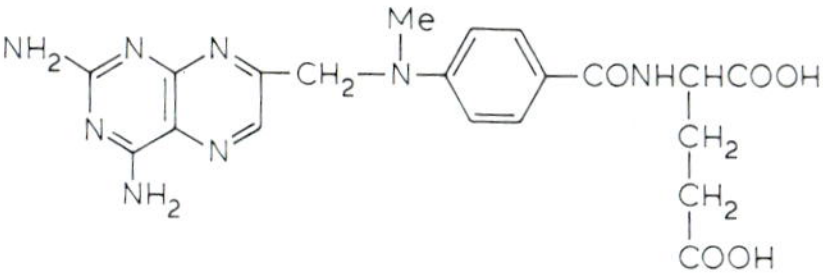

(19)

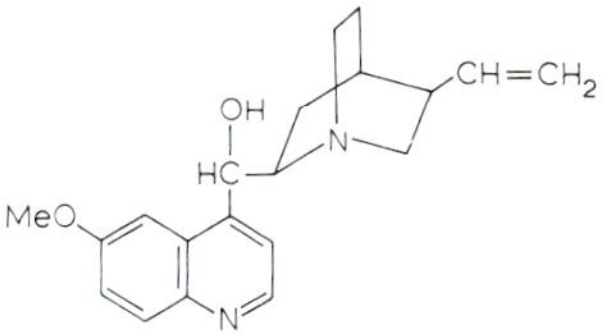

(20) 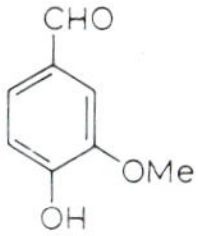 (21)

phthalazine (1), with the lung and small intestine containing considerably less activity [91, 105]. Values for human tissues are rather low compared with those from rabbit and guinea-pig but, again, only one substrate, 6-methylpurine (18), was used in the analysis [92]. Furthermore, human aldehyde oxidase is known to be very unstable following surgical excision, and post mortem samples often appear to be devoid of enzyme activity [74]. In our laboratories, using substrates based on the phthalazine nucleus, human liver aldehyde oxidase has been shown to have activity comparable to that of the guinea-pig enzyme [106].

It is easier to compare studies on xanthine oxidase, because one substrate, xanthine (3), is generally used to measure enzyme activity. In contrast to aldehyde oxidase, the mammary gland and the intestine have a higher xanthine oxidase content than the liver in most species [78, 92]; cats and humans are exceptions in that the intestinal and hepatic xanthine oxidase concentrations are approximately equal [92, 107]. Both molybdenum hydroxylases are found in the kidney, lung, spleen, stomach, muscle and heart [101, 102, 108, 109]. Little is known about the location of aldehyde oxidase within particular tissues, but xanthine oxidase is localized exclusively in sinusoid and capillary endothelial cells of hepatic, cardiac, pulmonary and adipose tissue [75, 110]. In mammary gland a positive reaction with antibodies to xanthine oxidase is seen in both the lactating epithelial cells and capillary endothelial cells [75]. The intestine also contains xanthine oxidase located primarily in the mucosal epithelium, with neither the lymph nodes nor the underlying muscle layer showing any significant enzyme activity [111]. In both the hamster and guinea-pig intestine there is a gradual decrease (approximately 50%) in xanthine oxidase activity from the proximal jejunum to distal ileum [111, 112]. Not only is intestinal xanthine oxidase responsible for the conversion of hypoxanthine and xanthine to uric acid, but Berlin and Hawkes have also found evidence for the active secretion of the hydroxylated purines into the intestinal lumen and have suggested that this secretory mechanism may be a pathway for the extrarenal excretion of urate [111]. However, other investigators have been unable to show a concentration gradient for purines across rodent intestinal mucosa [113]. Alternatively, intestinal xanthine oxidase, may be present in the mucosal cell to facilitate the transcellular transport of ionic iron by promoting its oxidative incorporation into mucosal transferrin [114, 115].

High concentrations of intestinal xanthine oxidase also have important implications in developing suitable dosage forms for drugs which are substrates for the enzyme; thus 6-mercaptopurine, which is administered almost exclusively in tablet form, undergoes substantial biotransformation during absorption by rat intestine [116] and also has a low oral bioavailability in man [117, 118]. Another possible site for extrahepatic metabolism is the skin. In those species tested, skin was found to have xanthine oxidase activity comparable to that of the kidney [92, 107]. If aldehyde oxidase is present in amounts similar to xanthine oxidase, drugs such as acyclovir, which is known to be oxidized by hepatic aldehyde oxidase [119], may be inactivated during passage through the skin from topical preparations.

Whole brain homogenates have been found to contain very low levels or to be devoid of either of the molybdenum hydroxylases [92, 107, 108, 120]. However, more specific assays for xanthine oxidase gave values of 2–20 nmol xanthine transformed/mg per h for homogenates of cortex or whole brain from mouse, rat, guinea-pig, rabbit and cow [46, 121]. Less activity was detected in the cerebellum [121] and, again, the cranial capillary endothelial cells were found to be enriched in xanthine oxidase [46]. Brain hypoxanthine concentrations are reported to rise during ischaemia due to increased ATP breakdown, and Betz [46] proposed that brain capillaries may be susceptible to damage from the xanthine-oxidase-derived oxygen radicals arising from increased substrate availability. There is also evidence [122] that the vascular injury associated with intestinal ischaemia is a result of oxygen radicals produced *via* xanthine oxidase. Rat brain xanthine oxidase is completely inhibited by subcutaneous injection (30 mg/100 g) of barbiturate, a typical inducer of the cytochrome *P*-450 monoxygenase system [123]. The concentrations used in this study are rather higher than those usually employed in induction studies, although inhibition of both brain and serum xanthine oxidase was also observed *in vitro* [123].

It has been proposed that the appearance of xanthine oxidase in the serum may be a useful indication of liver injury or, more specifically, capillary endothelial cell damage [107, 124, 125]. However, some mammals have reasonably high levels of the enzyme under normal conditions; these include rat, cow, donkey, badger, rabbit, guinea-pig, mouse and dog [107, 126]. Al-Khalidi and Chaglassian found the concentration of xanthine oxidase in whole blood to be about half of that contained in the serum in most mammals studied [107]. On the other hand, xanthine oxidase has been identified in mouse leucocytes, the levels of which are significantly increased in bacterial- or protazoally infected animals [127]. An analogous situation, but with a more marked effect, also occurs in the liver of similarly treated animals [128]. The authors have postulated that the increase in xanthine oxidase activity in leucocytes may represent a prompt defence mechanism to infection, probably acting *via* the production of superoxide anion [47]. Rat plasma is reported to be devoid of aldehyde oxidase activity [125] but guinea-pig leucocytes contain a molybdenum hydroxylase with a specificity similar to that of guinea-pig liver aldehyde oxidase which is inactive towards xanthine [69, 129].

Although attempts to identify either enzyme in human sera have so far proved unsuccessful, exceptionally high concentrations of antibodies to xanthine oxidase have been found in human sera as well as in that from various animals [124]. Despite these unusually high levels, it is proposed that the antibodies result from self-immunization to the xanthine oxidase antigen present in the endothelial cells of capillaries, although the authors also acknowledge that the antibodies in human sera may be induced by bovine milk xanthine oxidase ingested in dairy products [124].

The thyroid also contains xanthine oxidase [48, 107]. In rats the thyroid has about 30% of liver enzyme activity, with a ratio of dehydrogenase : oxidase levels of approximately 3 : 1 [130]. As a result of *in vitro* experiments, Lee and Fischer reported that the hydrogen peroxide produced by the oxidase form of xanthine oxidase is required for the biosynthesis of thyroid hormones [48]. However, Japanese workers [130] who did find a suppression of thyroid hormone biosynthesis *in vivo* by the xanthine oxidase inhibitor allopurinol could not produce any inhibitory effect when the enzyme was inactivated by the administration of tungsten. They concluded that the biosynthesis of thyroid hormones was not mediated by xanthine oxidase.

SPECIES DISTRIBUTION

Although a comparison of species variation is complicated by the lack of a common analytical method, it is clear that the molybdenum hydroxylases are widely distributed throughout the animal kingdom. Monocellular organisms such as bacteria have been shown to contain xanthine oxidase [131]. This enzyme has also been found in lentil seedlings; the activity of the enzyme increases from germination to a maximum in 24 h, after which time the activity gradually decreases [132]. Two isoenzymes of aldehyde oxidase have been identified in potato tubers, but this was tested only with aldehyde substrates [70]. In two extensive surveys comparing molybdenum hydroxylase distribution in animals, both aldehyde oxidase and xanthine oxidase activities were measured with different substrates, the former enzyme with 6-methylpurine (18) and vanillin (21) and the latter enzyme with xanthine (3) and hypoxanthine (22), which may account for any discrepancies between the two reports [92, 104]. The results can be summarized as follows: both enzymes were detected in some multicellular phylogenetically primitive species such as Coelenterata (e.g., sea anemone) [92] and aldehyde oxidase was also found to be prevalent amongst molluscs, crustaceans and insects [104]. Two insects appeared to display xanthine dehydrogenase activity as larvae which changed to aldehyde oxidase activity in their adult stage [104]. The species difference in aldehyde oxidase activity appears to be more pronounced than that in xanthine oxidase, and Wurzinger and Hartenstein [104] concluded that the former enzyme is phylogenetically more primitive to xanthine oxidase. With regard to birds, xanthine oxidase was present in four out of six species tested in one study, whereas none of these displayed aldehyde oxidase activity [104]. Krenitsky, Tuttle, Cattau and Wang also found very little aldehyde oxidase activity in avian liver, although higher levels of the enzyme were found in avian kidney [92].

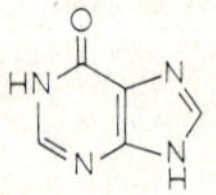

(22)

Amongst mammals, xanthine oxidase levels appear to be less variable than aldehyde oxidase, with the highest activity for the former enzyme observed in dog, rat and cat (see *Table 3.2*). This is convenient for those drugs which are potential substrates for the enzyme, as dog and rat are the species routinely employed for metabolism and toxicity studies. For example, 6-mercaptopurine undergoes substantial biotransformation to 6-thiouric acid in both man and rat [116–118].

Unfortunately, it is more difficult to choose an animal model for human aldehyde oxidase because of the different substrate variation for each species. *Table 3.3* compares the *in vivo* and *in vitro* activity of the enzyme from some

Table 3.3. ACTIVITY OF HEPATIC ALDEHYDE OXIDASE TOWARDS DIFFERENT DRUGS IN SOME COMMON LABORATORY SPECIES [59, 103, 119, 133–135]

		In vitro *activity*[a]		In vivo *activity*
Species	*Quinine*[b]	*N^1-Methyl-nicotinamide*[c]	*Methotrexate*[d]	*Acyclovir (% excreted as 8-hydroxyacyclovir)*
Rabbit	100	100	100	3.5
Rat	50	3.3	0.2	0.6
Dog	0	n.d.[e]	n.d.	0.6
Mouse	10	14.8	0.1	1.1
Guinea-pig	0	55	2.2	3.1
Monkey	9	9.8[f]	n.d.	15.1

[a] All *in vitro* activities expressed relative to rabbit enzyme activity (100%).
[b] % Destruction quinine/0.2 g liver per h.
[c] Rabbit N^1-methylnicotinamide oxidase activity = 0.121 μmol/mg per 10 min.
[d] Rabbit methotrexate oxidase activity = 0.218 μmol/mg per h.
[e] n.d., not determined.
[f] Homogenate prepared from frozen liver.

common laboratory species (see also *Table 3.2*). The amount of 8-hydroxy-9-(2-hydroxyethoxymethyl)guanine (8-hydroxyacyclovir) excreted in man has not been determined, but oral administration of acyclovir (24, *enol* form) gives a mean urinary recovery of the parent drug of only 14.4% [136]. The aldehyde-oxidase-mediated products of both methotrexate (19) and quinine (20) account for up to 33% and 40%, respectively, of the administered dose in man [137–140]. Dog consistently shows minimal aldehyde oxidase activity, both *in vivo* and *in vitro* [92, 103, 104, 141], which can lead to a marked species difference between dog and man in the bioavailability of certain drugs. Thus carbazeran (25), a potent cardiac stimulant in dogs, is rapidly inactivated in man and baboon, *via* 4-hydroxylation catalysed by aldehyde oxidase, whereas in the dog this metabolic route is absent [142, 143]. Rat liver exhibits marked variation in aldehyde oxidase activity between different animals [77, 144, 145]; even within the same inbred strain of rat, enzyme activity can vary up to 20-fold [144, 145]. This means that neither of these species is really suitable for testing drugs which may be oxidized by the enzyme.

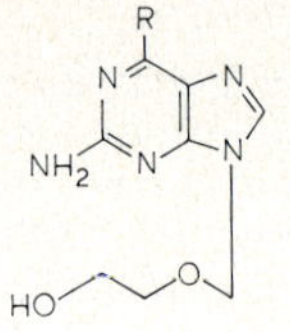

(23) R = H
(24) R = OH

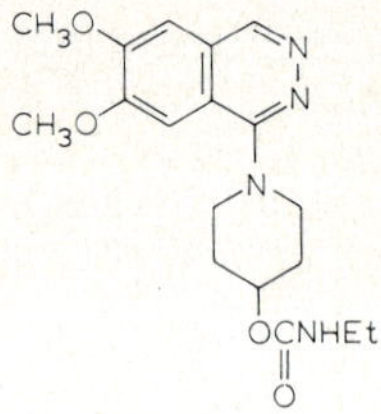

(25)

On the other hand, most workers prepare aldehyde oxidase from rabbit liver on the assumption that rabbit enzyme is able to catalyse a wider range of heterocyclic substrates than that from other mammalian species [3, 33, 146]. With some drugs it is possible to extrapolate from *in vitro* rabbit studies to metabolic biotransformation in humans. Thus, methotrexate (19) is rapidly oxidized by rabbit liver aldehyde oxidase, whereas rat and mouse liver are much less active in this respect [6, 133]. Oxidation to the inactive 7-hydroxy analogue accounts for approximately 30% of the dose excreted in rabbits [103, 147], but no metabolites were found in rats [148]. In high-, medium- and low-dose cancer therapy, 7-hydroxymethotrexate is the major metabolite identified and plasma concentrations of the metabolite exceed those of the parent drug after the end of dosing [147, 149–154]. However, rabbit and human hepatic aldehyde oxidase do not always have the same substrate specificity; thus, carbazeran (25), which is rapidly oxidized by baboon and human aldehyde oxidase, is not a substrate for the rabbit liver enzyme [155]. Furthermore, some quaternary substrates can be oxidized simultaneously at two alternative positions by hepatic aldehyde oxidase [57, 59, 61] and the ratio of the oxidation products is species-dependent [57, 59]. We have demonstrated that the major metabolite of *N*-phenylquinolinium perchlorate (8b) changes from the 2-quinolone with rabbit aldehyde oxidase to the 4-quinolone with guinea-pig enzyme [57]. Under identical conditions, 80% of *N*-phenylquinolinium perchlorate is converted exclusively to the 4-quinolone with human liver aldehyde oxidase [156]. As carbazeran is also an efficient substrate of guinea-pig enzyme [155], the results suggest that in respect to aldehyde-oxidase-catalysed oxidation the guinea-pig, rather than the rabbit, more closely resembles man.

FACTORS AFFECTING SUBSTRATE AND INHIBITOR SPECIFICITY

In this section are described the important chemical features of those substrates which are oxidized by the molybdenum hydroxylases. Although these enzymes, particularly aldehyde oxidase, also catalyse numerous reductive reactions under anaerobic conditions *in vitro*, it has not yet been established whether they occur under physiological conditions and there are as yet insufficient examples of any one reduction reaction to permit any conclusions regarding the structure of substrates. Thus, such reactions will not be discussed here (see [11] and references therein). Properties of those inhibitors which bind at the Mo centre and are also substrate analogues will also be included. However, the interaction of inhibitors such as cyanide and arsenite with the molybdenum hydroxylases and the mechanism of action of the specific xanthine oxidase inhibitor, allopurinol, have been comprehensively described elsewhere [8, 12, 14, 157].

ALDEHYDES

Aliphatic aldehydes are oxidized by the molybdenum hydroxylases; the K_m values for the same substrate are much lower with aldehyde oxidase than xanthine oxidase (*Table 3.4*) [33, 158–160], but even these are some 100-times higher than the corresponding value with an aldehyde dehydrogenase [161]. V_{max} values obtained with xanthine oxidase decrease with increasing carbon chain length, while the K_m values do not alter significantly (*Table 3.4*) [160, 162]. Optimum activity, towards aldehyde oxidase occurs with a three- or four-carbon unbranched chain [33, 163, 164]. Longer aliphatic aldehydes (up to sixteen C) have decreasing V_{max} values, but these are achieved at lower substrate concentrations [163–165]. The importance of a lipophilic binding site near the Mo centre is indicated by the lower K_m values for aromatic aldehydes, although once again these substrates have a greater affinity for aldehyde oxidase than xanthine oxidase [33, 158–160]. In this aspect, aldehyde oxidase would appear to be aptly named. In addition, an *N*-heteroaromatic ring appears to facilitate binding of the substrate but does not necessarily enhance the rate of oxidation [160]. However, if there is a vacant α-position in the ring system, e.g.,

Table 3.4. MOLYBDENUM HYDROXYLASE ACTIVITY WITH ALIPHATIC AND AROMATIC ALDEHYDES [33, 159, 160, 163]

	Bovine milk xanthine oxidase		*Hepatic aldehyde oxidase*			
			Hog		*Man*	*Rabbit*
Substrate	K_m *(mM)*	V_{max}	K_m *(mM)*	V_{max}	V_{max}	K_m *(mM)*
Acetaldehyde	130	100[a]	100	100[b]	100[c]	1
Proprionaldehyde	430	23.3	30	110	132	–[d]
Butyraldehyde	142	2.4	25	261	–	–
Pentanaldehyde	–	–	1.25	52	–	–
Heptaldehyde	–	–	1.3	7	–	–
Salicylaldehyde	1	7.7	–	–	–	0.1
Pyridine-2-aldehyde	0.36	3.4	–	–	–	–
Pyridine-3-aldehyde	0.05	2.7	–	–	–	–
Pyridoxal	–	0	–	–	25	–

[a] Bovine milk oxidase activity with acetaldehyde = 100/s.
[b] Hog liver aldehyde oxidase activity with acetaldehyde = 7.9 μmol/mg per min.
[c] Human aldehyde oxidase activity, comparative rates only.
[d] Not determined.

quinoline-6-aldehyde, oxidation of the nucleus may precede that of the aldehyde group [166].

The oxidation of substituted benzaldehydes by xanthine oxidase is sterically hindered by bulky substituents at the *ortho* (*o*) position (*Table 3.5*) [167]. Increasing the size of the halo-substituent dramatically decreases the oxidation of the *o*-substituted compound, whereas that of the *p*-halobenzaldehyde increases due to the increased inductive effect. The positional specificity was not due to electronic effects, because the oxidation rate was also decreased with electron-donating *o*-substituents. Although the substrates of aldehyde oxidase have not been so rigourously examined, the enzyme does appear to be subject to similar steric considerations, as *o*-chloro- and *o*-nitrobenzaldehyde are oxidized at much lower rates than benzaldehyde itself [33].

Phosphorylated aldehydes (26), which are structurally related to aldophosphamide (14), the aldehyde metabolite of cyclophosphamide, are also substrates for aldehyde oxidase [168]. A good correlation was found between the ability of the compounds (26) to inhibit N^1-methylnicotinamide oxidation competitively *in vitro* (as a competitive substrate) and the *in vivo* potentiation of cyclophosphamide efficacy in mice.

Table 3.5. SUBSTITUENT EFFECTS ON THE XANTHINE-OXIDASE-CATALYSED OXIDATION OF BENZALDEHYDE [167]

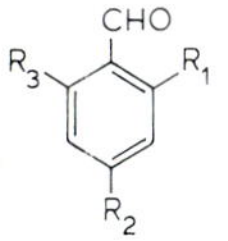

Substituent			
R_1	R_2	R_3	*Relative oxidation rate*
H	H	H	100[a]
Cl	H	H	70
Br	H	H	25
I	H	H	3
H	Cl	H	245
H	Br	H	264
H	I	H	453
Cl	H	Cl	1.5
Me	H	H	43
$Me(CH_2)_3$	H	H	5
MeO	H	H	26
H	Me	H	132
H	$Me(CH_2)_3$	H	108
H	MeO	H	53

[a] Bovine milk xanthine oxidase activity with benzaldehyde = 0.067 μmol/mg per min.

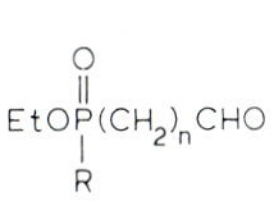

(26)

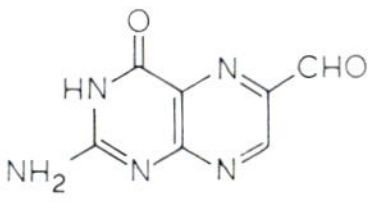

(27)

There are few reports of aldehydes acting solely as inhibitors of either molybdenum hydroxylase. With most of the aldehydes listed in *Table 3.4*, xanthine oxidase is reversibly inactivated at high substrate concentrations as it is turning over [169]. With the exception of formaldehyde inactivation, enzyme activity is completely recovered by incubation of the samples with oxygen [169]. The irreversible loss of activity observed with formaldehyde may be ascribed to the compound reacting at other sites in the enzyme as well as the Mo centre [160, 169, 170]. 2-Amino-4-oxopteridin-6-aldehyde (27) also progressively inactivates xanthine oxidase and shows a high affinity for the enzyme [4, 171]. However, this aldehyde (27) is probably also slowly oxidized by xanthine oxidase to the pterin carboxylic acid, as inhibitory activity gradually decreases when the aldehyde is incubated with enzyme [4, 171, 172]. In contrast, compound (27) shows substrate activity only with aldehyde oxidase [159].

N-HETEROCYCLES

Lipophilicity and the size of substrate-binding site

Although aldehyde oxidase and xanthine oxidase differ markedly in their substrate and inhibitor specificities, there is considerable evidence to suggest that binding of both substrates and inhibitors to either enzyme is facilitated by hydrophobic interaction in the enzyme active site.

Simple unsubstituted monocycles such as pyridine, pyrazine, pyrimidine, pyridazine and pyrrole show little or negligible binding to either enzyme [3, 10, 173]. However, we have shown that fusion or substitution of a phenyl group significantly increases the affinity of the compounds towards aldehyde oxidase, and in some cases the heterocycles are substrates for xanthine oxidase (*Table 3.6*) [50, 173]. Furthermore, studies with compounds containing a

Table 3.6. SPECIFICITY OF THE MOLYBDENUM HYDROXYLASES TOWARDS UNCHARGED *N*-HETEROCYCLES [50, 173–175]

	K_m *(mM)*	
Substrate	*Aldehyde oxidase*[a]	*Xanthine oxidase*[b]
Quinoline (1-azanaphthalene)	3	–
Isoquinoline (2-azanaphthalene)	0.2	–
Cinnoline (1,2-diazanaphthalene)	0.14	–
Quinazoline (1,3-diazanaphthalene)	0.015	0.25
Quinoxaline (1,4-diazanaphthalene)	0.18	–
Phthalazine (2,3-diazanaphthalene)	0.099	2.6
4-Phenylpyrimidine	0.76	–
Acridine (2,3-benzoquinoline)	0.5[c]	–
Phenanthridine (3,4-benzoquinoline)	<0.001	–
5,6-Benzoquinoline	–	–
7,8-Benzoquinoline	–	–

[a] Rabbit liver aldehyde oxidase.
[b] Bovine milk xanthine oxidase.
[c] Rat liver aldehyde oxidase.

second benzene ring fused on to the quinoline nucleus provide additional information on the limits of the hydrophobic binding site for uncharged substrates of aldehyde oxidase. Thus, 3,4-benzoquinoline, phenanthridine (28a), is one of the most efficient substrates of aldehyde oxidase, with a $K_m < 10^{-6}$ M, and the 2,3-isomer, acridine (29a), is rapidly oxidized by rat liver aldehyde oxidase [173, 174]. On the other hand, 5,6-benzoquinoline (30a) and 7,8-benzoquinoline (31a) do not bind to the enzyme [173], nor is a lactam metabolite isolated from rat liver incubations [176]. This indicates that the limit of the hydrophobic area is within 2.4–4.8 Å (i.e., 1–2 benzene rings) of the heterocyclic ring [177]. The lipophilic regions may also be important in the binding of the phenothiazines (32), which are potent competitive inhibitors of rabbit and human liver aldehyde oxidase [33].

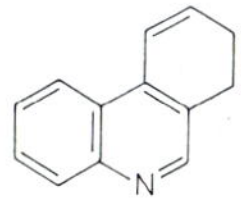

(28a)
(28b) 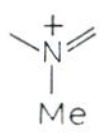 $\overset{+}{N}$–Me

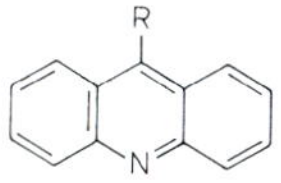

(29a) R = H
(29b) R = NH–(2-OMe-4-$NHSO_2Me$-phenyl)

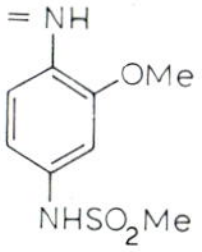

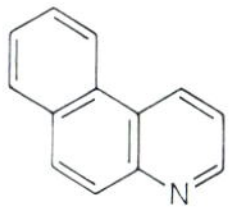

(30a)
(30b) 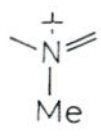 $\overset{+}{N}$–Me

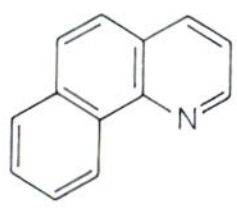

(31a)
(31b) 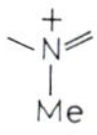 $\overset{+}{N}$–Me

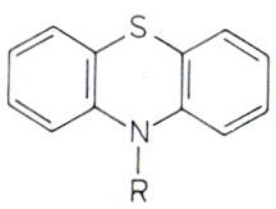

(32)

NH_2 NH_2 $\overset{+}{N}$ Br^- Et

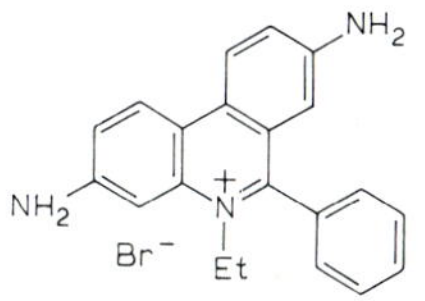

(33)

Despite the charge on *N*-heterocyclic cations, lipophilicity is also a determining factor in the binding of such compounds to aldehyde oxidase; thus, *N*-methylpyridinium does not serve as a substrate for aldehyde oxidase but only functions as a poor competitive inhibitor [159], whereas more lipophilic analogues are excellent substrates of the enzyme (*Table 3.7*) [57]. However,

Table 3.7. APPARENT MICHAELIS-MENTEN CONSTANTS FOR CATIONIC SUBSTRATES OF MOLYBDENUM HYDROXYLASES [57, 58, 62, 159]

	K_m *(mM)*	
Substrate	*Aldehyde oxidase*[a]	*Xanthine oxidase*[b]
N-Methylpyridinium	19[c]	–
N-Phenylpyridinium	n.d.	>70
N-Methylquinolinium	1.6	22
N-Phenylquinolinium	0.07	n.d.
N-Methylisoquinolinium	2[d]	–
N-Methylphenanthridinium	0.12	0.4
N-Methyl-5,6-benzoquinolinium	0.25	1.2
N-Methyl-7,8-benzoquinolinium	0.23	1.8

[a] K_m values determined at pH 7 and 30 °C with rabbit liver aldehyde oxidase.
[b] K_m values determined at pH 10.6 and 25 °C with bovine milk xanthine oxidase.
[c] K_i as competitive inhibitor of N^1-methylnicotinamide.
[d] K_i as competitive inhibitor of 6-methylpurine.

enzyme activity towards quaternized substrates does not necessarily reflect that observed with their uncharged counterparts, as can be seen from a comparison of *Tables 3.6* and *3.7*. Quaternization of a ring nitrogen normally activates a heterocyclic ring towards nucleophilic attack, yet *N*-methylisoquinolinium only binds as a weak competitive inhibitor, whereas isoquinoline itself is a reasonable substrate [50, 57, 62]. Furthermore, it is perhaps necessary to postulate a larger hydrophobic binding area to accommodate quaternary substrates such as the *N*-methylated benzoquinolines, which do not bind in the uncharged form [57, 173]. An extended lipophilic site is also required to bind to the extremely potent cationic competitive inhibitors, ethidium bromide (33) and 4′-(9-acridinylamino)methansulphon-*m*-anisidine (*N*-[4-(9-acridinylamino)-2-methoxyphenyl]methanesulphonamide) (29b) [62, 178]. On the other hand, both quaternary and uncharged substrates appear to act at the same site on aldehyde oxidase, as the oxidation of non-quaternized substrates is competitively inhibited by a number of cationic substrates or inhibitors [62].

Inhibition studies using xanthine oxidase also suggest that cationic substrates are oxidized at the same enzymic site as xanthine [179]. In this case, however, initial dissociation of an essential amino acid within the active site as a prerequisite is indicated before quaternary compounds bind, as these substrates are only very slowly oxidized at pH 7 and investigations have to be performed at pH > 9.6 [58, 179, 180]. Nevertheless, analogous substituent effects are observed with both molybdenum hydroxylases (*Table 3.7*) and a relatively large hydrophobic binding site is again indicated with xanthine oxidase.

Conventional substrates of xanthine oxidase based on the purine nucleus can undergo oxidation in both heterocyclic rings at positions 2,6 and/or 8. Substitution of an aryl group at carbon 8 in hypoxanthine (22) hinders oxidation at carbon 6, although such compounds are very strongly bound to the enzyme [181]. An alternative approach, used by Leonard's group in the United States to study the spatial restrictions of the binding site in xanthine oxidase, is to laterally extend the purine ring, keeping the peripheral rings intact [177, 182]. The lipophilic heterocycles used in these studies were based on either hypoxanthine (22) or xanthine (3), which would normally undergo oxidative attack at carbon 6 or 8, respectively. Xanthine oxidase readily oxidizes both *lin*-benzohypoxanthine (34a) and *lin*-benzoxanthine (34b), but the K_m value for (34a) is 5–10-times higher than that obtained for hypoxanthine [177, 182]. Binding of the angular benzopurines is obviously sterically hindered to a greater or lesser extent; thus, the *prox*-benzoanalogues (35a, b) are both oxidized at much lower rates than xanthine, whereas *dist*-benzoxanthine (36b) is not a substrate, although oxidation in the pyrimidine ring of *dist*-benzohypoxanthine

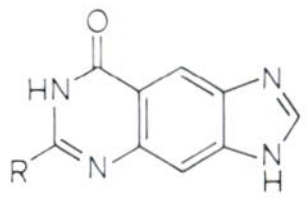

(34a) R = H
(34b) R = OH as lactam

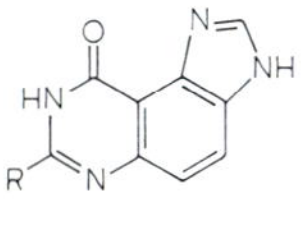

(35a) R = H
(35b) R = OH as lactam

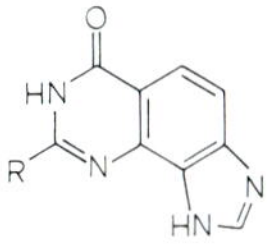

(36a) R = H
(36b) R = OH as lactam

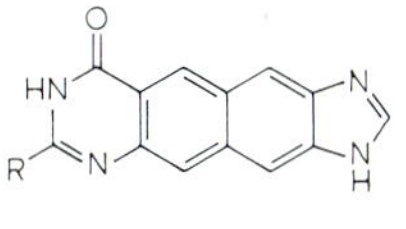

(37a) R = H
(37b) R = OH as lactam

(36a) is observed. A further extension to give *lin*-naphthohypoxanthine (37a) and *lin*-naphthoxanthine (37b) provides additional information on the dimensions of the binding site. The former compound (37a) is oxidized at carbon 7 to (37b) with a K_m value similar to that for *lin*-benzohypoxanthine (34a), whereas the V_{max} values increase in the order hypoxanthine < *lin*-benzohypoxanthine < *lin*-naphthohypoxanthine. In contrast, *lin*-naphthoxanthine (37b) is refractory to oxidation and functions only as a non-competitive inhibitor.

The widespread use of allopurinol (38), a xanthine oxidase inhibitor, in the treatment of gout has led to the synthesis and testing of numerous pyrazolopyrimidine analogues, and it is from these extensive studies that the increased binding due to hydrophobic substituents is most apparent. In the studies originated by Baker, around 60 different purines and pyrazolo[3,4-*d*]-pyrimidines were tested as inhibitors of hypoxanthine oxidation [183–185]. Enhanced binding due to a hydrophobic group could be demonstrated in the hypoxanthine, adenine guanine and pyrazolo[3,4-*d*]pyrimidine series, although

Table 3.8. POTENT INHIBITORS OF XANTHINE OXIDASE

Compound	I_{50}[a] *(μM)*	*Reference*
8-Phenylhypoxanthine	0.062	184
9-Phenylhypoxanthine	13	184
8-(3-Nitrophenyl)adenine	0.016	184
9-Phenyladenine	900	184
8-Phenylguanine	7.4	183
9-Methylguanine	58	183
9-Phenylguanine	0.41	183
Allopurinol	0.87	184
6-(3-Nitrophenyl)allopurinol	0.70	184
3-Hydroxy-1-(2′,4′-dinitrophenyl)-1*H*-pyrazolo[4,3-*c*]pyridine	52	189
3-Hydroxy-1-(2-nitrophenyl)-1*H*-pyrazolo[3,4-*b*]pyridine	250	190
2-Methyl-3-(2-pyridyl)-pyrido[2,3-*d*]pyrimidine-4-one	100	191
Pyrazolo[1,5-*a*]pyrimidin-7-one	11	194
2-Phenylpyrazolo[1,5-*a*]pyrimidin-7-one	50	194
3-Phenylpyrazolo[1,5-*a*]pyrimidin-7-one	0.4	194
3-(3-Tolyl)pyrazolo[1,5-*a*]pyrimidin-7-one	0.06	194
7-Hydroxypyrazolo[1,5-*a*]pyrimidin-7-one	4.5	194
3-(3-Tolyl)-7-hydroxypyrazolo[1,5-*a*]pyrimidin-7-one	0.025	194
3-(2-Furyl)-5-(4-pyridyl)-1,2,4-triazole	15	195
3-(3-Pyridyl)-5-(4-pyridyl)-1,2,4-triazole	0.02	195
3-(4-Pyrimidyl)-5-(4-pyridyl)-1,2,4-triazole	0.04	195

[a] I_{50} value is the concentration of compound to produce a 50% inhibition of control reaction using either hypoxanthine or xanthine as substrate.

the position of the aromatic substituent was critical (*Table 3.8*) [183–185]: maximum reversible inhibition was obtained with 8-phenylhypoxanthine (39), and 8-(3-nitrophenyl)adenine (40) but not with 9-phenylguanine (41) and 6-(3-nitrophenyl)allopurinol (42). All these compounds have I_{50} values against hypoxanthine of less than 1 μM, which are 12–54-fold better than allopurinol itself [185]. A number of 5-methyl-6-substituted-pyrrolo[2,3-*d*]pyrimidine-2,4-diones (43) have also been tested as xanthine oxidase inhibitors, but only

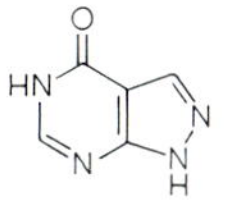

(38)

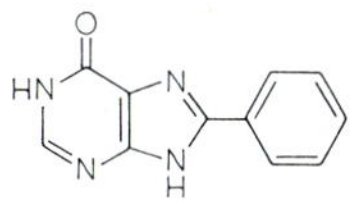

(39) 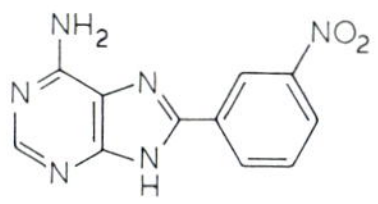 (40)

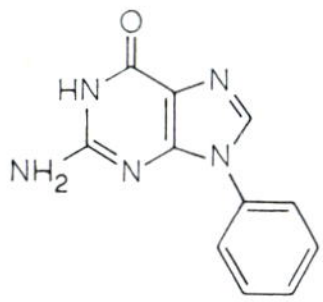

(41) 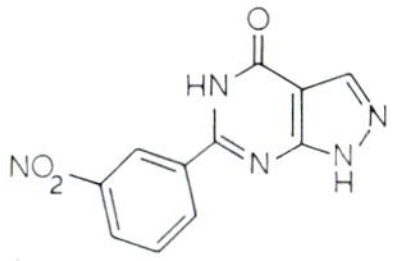 (42)

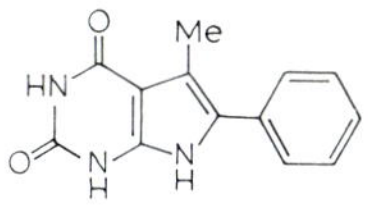

(43) 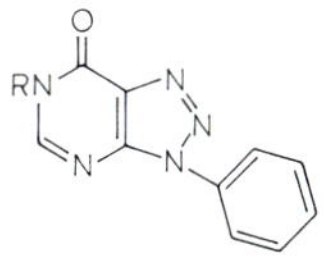 (44)

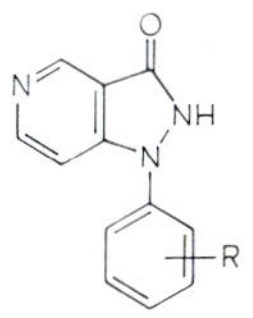

(45) 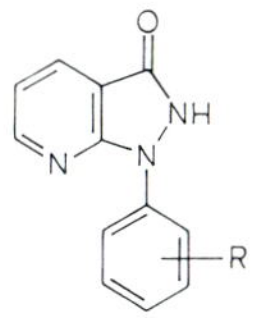 (46) 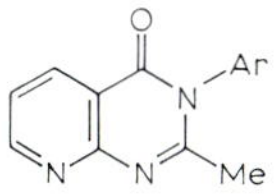 (47)

the 6-phenyl analogue showed significant inhibitory activity [186]. Analogues based on the purine or allopurinol nucleus are so structurally similar to endogenous molecules that they also interfere with other purine-metabolizing enzymes; for example, 9-phenylguanine (41) also inhibits guanine deaminase [184] and allopurinol can be converted to unnatural nucleosides and nucleotides [187]. Even 8-azapurines (44), which contain an additional nitrogen atom in the imidazole ring, inhibit xanthine oxidase and two purine deaminases, although the effect of 9-aryl substitution is different in each case [188]. However, pyrazolopyrimidines lacking the nitrogen atom at position 5 (pyrazolo-[4,3-*c*]pyridines) (45) or position 7 (pyrazolo[3,4-*b*]pyridines) (46) and pyrido-[2,3-*d*]pyrimidines (47) all substituted with hydrophobic groups were found to selectively inhibit xanthine oxidase but to be inactive towards guanine deaminase, guanosine deaminase and adenosine deaminase [189–191]. The inhibition exhibited by the latter group of compounds was found to be competitive [191], although the I_{50} values in each case were some 1000-times higher than those for allopurinol (*Table 3.8*).

Two groups of compounds appear to show the most promise as selective xanthine oxidase inhibitors: those based on a pyrazolo[1,5-*a*]pyrimidine nucleus, which show mixed or non-competitive inhibition; and substituted triazoles, which are competitive inhibitors [192, 193]. The 3-position of the pyrazolo[1,5-*a*]pyrimidines (48) is spatially equivalent to the 9-position of purine, and 3-aryl-substituted compounds were found to be 30–160-times better inhibitors than allopurinol (*Table 3.8*) [194]. In contrast, the hetero-rings in the potent substituted triazole inhibitors (49) are no longer fused, but those compounds with substituted aryl groups in the 3-position have the highest levels of intrinsic activity (*Table 3.8*) [195].

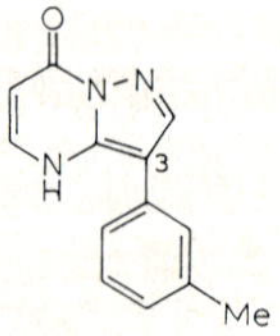

(48)

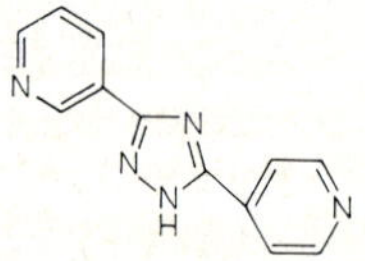

(49)

Steric factors and position of oxidation

In spite of the close structural relationship of the molybdenum hydroxylases, including a tendency for hydrophobic substrate/enzyme interaction, there is a very significant difference in the substrate specificity of the two enzymes. Not only is there considerable variation in the affinities for substrates and inhibitors, but there is often a difference in the position of oxidative attack. As both enzymes catalyse apparently similar nucleophilic reactions, this difference cannot be explained solely from electronic considerations and is probably due, to a great extent, to the differential response of each enzyme to steric factors.

N^1-Methylnicotinamide (7a) is oxidized predominantly at position 6 by rabbit liver aldehyde oxidase [60, 196, 197]. As the size of the alkyl substituent increases (7b–7d), oxidation at carbon 4 becomes more favourable, and with the *t*-butyl analogue (7e) only 4-pyridone is produced (*Table 3.1*) [60, 197]. The K_m for compounds (7a–7e) decreases with increasing size of the alkyl group, pointing to a higher affinity for those compounds with more hydrophobic substituents. However, the larger branched alkyl groups not only sterically hinder oxidation at carbon 6 but also decrease the overall oxidation rate. The major products of aryl or arylalkyl-substituted 3-carbamoylpyridinium chlorides (7f–7n) with rabbit liver aldehyde are also 6-pyridones [51].

N-Methyl and *N*-ethyl analogues of (7) and (8) are converted exclusively to the 6-pyridones and 2-quinolones, respectively, by xanthine oxidase [58, 179], although in the latter case substitution with the longer alkyl group causes an increase in K_m and a sharp drop in maximal oxidation rate for the quinolinium derivative [58]. The unfavourable steric reaction with ethyl as opposed to methyl substituents suggests that longer-chain analogues would not react at all. With *N*-aryl-3-carbamoylpyridinium chlorides, the main product formed by aldehyde oxidase, the 2-pyridone, is the minor product in the xanthine oxidase reaction and *vice versa* [51].

Purines make up the largest group of substrates studied for xanthine oxidase; in general, these compounds are better substrates for this enzyme than for aldehyde oxidase. The parent base is sequentially oxidized *via* hypoxanthine (22) and xanthine (3) to purine-2,6,8-trione (uric acid) by xanthine oxidase, whereas aldehyde-oxidase-catalysed oxidation of purine occurs exclusively at carbon 8. Introduction of a methyl group at carbon 6 (18) completely abolishes activity with the former enzyme but does not significantly affect oxidation by aldehyde oxidase [10]. However, a larger *n*-propyl substituent lowers the oxidation rate with the latter enzyme by about 60%. The contrasting results obtained from 6-substitution are not unexpected in view of the different initial sites of purine oxidation. However, substitution of a methyl group at position

7, 8 or 9 also drastically diminishes the rate of oxidation with xanthine oxidase [10, 198], which indicates that binding of the imidazole ring is also critical for oxidation at carbon 6. Surprisingly, 8-methylpurine is slowly oxidized by aldehyde oxidase, although in this case attack must occur in the pyrimidine ring [10]. It is also interesting to note that 6-mercaptopurine, a widely used cytotoxic drug, is attacked by both enzymes at a relatively low oxidation rate. In both cases, initial oxidation is at carbon 8, followed by attack at carbon 2 to yield 6-thiouric acid for xanthine oxidase [10, 199].

Hypoxanthine is oxidized at carbon 2 by both molybdenum hydroxylases, although xanthine oxidase is much more effective as a catalyst in this reaction [10]. A methyl substituent in this position prevents oxidation by either enzyme. Introduction of *N*-methyl substituents into the hypoxanthine nucleus produces dramatic effects on enzymic oxidation rates and also gives some insight into the productive modes of binding to each enzyme. Thus, it has been proposed that hypoxanthine tautomerizes in the xanthine oxidase-substrate complex to the 3-NH-form with a simultaneous shift of the NH-group in the imidazole ring from position 9 to 7 [198, 200]. In support of this hypothesis, when tautomerism in the imidazole ring is prevented by substitution at N-7 or N-9, such compounds are almost refractory to oxidation (see *Table 3.9*)

Table 3.9. EFFECT OF METHYLATION ON OXIDATION OF HYPOXANTHINE AND 7-DEAZAHYPOXANTHINE (10, 198, 200]

	Relative oxidation rate	
Compound	*Xanthine oxidase*	*Aldehyde oxidase*
Hypoxanthine	108	3
1-Methylhypoxanthine	–[a]	84
3-Methylhypoxanthine	0.6	710
7-Methylhypoxanthine	0	27
8-Methylhypoxanthine	94	–
9-Methylhypoxanthine	0	2
7-Deazahypoxanthine	9.3	–
1-Methyl-7-deazahypoxanthine	0	–
3-Methyl-7-deazahypoxanthine	20	–
7-Methyl-7-deazahypoxanthine	10	–
9-Methyl-7-deazahypoxanthine	0[b]	–
6-Methoxy-7-deazapurine	0[c]	–

[a] 1-Methylhypoxanthine has an oxidation rate 15% that of hypoxanthine [10].
[b] Competitive inhibitor, K_i = 100 μM.
[c] Competitive inhibitor, K_i = 180 μM.

[10, 198, 200]. On the other hand, the 8-position does not appear to play any rôle in the formation of the enzyme-substrate complex, as 8-methylhypoxanthine is oxidized almost as efficiently as hypoxanthine itself [200]. Furthermore, where tautomerism in the pyrimidine ring is blocked, as in 1-methylhypoxanthine (51a), the compound is not a substrate [198, 200], but if the double bond is already fixed between positions 1 and 2, which is the case with the 3-methylisomer (51b), it is slowly oxidized by xanthine oxidase. Although the latter compound has the correct tautomeric form for the most productive binding, the bulky methyl group appears to sterically hinder attack at carbon 2, which accounts for the lower oxidation rate. Analogous results have also been obtained by Rosemeyer and Seela using monomethyl derivatives of 7-deazahypoxanthine (52a, b) [201]. Unlike the hypoxanthines, which can undergo two sequential oxidation steps, 7-deaza analogues of hypoxanthine are oxidized exclusively at carbon 2 and therefore the regiochemical requirements for the single oxidation step can be more accurately obtained. In the deazapurine series, methylation of N-1 results in a complete loss of substrate activity, whereas the 3-methyl isomer is a slightly poorer substrate than 7-deazahypoxanthine itself [201]. It is thus assumed that this position, in the tautomeric form (52b), is necessary for binding and oxidation at C-2. In addition, a free carbonyl group at C-6 is also essential, since in the *O*-methyl derivative, binding is decreased significantly, leading to weak inhibition properties. A similar dependence on the correct tautomeric form of the pyrimidine ring for binding to xanthine oxidase can also be demonstrated in *N*-methylated allopurinol series (54a, b) and the isomeric *N*-methyl-7-deazaguanines (53a,b) [202, 203].

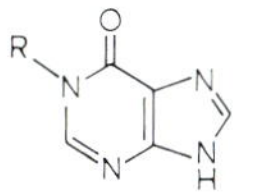

(50a) R = H
(51a) R = Me

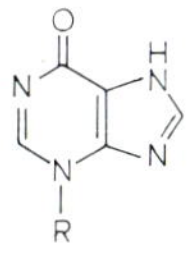

(50b)
(51b)

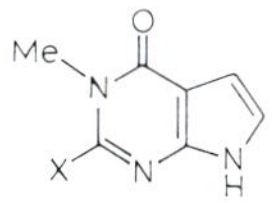

(52a) X = H
(53a) X = NH_2

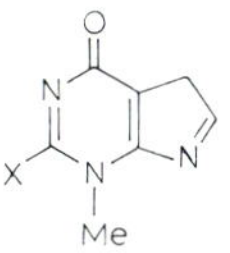

(52b)
(53b)

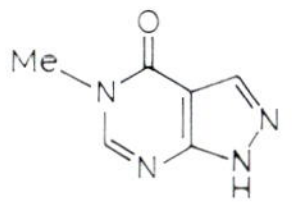

(54a)

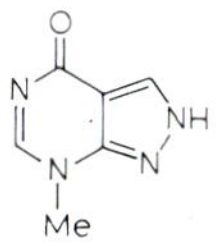

(54b)

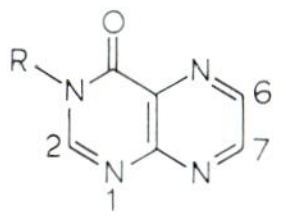

(55a) R = Me
(56a) R = H

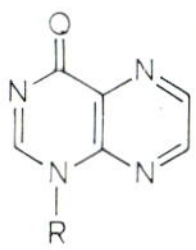

(55b)
(56b)

The effect of methylation on aldehyde-oxidase-catalysed oxidation of hypoxanthine has not been so rigourously analysed. However, a similar requirement for the 3-*H* tautomer is indicated, as 3-methylhypoxanthine (51b) is a much more efficient substrate than hypoxanthine for this enzyme [10].

In contrast, pteridines are excellent substrates for both molybdenum hydroxylases and each enzyme can catalyse oxidation at carbon 2, 4 or 7, although not necessarily in the same molecule [10, 204–206]. For a comparison of the position of attack by the enzymes, the reader is referred to a recent review by the author [11]. Unfortunately, the sequence of oxidation in the multistep pathways is not always clear and therefore the effects of substitution are difficult to interpret. 1-Methylpteridin-4-one (55b), which is apparently oxidized at carbon 2, is converted much more rapidly than the 3-methyl isomer (55a) [206]. These derivatives correspond to the 3-methyl and 1-methyl analogues of hypoxanthine, respectively. Nevertheless, it is not possible to make direct comparisons of the *N*-methyl derivatives with unsubstituted pteridin-4-one (56b) because the site of oxidation may differ [206, 207].

Effect of charge and electron density

In addition to lipophilic groups increasing the affinity of heterocyclic compounds for the molybdenum hydroxylases, a similar effect can be observed by decreasing the electron density at the carbon undergoing oxidation. This may be achieved in a number of different ways. Quaternization of a ring nitrogen atom may activate a molecule towards enzymic attack; thus, *N*-methylquinolinium (8a) is oxidized much more rapidly by aldehyde oxidase than the parent base, quinoline [57]. However, in monocyclic systems, quaternization alone may not be sufficient to promote molybdenum-hydroxylase-catalysed oxidation; *N*-methylpyridinium cations are not substrates for either enzyme unless an electron-withdrawing substituent such as $CONH_2$ is present at carbon 3 (7a–7n) [10, 58, 159]. The presence of the electron-withdrawing group itself does not confer substrate activity on a molecule; thus nicotinamide is refractory to enzymic oxidation [10]. Bunting and Gunasekara [180] have therefore postulated that in cationic substrates there is an important interaction between an electronegative atom *X* (X = O, N or S) in the substituent and a functional group in the active site of xanthine oxidase [180]. Although such an interaction has a favourable effect on the oxidation of pyridinium compounds, 3-substituted quinolinium salts are thought to be oriented in such a way in the active site of aldehyde oxidase that oxidation is directed completely at carbon 4 and the reaction velocity is very low, despite the electron-withdrawing $CONH_2$ group [61].

Substrates have a minimum requirement for a heterocyclic nitrogen atom in an aromatic system although Banks and Barnett have shown recently that the stabilized carbonium ion salt, tropylium tetrafluoroborate, is oxidized by aldehyde oxidase to cycloheptatrienone (tropone) [207a]. Additional nitrogen atoms have a further activating effect towards molybdenum hydroxylase activity; thus all four diazanaphthalenes are better substrates for aldehyde oxidase than the monoazanaphthalenes, quinoline and isoquinoline (see *Table 3.6*) [50] and 1,3,5,8-tetraazanaphthalenes (pteridines) are rapidly oxidized by both enzymes [10, 49, 204]. A low electron density at a carbon atom appears to be more important in influencing the position of aldehyde-oxidase-catalysed attack than that of xanthine oxidase. In the latter case, the relative position of the electronegative nitrogen atoms, perhaps to bind to an electropositive centre in the enzyme active site, seems to be more significant. In purine, for instance, carbon 8 is the most electron-deficient and is also the position of aldehyde oxidase attack, whereas the compound is oxidized at carbon 6 by xanthine oxidase [10, 208]. Furthermore, cinnoline (6), despite the two adjacent nitrogen atoms, is refractory to oxidation by the latter enzyme but is converted to cinnolin-4-one by aldehyde oxidase [50]. In this case, oxidation does not occur adjacent to a nitrogen atom but at the position of lowest electron density. Although attack is normally in the heterocyclic ring, some compounds, such as acridine (29a) and phenazine methosulphate (57), are oxidized by aldehyde oxidase at an electron-deficient carbon in a carbocyclic ring [174, 209]. A low

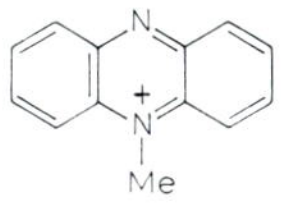

(57)

electron density may also affect reactions catalysed by xanthine oxidase; out of the four oxidizable carbon atoms in the pteridine ring system, only the 6-position is not oxidized by either molybdenum hydroxylase [10, 204, 205]. Substitution of a third nitrogen atom into the pyrazine ring at position 7 means that in 7-azapteridine-2,4-dione carbon 6 carries a greater positive charge than the corresponding position in pteridine-2,4-dione and the former compound is oxidized, albeit at a low rate, at carbon 6 [210]. However, it is also possible that oxidation may be enhanced at this position because the additional nitrogen assists in binding to the active site rather than a lowered electron density at the adjacent carbon.

Both molybdenum hydroxylases can oxidize uncharged and cationic substrates, although at physiological pH aldehyde oxidase is much more active towards the latter group of compounds. It does not necessarily follow, however, that when compounds have a pK_a value around 7, the neutral and protonated forms both bind to the enzyme. Kinetic studies at different pH values have indicated that it is the unprotonated form of 2-aminophthalazine that reacts with the molybdenum hydroxylases, either as a substrate for aldehyde oxidase or a competitive inhibitor of xanthine oxidase [211]. In contrast, at pH 7–8 many of the purine and pteridine substrates are present as mixtures of uncharged molecules and monoanions, both of which are rapidly oxidized by xanthine oxidase [181, 200, 206, 210, 212].

Anions may not be substrates at all for aldehyde oxidase. The cytotoxic drug, methotrexate (19), is a substituted pteridine with both basic and acidic ionizable groups present; the carboxylic acid groups, part of a glutamic acid residue, are predominantly ionized at pH 7 [213]. Uncharged di-*n*-alkyl esters of methotrexate show significantly greater activity towards aldehyde oxidase than the parent drug [214]. This could be due to increased lipophilicity, but since the K_m values for methotrexate decrease as the pH value is lowered from 8 to 6, it has been interpreted to indicate that substrate activity for aldehyde oxidase is a property of the undissociated carboxylic acid form of the drug [214]. In support of this proposal, 4-amino-4-deoxy-N^{10}-methylpteroic acid (APA), which lacks the dibasic glutamic acid group, is oxidized 2–3-times more rapidly by rabbit liver aldehyde oxidase than is the parent base [215]. Furthermore, there is an inverse relationship between the affinity of methotrexate polyglutamates for the enzyme and the number of glutamic acid residues (n = 1–4) [146, 216, 217].

Neither methotrexate nor its microbial breakdown product, APA, is a substrate for xanthine oxidase [6, 215], although this may be more a function of the 2,4-diaminopteridine moiety, which itself is refractory to xanthine oxidase [204], rather than the glutamic acid residue. In fact, methotrexate is a potent competitive inhibitor of this enzyme, with a K_i value of around 25 μM [218, 219]. There is considerable controversy as to whether folic acid, a substituted 2-aminopteridin-4-one, is also an inhibitor of xanthine oxidase. It is not oxidized at carbon 7, unlike the parent compound, which is a poor substrate [204]. However, some workers have shown that folic acid is an extremely potent competitive inhibitor of xanthine oxidase, some 10-times more effective *in vitro* than allopurinol, whereas other reports claim that the inhibition is due to the contaminant 2-amino-4-oxopteridine-6-aldehyde (27), which is a photolytic breakdown product of folic acid [4, 171, 172, 218–220].

CONCLUSION

The previous sections serve to illustrate the wide range of compounds that can interact with the molybdenum hydroxylases, although the survey is by no means exhaustive. Substrate and inhibitor specificity is governed by an interplay of steric, lipophilic and electronic factors. With simple molecules, it may be possible to demonstrate a correlation between enzymic activity and electron density [50] or partition coefficient, which is indicative of lipophilicity [175], but with more complex substrates it is often difficult to differentiate between such effects.

Because the enzymes have such an extensive specificity and are so widely distributed, it is perhaps inappropriate to select a single 'physiological' substrate, although there are a number of possible candidates. In spite of the fact that all purines arising from nucleic acid breakdown are normally channelled *via* hypoxanthine for xanthine oxidase oxidation [221–223], individuals lacking the enzyme do not seem to suffer any severe clinical effects [224, 225]. However, xanthinuria is not necessarily linked with a lack of aldehyde oxidase, as in some patients endogenous purines and allopurinol still undergo oxidation [226]. Thus, even in normal subjects, both molybdenum hydroxylases may be involved in purine oxidation. Both xanthine oxidase and aldehyde oxidase can catalyse the oxidation of vitamin A aldehyde to the corresponding acid and it has therefore been suggested that aldehyde oxidase may play a pivotal rôle in vitamin A metabolism [227, 228]. The latter enzyme is also active in the degradation of N^1-methylnicotinamide and vitamin B-6 (pyridoxal, (17)) [1, 229]. In humans, pyridoxic acid, the dead-end metabolite of pyridoxal, rapidly appears in the urine after administration of the vitamin [230]. Nevertheless, it seems unlikely that a key regulatory enzyme in the catabolism of the above compounds would have such a wide substrate specificity.

Conversely, it has been argued that the main function of xanthine oxidase may not be the oxidative half-reaction, but may rather reflect the reductive capacity of the enzyme to produce hydrogen peroxide and superoxide radical [231]. These 'active' forms of oxygen could act as general oxidizing agents for compounds such as ethanol, tryptophan and adrenaline [231], or have a specific rôle in those tissues previously mentioned [46–48, 114, 115, 122]. Such arguments are based on the utilization of oxygen as the enzymic electron acceptor whereas, in fact, NAD^+ is involved in xanthine oxidase reoxidation *in vivo* and thus these views, at present, must be regarded as speculative. Indeed, the wide distribution of both molybdenum hydroxylases suggests that their primary metabolic function is of a more fundamental nature rather than a highly specialized one [2]. In this respect, the molybdenum hydroxylases resemble the cytochrome *P*-450 system [11] and the author believes that aldehyde oxidase and xanthine oxidase have an important rôle in the detoxification of xenobiotics, not only in oxidation but also in dehydrogenation and reduction reactions.

Drug metabolism studies are usually performed *in vivo* or with 10 000 × *g* supernatant fractions. Although it is possible to differentiate between nucleophilic molybdenum hydroxylase oxidation to produce lactam metabolites, and electrophilic microsomal oxidative attack, which results in the formation of phenols, the participation of the former enzymes is often ignored. As aldehyde oxidase has recently been identified in microsomal fractions [72–75], the use of the 100 000 × *g* pellet does not necessarily preclude the involvement of the enzyme in metabolic oxidation, either as a soluble fraction contaminant or a true membrane-bound enzyme. For example, we have shown phenanthridine (30a) to be rapidly converted to 6-phenanthridone by partially purified aldehyde oxidase [173], devoid of microsomal activity, yet the same reaction has been attributed to the microsomal enzyme system [232, 233], even though it is unlikely that electrophilic attack would occur at such an electropositive carbon adjacent to a ring nitrogen atom. Firm evidence of microsomal attack could be afforded by including specific inhibitors of each system such as menadione and SKF-525A in incubation mixtures. In addition, much more information regarding the induction and effects of *in vivo* inhibition of the molybdenum hydroxylases is required. These enzymes can be induced by substrates such as phthalazine (1) or xanthine (3) [234, 235], but relatively little is known about the effect of classical microsomal inducers. Although allopurinol (38) is prescribed frequently to inhibit xanthine oxidase in gout [236, 237] or to potentiate the action of 6-mercaptopurine in cancer chemotherapy [238], there is still considerable potential for the use of aldehyde oxidase inhibitors such as 1-hydrazinophthalazine [239] in drug therapy. A further development utilizing molybdenum hydroxylase action is in the activation of prodrugs. 6-Deoxyacyclovir (23), a congener of acyclovir (24), is readily oxidized *in vitro* to the parent drug by xanthine oxidase, but is also deactivated by aldehyde oxidase [240]. However, the compound is extensively metabolized to acyclovir (24) in humans, which suggests that the prodrug, 6-deoxyacyclovir, might be superior to acyclovir itself for oral administration [240].

ACKNOWLEDGEMENTS

I would like to thank Dr. J.A. Smith for encouragement and many helpful discussions and to express my appreciation to Mrs. J. Joyce and Ms. P. Moon for their considerable efforts in typing the manuscript.

REFERENCES

1 M. Stanlovic and S. Chaykin, Arch. Biochem. Biophys., 145 (1971) 35.
2 T.A. Krenitsky, Adv. Exp. Med. Biol., 41A (1974) 57.
3 W.E. Knox, J. Biol. Chem., 163 (1946) 699.
4 O.H. Lowry, O.A. Bessey and E.J. Crawford, J. Biol. Chem., 180 (1949) 399.
5 Y. Hashimoto, K. Shibata and H. Sakamoto, Nihon Univ. J. Med., 2 (1960) 229.
6 D.G. Johns, A.T. Iannotti, A.C. Sartorelli, B.A. Booth and J.R. Bertino, Biochim. Biophys. Acta, 105 (1965) 380.
7 R.L. Felsted and S. Chaykin, Methods Enzymol., 18B (1971) 216–222.
8 R.C. Bray, in The Enzymes, ed. P.D. Boyer (Academic Press, London) 3rd Edn. Vol. 12, Part B (1975) pp. 300–419.
9 E.W. Holmes and W.N. Kelly, in Handbuch der inneren Medizin, eds. N. Zollner and W. Grobner (Springer-Verlag, Berlin, 1976), p. 627.
10 T.A. Krenitsky, S.M. Neil, G.B. Elion and G.C. Hitchings Arch. Biochem. Biophys., 150 (1972) 585.
11 C. Beedham, Drug Metab. Rev., 16 (1985) 119.
12 M.P. Coughlan, in Molybdenum and Molybdenum-Containing Enzymes, ed. M.P. Coughlan (Pergamon Press, Oxford, 1980) pp. 119–185.
13 K.V. Rajagopalan, in Enzymatic Basis of Detoxification, ed. W.B. Jakoby (Academic Press, London) Vol. 1 (1980) pp. 295–309.
14 R. Hille and V. Massey, in Molybdenum Enzymes, ed. T.P. Spiro (Academic Press, New York, 1985) pp. 443–518.
15 H. Beinert and P. Hemmerich, Biochem. Biophys. Res. Commun., 18 (1965) 212.
16 K.V. Rajagopalan, P. Handler, G. Palmer and H. Beinert, J. Biol. Chem., 243 (1968) 3784.
17 D.J. Lowe, R.M. Lynden-Bell and R.C. Bray, Biochem. J., 130 (1972) 239.
18 D.J. Lowe and R.C. Bray, Biochem. J., 169 (1978) 471.
19 M.J. Barber, M.P. Coughlan, K.V. Rajagopalan and L.M. Siegel, Biochemistry, 21 (1982) 3561.
20 M.J. Barber, J.C. Salerno and L.M. Siegel, Biochemistry, 21 (1982) 1648.
21 G.N. George, in Flavins and Flavoproteins, eds. R.C. Bray, P.C. Engel and S.G. Mayhew (Walter de Gruyter, Berlin, 1984) pp. 325–330.
22 M.P. Coughlan and I. Ni Fhaolain, Proc. R. Irish Acad., 79B (1979) 13.
23 R.C. Bray in Ref. 21, pp. 707–722.
24 J.T. Spence, M.J. Barber and L.M. Siegel, Biochemistry, 21 (1982) 1656.
25 J.S. Olsen, D.P. Ballou, G. Palmer and V. Massey, J. Biol. Chem., 249 (1974) 4363.
26 R. Hille in Ref. 21, pp. 683–686.
27 R. Hille and V. Massey, J. Biol. Chem., 261 (1986) 1241.
28 A. Battacharya, G. Tollin, M. Davis and D.E. Edmondson, Biochemistry, 22 (1983) 5270.
29 H.L. Gurtoo and D.G. Johns, J. Biol. Chem., 246 (1971) 286.
30 U. Branzoli and V. Massey, J. Biol. Chem., 244 (1969) 1692.
31 H. Komai, V. Massey and G. Palmer, J. Biol. Chem., 244 (1969) 1692.
32 K.V. Rajagopalan, I. Fridovich and P. Handler, J. Biol. Chem., 237 (1962) 922.
33 D.G. Johns, J. Clin. Invest., 46 (1967) 1492.
34 A.G. Porras, J.S. Olsen and G. Palmer, J. Biol. Chem., 256 (1981) 9096.
35 F. Stirpe and E. Della Corte, J. Biol. Chem., 244 (1969) 3855.
36 W.R. Waud and K.V. Rajagopalan, Arch. Biochem. Biophys., 172 (1976) 354.
37 S.-W. Kimm and Y.-S. Juhnn, Korean J. Biochem., 15 (1983) 71.

38 E. Della Corte and F. Stirpe, Biochem. J., 126 (1972) 739.
39 T.A. Krenitsky and J.V. Tuttle, Arch. Biochem. Biophys., 185 (1978) 370.
40 M. Nakamura and I. Yamazaki, J. Biochem. (Tokyo), 92 (1982) 1279.
41 W.R. Waud and K.V. Rajagopalan, Arch. Biochem. Biophys., 172 (1976) 365.
42 M.G. Battelli, Febs. Lett., 113 (1980) 47.
43 Z.W. Kaminski and M.M. Jezewska, Biochem. J., 207 (1982) 341.
44 Z.W. Kaminski and M.M. Jezewska, Biochem. J., 181 (1979) 177.
45 R.S. Roy and J.M. McCord, Fed. Proc., 41 (1982) 767.
46 L. Betz, J. Neurochem., 44 (1985) 574.
47 E. Tubaro, B. Lotti, C. Santiangelli and G. Cavallo, Biochem. Pharmacol., 29 (1980) 3018.
48 H.S. Lee and A.G. Fischer, Int. J. Biochem., 9 (1978) 559.
49 J.J. McCormack, B.A. Allen and C.N. Hodnett, J. Heterocycl. Chem., 15 (1978) 1249.
50 C. Stubley, J.G.P. Stell and D.W. Mathieson, Xenobiotica, 9 (1979) 475.
51 S.A.G.F. Angelino, D.J. Buurman, H.C. van der Plas and F. Müller, Recl. Trav. Chim. Pays-Bas, 102 (1983) 331.
52 G. Palmer and J.S. Olsen in Ref. 12, pp. 189–220.
53 D.E. Edmondson, D.P. Ballou, A. van Heuvelen, G. Palmer and V. Massey, J. Biol. Chem., 248 (1973) 6135.
54 H.C. van der Plas, J. Heterocycl. Chem., 19, Suppl. Vol VI (1982) S-1.
55 P.C. Ruenitz and H.G. Thomas, Arch. Biochem. Biophys., 239 (1985) 270.
56 H.G. Thomas and D.N. Silverman, Arch. Biochem. Biophys., 241 (1985) 649.
57 S.M. Taylor, C. Stubley-Beedham and J.G.P. Stell, Biochem. J., 220 (1984) 67.
58 J.W. Bunting, K.R. Laderoute and D.J. Norris, Can. J. Biochem., 58 (1980) 49.
59 R.L. Felsted and S. Chaykin, Biochim. Biophys. Acta, 485 (1977) 489.
60 S.A.G.F. Angelino, D.J. Buurman, H.C. van der Plas and F. Müller, Recl. Trav. Chim. Pays-Bas, 98 (1982) 342.
61 S.A.G.F. Angelino, B.H. van Valkengoed, D.J. Buurman, H.C. van der Plas and F. Müller, J. Heterocycl. Chem., 21 (1984) 107.
62 S.M. Taylor, Specificity of aldehyde oxidase towards *N*-heterocyclic cations, Ph.D. Thesis, Bradford University (1984).
63 P.C. Ruenitz, T.P. McClennon and L.A. Sternson, Drug Metab. Dispos., 7 (1979) 204.
64 H.G. Thomas and P.C. Reunitz, J. Heterocycl. Chem., 21 (1984) 1057.
65 S. Brandänge and L. Lindblom, Biochem. Biophys. Res. Commun., 91 (1979) 991.
66 J.W. Gorrod and A.R. Hibberd, Eur. J. Drug Metab. Pharmacokin., 7 (1982) 293.
67 D.L. Hill, W.R. Laster and R.F. Struck, Cancer Res., 32 (1972) 658.
68 Y. Hashimoto and C. Shimizu, Nihon Univ. J. Med., 1 (1959) 175.
69 J.A. Badwey, J.M. Robinson, M.J. Karnovsky and M.L. Karnovsky, J. Biol. Chem., 256 (1981) 3479.
70 G.M. Rothe, Biochem. Physiol. Pflazen, 167 (1975) 411.
71 E. Della Corte, G. Gozetti, F. Novello and F. Stirpe, Biochim. Biophys. Acta, 191 (1969) 164.
72 T.L. Seely, P.B. Mather and R.S. Holmes, Comp. Biochem. Physiol., 78B (1984) 131.
73 B. Mousson, P. Desjacques and P. Baltassat, Enzyme, 29 (1983) 32.
74 P. Duley, O. Harris and R.S. Holmes, Clin. Exp. Res., 9 (1985) 263.
75 G. Bruder, H. Heid, E.-D. Jarasch, T.W. Keenan and I.H. Mather, Biochim. Biophys. Acta, 701 (1982) 357.
76 R.P. Igo, B. Mackler and H. Duncan, Arch. Biochem. Biophys., 93 (1961) 435.
77 B. Furnival, J.M. Harrison, J. Newman and D.G. Upshall, Xenobiotica, 13 (1983) 361.

78 G. Bruder, H.W. Heid, E.-D. Jarasch and I.H. Mather, Differentiation, 23 (1983) 218.
79 E.L. Bierman and R.E. Shank, J. Am. Med. Assoc., 234 (1975) 630.
80 R.K. Morton, Biochem. J., 57 (1954) 231.
81 L. Robert and J. Polonovski, Discuss. Faraday Soc., 20 (1955) 54.
82 G.R. Greenbank and M.J. Pallansch, J. Dairy Sci., 45 (1962) 958.
83 D.J. Ross, S.V. Sharnick and K.A. Oster, Proc. Soc. Exp. Biol. Med., 163 (1980) 141.
84 M.P.S. Gandhi and S.P. Ahuja, Zentrabl. Vet. Reihe A, 26 (1979) 635.
85 C.Y. Ho, R.T. Crane and A.J. Clifford, J. Nutr., 108 (1978) 55.
86 R.F. Volp and G.L. Lage, Proc. Soc. Exp. Biol. Med., 154 (1977) 488.
87 B. Schoutsen and J.W. De Jong, Arch. Int. Physiol. Biochim., 92 (1984) 379.
88 K.A. Oster, Med. Counterpoint, 5 (1973) 26.
89 K.A. Oster, Med. Counterpoint, 6 (1974) 39.
90 C.Y. Ho and A.J. Clifford, J. Nutr., 106 (1976) 1600.
91 H.C. Deeth, J. Dairy Sci., 66 (1983) 1419.
92 T.A. Krenitsky, J.V. Tuttle, E.L. Cattau and P. Wang, Comp. Biochem. Physiol., 49B (1974) 687.
93 J.P. Zikakis, M.A. Dressel and M.R. Silver, in Instrumental Analysis of Foods, eds. G. Charalambous and G. Inglett (Academic Press, New York) Vol. 2 (1983) pp. 243–303.
94 M.P.S. Gandhi and S.P. Ahuja, Zentrabl. Vet. Reihe A, 26 (1979) 763.
95 V.V. Modi, E.C. Owen and R. Proudfoot, Proc. Nutr. Soc., 18 (1959) i.
96 J. P. Zikakis, T.M. Dougherty and N.O. Biasotto, J. Food Sci., 41 (1976) 1408.
97 R.W. Topham, J.H. Woodruff and M.C. Walker, Biochemistry, 20 (1981) 319.
98 R.S. Holmes, Biochem. Genet., 17 (1979) 517.
99 M.G. Battelli, E. Lorenzoni and F. Stirpe, Biochem. J., 131 (1973) 191.
100 M.W. Bailey and R. Eisenthal, Biochem. J., 143 (1974) 149.
101 M.-L. Chen and W.L. Chiou, Drug Metab. Dispos., 10 (1982) 706.
102 K. Sasaki, R. Hosoya, Y.-M. Wang and G.L. Raulston, Biochem. Pharmacol., 33 (1983) 503.
103 F.E. Kelsey and F.K. Oldham, J. Pharmacol. Exp. Ther., 79 (1943) 77.
104 K-H. Wurzinger and R. Hartenstein, Comp. Biochem. Physiol., 49B (1974) 171.
105 C. Beedham, S.E. Bruce and D.J. Rance, Eur. J. Drug Metab. Pharmacokin. (in press)
106 C. Beedham, unpublished data.
107 U.A.S. al-Khalidi and T.H. Chaglassian, Biochem. J., 97 (1965) 318.
108 R.S. Holmes, Biochem. Physiol., 61B (1978) 339.
109 B. Schoutsen, P. de Tombe, E. Harmsen, E. Keijzer and J.W. de Jong, Adv. Exp. Med. Biol., 165 (1984) 497.
110 E.-D. Karasch, C. Grund, G. Bruder, H.W. Heid, T.W. Keenan and W.W. Franke, Cell, 25 (1981) 67.
111 R.D. Berlin and R.A. Hawkes, Am. J. Physiol., 251 (1968) 932.
112 S.E. Bruce and C. Beedham, unpublished observation.
113 A.H. Khan, S. Wilson and J.C. Crawhall, Can. J. Physiol. Pharmacol., 53 (1975) 113.
114 R.W. Topham, M.C. Walker, M.P. Calisch and R.W. Williams, Biochemistry, 21 (1982) 4529.
115 R.W. Topham, M.R. Jackson, S.A. Jaslin and M.C. Walker, Biochem. J., 235 (1986) 39.
116 W.R. Ravis, J.S. Wang and S. Feldman, Biochem. Pharmacol., 33 (1983) 443.
117 T.L. Loo, J.K. Luce, M.P. Sullivan and E. Frei, Clin. Pharmacol. Ther., 9 (1968) 180.
118 S. Zimm, J.M. Collins, R. Riccardi, D. O'Neill, P.K. Narang, B. Chabner and D.G. Poplack, N. Engl. J. Med., 308 (1983) 1005.

119 S.S. Good and P. De Miranda, Am. J. Med. Acyclovir Symp., 73 (1982) 91.
120 W.W. Westerfield and D.A. Richter, Proc. Soc. Exp. Biol. Med., 71 (1949) 181.
121 G.G. Villela, Experientia, 24 (1968) 1101.
122 D.A. Parks and D.N. Granger, Am. J. Physiol., 245 (1983) G285.
123 G.G. Villela, O.R. Affonso and E. Mitidieri, Biochem. Pharmacol., 15 (1966) 1894.
124 G. Bruder, E.-D. Jarasch and H.W. Heid, J. Clin. Invest., 74 (1984) 783.
125 A. McHale, H. Grimes and M.P. Coughlan, Int. J. Biochem., 10 (1979) 317.
126 Y. Yonetani and Y. Ogawa, Chem. Pharm. Bull., 25 (1977) 448.
127 E. Tubaro, B. Lotti, C. Santiangeli and G. Cavallo, Biochem. Pharmacol., 29 (1980) 1945.
128 E. Tubaro, B. Lotti, C. Croce, G. Borelli and G. Cavallo, Biochem. Pharmacol., 29 (1980) 1939.
129 C. Beedham, J. Pharm. Pharmacol., 38 (1986) 57.
130 J. Kawada, Y. Shirakawa, Y. Yashimura and M. Nishida, J. Endocrinol., 95 (1982) 117.
131 C.A. Woolfolk and J.S. Downard, J. Bacteriol., 130 (1977) 1175.
132 R. Kumar and V. Taneja, Biochim. Biophys. Acta, 485, (1977) 489.
133 D.G. Johns, A.T. Iannotti, A.C. Sartorelli and J.R. Bertino, Biochem. Pharmacol., 15 (1966) 555.
134 P. De Miranda, H.C. Krasney, D.A. Page and G.B. Elion, J. Pharmacol. Exp. Ther., 219 (1981) 309.
135 H.C. Krasney, P. De Miranda, M.R. Blum and G.B. Elion, J. Pharmacol. Exp. Ther., 216 (1981) 281.
136 P. De Miranda, R.J. Whitley, N. Barton, D. Page, T. Creagh-Kirk, S. Liao and R. Blum, Program and Abstracts of the 22nd Interscience Conference on Antimicrobial Agents and Chemotherapy (American Society for Microbiology, Washington, DC, 1982) p. 140.
137 S.A. Jacobs, R.G. Stoller, B.A. Chabner and D.G. Johns, Cancer Treat. Rep., 61 (1977) 651.
138 H. Briethaupt and E. Kuenzlen, Cancer Treat. Rep., 66 (1982) 1733.
139 M. Przbylski, J. Preiß, J. Dennebaum and J. Fischer, Biomed. Mass Spectrom., 9 (1982) 22.
140 B.B. Brodie, J.E. Baer and L.C. Craig, J. Biol. Chem., 188 (1951) 567.
141 P. De Miranda, H.C. Krasney, D.A. Page and G.B. Elion, Am. J. Med. Acyclovir Symp., 73 (1982) 31.
142 B. Kaye, J.L. Offerman, J.L. Reid, H.L. Elliott and W.S. Hillis, Xenobiotica, 14 (1984) 935.
143 B. Kaye, D.J. Rance and L. Waring, Xenobiotica, 15 (1985) 237.
144 M. Stanlovic and S. Chaykin, Arch. Biochem. Biophys., 145 (1971) 27.
145 D.W. Lundgren, H. Tevesz and E.A. Krill, Anal. Biochem., 148 (1985) 461.
146 G. Fabre, R. Seither and I.D. Goldman, Biochem. Pharmacol., 35 (1986) 1325.
147 M.-L. Chen, HPLC assay and pharmacokinetics of methotrexate, Ph.D. Thesis, University of Illinois, 1982.
148 S.M. El Dareer, K.F. Tillery and D.L. Hill, Cancer Treat. Rep., 65 (1981) 101.
149 S.A. Jacobs, R.G. Stoller, B.A. Chabner and D.G. Johns, J. Clin. Invest., 57 (1976) 534.
150 J. Lankelma, E. Van der Klein and F. Raemakers, Cancer Lett., 9 (1980) 133.
151 K.K. Chan, S.B. Nayer, J.L. Cohen, R.T. Chlebowski, H. Liebman, D. Stolinsky, M.S. Mitchell and D. Farquhar, Res. Commun. Chem. Pathol. Pharmacol., 28 (1980) 551.
152 M.-L. Chen, W.P. McGuire, T.E. Lad and W.L. Chiou, Int. J. Clin. Pharmacol. Ther. Toxicol., 22 (1984) 1.
153 A.L. Stewart, J.M. Margison, P.M. Wilkinson and S.B. Lucas, Cancer Chemother. Pharmacol., 14 (1985) 165.

154 R. Ertmann, S. Bielack and G. Landbeck, J. Cancer Res. Clin. Oncol., 109 (1985) 86.
155 C. Beedham and D.J. Rance, unpublished data.
156 C. Beedham, S.E. Bruce, D.C. Critchley, Y. Al-Tayib and D.J. Rance, Eur. J. Drug Metab. Pharmacokin. (in press)
157 R. Hille and V. Massey, Pharmacol. Ther., 14 (1981) 249.
158 V.H. Booth, Biochem. J., 32 (1938) 503.
159 K.V. Rajagopalan and P. Handler, J. Biol. Chem., 239 (1964) 2027.
160 F.F. Morpeth, Biochim. Biophys. Acta, 744 (1983) 328.
161 R.J. Feldman and H. Weiner, J. Biol. Chem., 247 (1972) 260.
162 V.H. Booth, Biochem. J., 32 (1938) 494.
163 G. Palmer, Biochim. Biophys. Acta, 56 (1962) 444.
164 J. Kurth, B. Neumann and H. Aurich, Acta Biol. Med. Ger., 37 (1978) 493.
165 J. Kurth and H. Aurich, Wiss. Z., Karl-Marx-Univ., Leipzig, Math.-Naturwiss. Reihe, 27 (1978) 47; Chem. Abstr., 89 (1978) 175426.
166 M. Hijikata, Nichidai Igaku Zasshi, 18 (1959) 1105.
167 G. Pelsey and A.M. Klibanov, Biochim. Biophys. Acta, 742 (1983) 352.
168 L.A. Cates, G.S. Jones, D.J. Good, H.V.-L. Tsai, V-S. Li, N. Caron, S-C. Tu and A.P. Kimball, J. Med. Chem., 23 (1980) 300.
169 F.F. Morpeth and R.C. Bray, Biochemistry, 23 (1984) 1332.
170 F. Ni Fhaolain and M.P. Coughlan, Int. J. Biochem., 8 (1977) 441.
171 T. Spector and R. Ferone, J. Biol. Chem., 259 (1984) 10784.
172 H.M. Kalckar, N.O. Kjeldgaard and H. Klenow, Biochim. Biophys. Acta, 5 (1950) 586.
173 C. Stubley and J.G.P. Stell, J. Pharm. Pharmacol., 32 (1980) 51P.
174 K.D. McMurtey and T.J. Knight, Mutat. Res., 140 (1984) 7.
175 M. Nightingale and C. Beedham, unpublished data.
176 E.J. LaVoie, E.A. Adams and D. Hoffman, Carcinogenesis, 4 (1983) 1133.
177 K.P. Moder and N.J. Leonard, J. Am. Chem. Soc., 104 (1982) 2613.
178 P.E. Gormley, E. Rossitch, M.E. D'Anna and R. Lysek, Biochem. Biophys. Res. Commun., 116 (1983) 759.
179 L. Greenlee and P. Handler, J. Biol. Chem., 239 (1964) 1090.
180 J.W. Bunting and A. Gunasekara, Biochim. Biophys. Acta, 704 (1982) 444.
181 F. Bergmann, L. Levene and H. Govrin, Biochim. Biophys. Acta, 484 (1977) 275.
182 N.J. Leonard, M.A. Sprecker and A.G. Morrice, J. Am. Chem. Soc., 98 (1976) 3987.
183 B.R. Baker, J. Pharm. Sci., 56 (1967) 959.
184 B.R. Baker and W.F. Wood, J. Med. Chem., 10 (1967) 1101.
185 B.R. Baker, W.F. Wood and J.A. Kozma, J. Med. Chem., 11 (1968) 661.
186 C.J. Betlach and J.W. Sowell, J. Pharm. Sci., 71 (1982) 269.
187 T.D. Beardmore and W.N. Kelly, Adv. Exp. Med. Biol., 41B (1974) 609.
188 A. Lucacchini, L. Bazzichi, G. Biagi, O. Livi and D. Bergini, Ital. J. Biochem., 31 (1982) 153.
189 S.S. Parmar, C. Dwivedi and B. Ali, J. Pharm. Sci., 61 (1972) 179.
190 S.S. Parmar, B.R. Pandey, C. Dwivedi and B. Ali, J. Med. Chem., 17 (1974) 1031.
191 S. Kaur, K. Raman, B.R. Pandey, J.P. Burthwal, K. Kishor, A. Kumar, K.P. Bhargava and B. Ali, Res. Commun. Chem. Pathol. Pharmacol., 21 (1978) 103.
192 R.H. Springer, M.K. Dimmitt, T. Novinson, D.E. O'Brien, R.K. Robins, L.N. Simon and J.P. Miller, J. Med. Chem., 19 (1976) 291.
193 D.E. Duggan, R.M. Noll, J.E. Baer, F.C. Novello and J.J. Baldwin, J. Med. Chem., 18 (1975) 900.
194 R.K. Robins, G.R. Revankar, D.E. O'Brien, R.H. Springer, T. Novinson, A. Albert, K. Senga, J.P. Miller and D.G. Streeter, J. Heterocycl. Chem., 22 (1985) 601.

195 J.J. Baldwin, P.A. Kasinger, F.C. Novello, J.M. Sprague and D.E. Duggan, J. Med. Chem., 18 (1975) 895.
196 W.E. Knox and W.I. Grossman, J. Am. Chem. Soc., 70 (1948) 2172.
197 R.L. Felsted, A.E.-Y. Chu and S. Chaykin, J. Biol. Chem., 248 (1973) 2580.
198 F. Bergmann, H.W. Kwietny, G. Levin and D.J. Brown, J. Am. Chem. Soc., 82 (1960) 598.
199 F. Bergmann and H. Ungar, J. Am. Chem. Soc., 82 (1960) 3957.
200 F. Bergmann and L. Levene, Biochim. Biophys. Acta, 429 (1976) 672.
201 H. Rosemeyer and F. Seela, Eur. J. Biochem., 134 (1983) 513.
202 F. Bergmann, A. Frank and H. Govrin, Biochim. Biophys. Acta, 570 (1979) 215.
203 F. Seela, W. Bussmann, A. Gotee and H. Rosemeyer, J. Med. Chem., 27 (1984) 981.
204 C.M. Hodnett, J.S. McCormack and J.A. Sabean, J. Pharm. Sci., 68 (1976) 1150.
205 F. Bergmann and H. Kwietny, Biochim. Biophys. Acta, 33 (1959) 29.
206 F. Bergmann, L. Levene, I. Tamir and M. Rahat, Biochim. Biophys. Acta, 480 (1977) 21.
207 H.S. Forrest, E.W. Hanly and J.M. Lagowski, Biochim. Biophys. Acta, 50 (1961) 596.
207a R.B. Banks and S.D. Barnett, Biochem. Biophys. Res. Commun. 140 (1986) 609.
208 Z. Nieman, Isr. J. Chem., 10 (1972) 819.
209 K.V. Rajagopalan and P. Handler, Biochem. Biophys. Res. Commun., 8 (1962) 43.
210 F. Bergmann, L. Levene, Z. Neiman and D.J. Brown, Biochim. Biophys. Acta, 222 (1970) 191.
211 C. Johnson, C. Beedham and J.G.P. Stell, Xenobiotica, 17 (1987) 17.
212 D. Lichtenberg, F. Bergmann and Z. Nieman, Isr. J. Chem., 10 (1972) 805.
213 D.G. Liegler, E.S. Henderson, M.A. Hahn and V.T. Oliverio, Clin. Pharmacol. Ther., 10 (1969) 849.
214 D.G. Johns, D. Farquhar, B.A. Chabner and J.J. McCormack, Biochem. Soc. Trans., 2 (1974) 602.
215 D.M. Valerino, D.G. Johns, D.S. Zaharko and V.T. Oliverio, Biochem. Pharmacol., 21 (1972) 821.
216 J.J. McGuire, P. Hsieh and J.R. Bertino, Biochem. Pharmacol., 33 (1984) 1355.
217 G. Fabre, I. Fabre, L.H. Matherly, J.P. Cano and I.D. Goldman, J. Biol. Chem., 259 (1984) 5066.
218 H.G. Kaplan, Biochem. Pharmacol., 29 (1980) 2135.
219 A.S. Lewis, L. Murphy, C. McCalla, M. Fleary and S. Purcell, J. Biol. Chem., 259 (1984) 12.
220 D.M. Valerino and J.K. McCormack, Biochim. Biophys. Acta, 184 (1969) 154.
221 R. Pomales, G.B. Elion and G.H. Hitchings, Biochim. Biophys. Acta, 95 (1965) 505.
222 R. Pomales, S. Bieber, R. Friedman and G.H. Hitchings, Biochim. Biophys. Acta, 72 (1963) 119.
223 G.H. Hitchings, in Biochemical Aspects of Antimetabolites and of Drug Hydroxylation, ed. D. Shugar (Academic Press, London, 1969) pp. 11–22.
224 J.E. Seegmiller, in Hereditary Xanthinuria. Metabolic Control and Disease, eds. P.K. Boney and L.E. Rosenberg (E.B. Saunders, Philadelphia, 1980) p. 842.
225 K. Nishioka, H. Yamanaka, T. Nishina, T. Hosoya and K. Mikanagi, Adv. Exp. Med. Biol., 165 (1984) 73.
226 H. Yamanaka, K. Nishioka, T. Sueuki and K. Kohno, Ann. Rheum. Dis., 42 (1983) 684.
227 S. Futterman, J. Biol. Chem., (1962) 677.
228 M.R. Lakshaman, C.S. Vaidyanathan and H.R. Cama, Biochem. J., 90 (1964) 569.
229 A.H. Merrill, Jr., J.M. Henderson, E. Wang, B.W. McDonald and W.J. Millikan, J. Nutr., 114 (1984) 1664.

230 J.R. Wozenski, J.E. Leklem and L.T. Miller, J. Nutr., 110 (1980) 275.
231 R. Fried, L.W. Fried and D.R. Babin, Eur. J. Biochem., 33 (1973) 439.
232 J.M. Benson, R.E. Royer, J.B. Galvin and R.W. Shimizu, Toxicol. Appl. Pharmacol., 68 (1983) 36.
233 E.J. Lavoie, E.A. Adams, A. Shigematsu and D. Hoffman, Drug Metab. Dispos., 13 (1985) 71.
234 C. Johnson, C. Stubley-Beedham and J.G.P. Stell, Biochem. Pharmacol., 33 (1984) 3699.
235 L.S. Dietrich, J. Biol. Chem., 211 (1954) 79.
236 G.B. Elion, A. Kovensky, G.H. Hitchings, E. Metz and R.W. Rundles, 15 (1966) 863.
237 V.H. Fenner, O. Schiemann and I. Gikalov, Arzneim-Forsch., 35 (1985) 1093.
238 G.B. Elion, S. Callahan, H. Nathan, S. Bieber, R.W. Rundles and G.H. Hitchings, Biochem. Pharmacol., 12 (1963) 85.
239 C. Johnson, C. Stubley-Beedham and J.G.P. Stell, Biochem. Pharmacol., 34 (1985) 4251.
240 T.A. Krenitsky, W.W. Hall, P. De Miranda, L.M. Beauchamp, H.J. Schaeffer and P.D. Whiteman, Proc. Natl. Acad. Sci. USA., 81 (1984) 3209.

Progress in Medicinal Chemistry – Vol. 24, edited by G.P. Ellis and G.B. West

4 Platinum Antitumour Agents

JAMES C. DABROWIAK, Ph.D. and WILLIAM T. BRADNER, Ph.D.[b]

[a]*Department of Chemistry, Syracuse University, Syracuse, NY 13244-1200 and* [b]*The Bristol Myers Company, Preclinical Anticancer Research, P.O. Box 4755, Syracuse, NY 13221-4755, U.S.A.*

INTRODUCTION

The discovery of the anticancer properties of the simple coordination compound *cis*-diamminedichloroplatinum(II), (1), and the successful introduction of the complex into the clinic represents an important advance for medicinal chemistry. Aside from the direct beneficial aspects of the compound as an anticancer agent, its success has served to focus attention on a hitherto largely unexplored area of medicinal chemistry, namely, the identification of inorganic substances as useful chemotherapeutic agents. Today, well over 1000 analogues of (1) have been synthesized and examined for their antitumour properties and a number of those have shown sufficient promise to be entered into clinical trials. In this review we concentrate on the parent compound and a limited number of its analogues which collectively demonstrate the breadth of structure-activity relationships which are known for the platinum-based antitumour agents. In addition to discussing the parent compound, we examine how the presence of optical activity in the complex and the ability of the compound to undergo substitution reactions affect its DNA binding and antitumour properties. Since anticancer activity is not confined to the divalent oxidation state of the metal ion, we also examine complexes of Pt(IV) which show promise as successors to (1). Finally, in light of the fact that biologically active *cis*-diamineplatinum(II) compounds are able to undergo oligomerization, the antitumour properties of a group of dimers and trimers are also presented and discussed.

CIS-DIAMMINEDICHLOROPLATINUM(II) (CISPLATIN) (1)

The most extensively studied platinum-containing antitumour agent is *cis*-diamminedichloroplatinum(II) (cisplatin) (1) [1, 2]. As reviews covering the chemistry of the compound, the chemical, biochemical and biological events pertaining to its mechanism of action as well as its antitumour properties have already appeared [3–9], we will focus on the basic elements of the drug's solution chemistry and present some of the more recent findings pertaining to its interactions with DNA. In a following section, we summarize the important antitumour effects of the compound.

CHEMISTRY AND MECHANISM

Dissolution of (1) in water results in the slow (hours) stepwise release of the bound chloride ions and the formation of mono and diaquo species which retain the *cis*-diammine geometry [10] (*Scheme 4.1*). In addition to aquation,

(1) cis-$[Pt(NH_3)_2Cl_2]$ (2) $trans$-$[Pt(NH_3)_2Cl_2]$

$$[Pt(NH_3)_2Cl_2] \underset{}{\overset{-Cl^-}{\rightleftharpoons}} [Pt(NH_3)_2Cl(OH_2)] \underset{}{\overset{-Cl^-}{\rightleftharpoons}} [Pt(NH_3)_2(OH_2)_2]$$

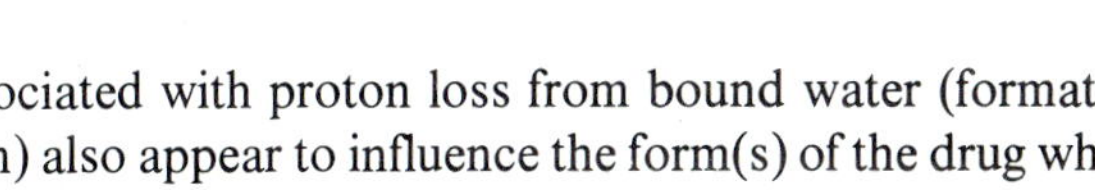

Scheme 4.1. The equilibria involving loss of chloride ion from cis-*diamminedichloroplatinum(II), (1), in aqueous media are shown.*

equilibra associated with proton loss from bound water (formation of bound hydroxide ion) also appear to influence the form(s) of the drug which is present at physiological values of pH [11]. It is believed that aquo (hydroxo) forms of the compound are the species which react with biological components and which ultimately give rise to the antitumour effects of the drug. Although substitution kinetics are somewhat faster than those observed for (1), analogous equilibra apply to the weakly biologically active *trans* isomer, (2), as well.

An additional complicating feature of the chemistry of cisplatin (1) is its ability to oligomerize in solution to form a μ-oxo-bridged dimer (3) and trimer (4) (*Scheme 4.2*). This process, which proceeds through the mono-deprotonated form of the diaquo complex, has been studied using ^{195}Pt-NMR spectroscopy [5, 12] and the products have been characterized *via* X-ray structural analysis [13–16]. Since oligomerization requires a relatively high

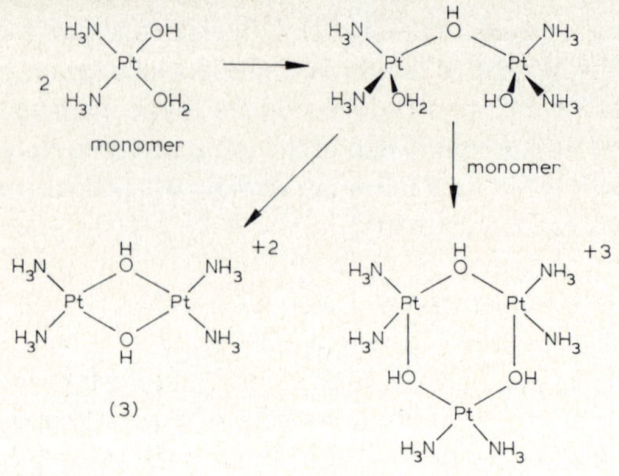

Scheme 4.2. The oligomerization of the aquo-hydroxo form of (1) in aqueous media is shown.

concentration of the drug in solution (at least about 10 mM), the events depicted in *Scheme 4.2* probably do not occur to any great extent with cisplatin *in vivo*. Due to the geometry of the complex, oligomerization is not possible for the *trans* isomer (2).

The important biological target for cisplatin is DNA [7, 17]. Both it and its weakly biologically active *trans* isomer are known to form DNA inter- and intrastrand cross-links as well as cross-links between proteins and DNA. Which of these lesions is important in the antitumour effects of cisplatin remains unclear, but recent work with SV-40 DNA in green monkey CV-1 cells [18] showed that inhibition of DNA replication, a property of the platinum antitumour agents, is probably not affected by the presence of protein-DNA cross-links. An analogue of (1), dichloro(ethylenediamine)platinum(II) (5), a compound which is incapable of forming protein-DNA cross-links, was found to inhibit replication as efficiently as (1) and (2). The facts that cisplatin is considerably more active than (2) as an antitumour agent and that it, unlike the *trans* compound, is capable of forming a cross-link with two consecutive nucleotides on the same strand of DNA, has focused considerable attention on the intrastrand cross-link as the important chemical event underlying the antitumour effects of cisplatin.

Recent studies [19–24] have shown that the intrastrand cross-link is the major adduct formed upon incubation of (1) or (5) with purified double-stranded DNA. Exhaustive digestion of platinated DNA *via* enzymatic or chemical means followed by analysis revealed that platinum binding most often occurred at the dinucleotide sequence GG, with lesser binding occurring at the sequences (all 5′ → 3′) AG and GNG, where N is any of the four DNA nucleotides. Interestingly, no evidence for platination at the sequence GA (opposite polarity to AG) has yet been obtained. Although the *trans* isomer (2) possesses a geometry which prevents it from forming nearest-neighbour intrastrand links, e.g., with GG or AG, the complex appears capable of reacting with the sequence GNG [25, 26]. Using a 'replicating mapping' assay, Pinto and Lippard [25] recently identified the DNA synthesis termination sequences induced by (1) and (2). By examining the 'stop-points' in the synthesis, it was

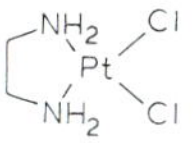

(5) 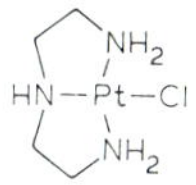 (6)

learned that the *trans* isomer inhibited at GNG sequences, while cisplatin stopped synthesis most often at $(dG)_n$ ($n \geqslant 2$) sites on the phage DNA. A compound which effectively models monoadducts between cisplatin and DNA, chlorodiethylenetriamineplatinum(II), (6), apparently bound to DNA but was incapable of inhibiting synthesis in the system.

In an effort to uncover the structures of DNA adducts of cisplatin, many reactions with the drug and synthetic oligonucleotides have also been carried out. Recently it has been shown that (1) and (2) can react in a cross-linking manner with the guanine residues in the trinucleotide d(GpTpG) [26]. In both cases, NMR measurements showed that the donor atoms bound to the platinum ions are two ammonia ligands and N-7 of guanine in positions 1 and 3 of the trinucleotide.

The structure of the major adduct between DNA and cisplatin has recently been studied *via* X-ray crystallography [27]. Reaction of the drug with d(GpG) yielded a complex having both N-7 atoms of the guanine residues of the dinucleotide bound to *cis* coordination sites on the metal ion. The distortions found in the sugar phosphate backbone in the complex suggested that platination of native double-helical DNA by (1) at a GG site would probably lead to a disruption of the Watson-Crick base-pairing or duplex kinking at the interaction site. Although this suggestion is consistent with others made earlier [28–31], recent NMR studies show that Watson-Crick base-pairing may not be disrupted by platination of DNA. For example, the octanucleotide d(GATCCGGC), which has been platinated at the guanine bases at positions 6 and 7 of the nucleotide by *cis*-$Pt(NH_3)_2^{2+}$, is able to form a stable duplex with its complementary DNA strand [32].

ANTITUMOUR EFFECTS

Preclinical studies

Spectrum of activity. The effects of platinum derivatives on tumour cells have been studied both *in vitro* and *in vivo*. Cytotoxicity testing *in vitro* has been popular as a prescreen for antitumour effects because of speed, efficiency and need for small amounts of compound [33]. However, platinum complexes present a number of problems, such as incompatibility with certain media and widely varying effects based on drug-cell contact time [34–36]. Thus, most antitumour testing has concentrated on *in vivo* systems, principally transplanted murine tumours. Early studies relied on sarcoma 180, the tumour system first revealing the *in vivo* effects of cisplatin [37, 38]. This tumour is still being used [39] but has largely been replaced by the lymphatic leukaemia L-1210 as a primary screen. The latter is the mainstay of analogue programmes at the National Cancer Institute [40], the Bristol-Myers Company [41], and the Wadley Institute [42]. As it is generally used in the ascitic form with therapy administered intraperitoneally (i.p.) to i.p. implanted tumour cells, L-1210 is highly responsive to the more reactive platinum complexes, but much less so to the stable compounds. Thus, the data generated on L-1210, if considered alone, could be misleading with respect to a number of derivatives which have gone into clinical trial, as will be shown in discussions of the individual agents.

Cisplatin is strongly inhibitory to many other tumours of rats and mice [43] and is not schedule-dependent, since a single dose day 1 after tumour implantation, or daily injection for 9 days, are equally effective on L-1210 [44]. Recently the M-5076 sarcoma has been shown to have an interesting pattern of response to cisplatin and selected analogues, suggesting its utility as a screen for these agents [45].

Potency. In biological systems, potency is defined as the dose required to achieve a given effect. In terms of *in vivo* tumour testing, a common approach is to identify the dose at which the optimum antitumour effects occur (optimum dose, O.D.). This is generally the same as the maximum tolerated dose. The single figure generated can be compared with that of related compounds and simple ratios of relative potency can be established. Cisplatin has a single-dose O.D. of about 8 mg/kg. Since cisplatin is the simplest compound, it is not surprising that only a few derivatives have slightly greater potency and a large number of analogues are considerably less potent (upwards of 20-times), though many have comparable antitumour effects.

Toxicity. The toxicity of cisplatin has been studied both in large animals such as dogs and monkeys [46] and in small animals as a basis for screening and comparing analogues [47, 48]. In general, many of the toxicities observed in the clinic can be duplicated or predicted in one or more laboratory animal species. Renal toxicity, which is the principal hazard of cisplatin therapy, can be accurately measured, and vomiting and myelosuppression can be demonstrated, but in a less quantitative manner.

Pharmacology. The pharmacokinetics of cisplatin have been extensively studied in both animals and humans [49, 50]. Briefly, cisplatin and its monoaquo and diaquo reaction products become 90% protein bound within 3 h of administration. Disappearance of cisplatin from the blood of animals is biphasic, with $t_{1/2}\,\alpha$ of under 1 h and $t_{1/2}\,\beta$ of 2 days in rats and 4–5 days in dogs.

Clinical studies

Cisplatin is approved in the United States for clinical treatment of metastatic testicular tumours, metastatic ovarian tumours and advanced bladder cancer. It has also demonstrated significant activity in head and neck cancer, cervical cancer, osteogenic sarcoma and prostate cancer [49]. It is generally used in combination with other active agents and is particularly effective against testicular tumours, where about 70% of patients are cured. Cisplatin is a very toxic drug, with renal toxicity being the most hazardous, though it can be reasonably controlled with hydration. Marked nausea and vomiting occur in most of the patients treated with the drug. Myelosuppression, ototoxicity and neurotoxicity are other side-effects which are encountered. It is obvious that an objective in preparing derivatives of cisplatin would be to achieve reductions in toxicity, particularly renal and gastric, which are most troublesome, while retaining or improving antitumour effects.

4-CARBOXYPHTHALATO(1,2-DIAMINOCYCLOHEXANE)-PLATINUM(II), (7), AND RELATED COMPOUNDS

CHEMISTRY AND MECHANISM

Metal complexes containing the bidentate amine ligand, 1,2-diaminocyclohexane (DACH), have been known for 50 years [51]. The ligand, when complexed to a metal ion, adopts a geometry which is highly dependent on the relative configurations about the two asymmetric carbon atoms in the compound. As is evident from *Figure 4.1*, complexation of the racemic form of the ligand (often denoted as the *trans* form) to a platinum ion causes the diamine to adopt a rigid cyclohexane chair-type conformation with the atoms of the

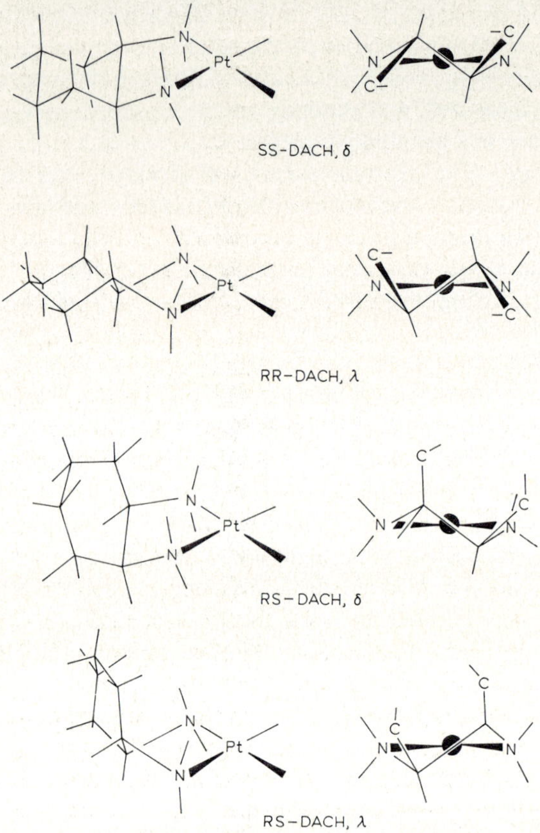

SS–DACH, δ

RR–DACH, λ

RS–DACH, δ

RS–DACH, λ

Figure 4.1. The structures of the various isomeric forms of 1,2-diaminocyclohexane (DACH) bound to platinum are shown. Stereochemical designations for DACH which are in common use are: trans-*(−)*-, R,R; trans- *(+)*- S,S; cis, R,S *or* S,R.

ligand lying close to the nitrogen donor plane. Depending on the absolute configurations of the two asymmetric centres in the compound, the five-membered chelate ring involving the metal ion can adopt either the so-called λ (carbon centres *R*,*R*) or δ (carbon centres *S*,*S*) conformation [52]. This stereochemistry has been confirmed by a number of X-ray structural analyses involving this form of the DACH ligand [53–57]. For the coordinated *meso* form of the ligand (often denoted as the *cis* form), the configurations about the two asymmetric carbon centres (carbon centres, *S*,*R*) force the atoms of the cyclohexane ring to lie on one side of the nitrogen-metal-nitrogen donor plane [39, 58]. Unlike the coordinated *R*,*R* and *S*,*S* forms of the ligand, which have locked chelate ring conformations, the λ and δ forms of the coordinated *meso* diamine easily interconvert in solution at room temperature [59].

The synthesis and antitumour properties of DACH platinum(II) complexes containing a variety of different types of leaving ligand have been reported [38, 39, 60–67]. Although early studies with the compounds were content to utilize mixtures of geometric and optical isomers, the realization that the stereochemistry of the DACH ligand influenced antitumour activity underscored the importance of working with isomerically pure forms of the diamine [68–70].

In attempting to uncover the structural origins of the antitumour effects of the various isomeric DACH complexes, a few studies have focused on the interactions of the compounds with DNA [71, 72]. Reaction of Pt(II)(*R*,*R*-DACH)Cl_2, Pt(II)(*S*,*S*-DACH)Cl_2 and Pt(II)(*R*,*S*-DACH)Cl_2 with the dinucleotide d(GpG) resulted in platinum binding to the N-7 atoms of both guanine residues of the dinucleotide [72]. Thus, the interaction is directly analogous to the major intrastrand cross-link produced by cisplatin in its interactions with DNA. Since Pt(II)(*R*,*R*-DACH)Cl_2 and Pt(II)(*S*,*S*-DACH)Cl_2 are both optically active, loss of the chloride ions and ligation to the optically active dinucleotide results in diastereomers having different HPLC and CD properties. Comparison of the CD properties of the products obtained from the Pt-DACH reaction with those obtained from reaction of cisplatin with [d(GpG)] suggested that, like the latter complex, the dinucleotide in the Pt-DACH adducts has a head-to-head arrangement with anti-anti configurations of the bases. Since the complex containing the *meso* form of the ligand (*R*,*S* carbon centres) has its cyclohexane moiety displaced to one side of the N-Pt-N donor plane, reaction with the dinucleotide results in two non-interconverting structures. Although both have the same nucleotide arrangement, i.e., head-to-head with and anti-anti configuration of bases, one has the cyclohexane group on the same side of the coordination plane of the complex with O(6) of guanine, Adduct A, while the other has these groups on opposite sides of the coordination plane, Adduct B. For the reaction with the simple dinucleotide d(GpG), both of these complexes are produced in equal amounts.

An illustration of how DNA structure can influence the product distribution in this case was obtained by reaction of Pt(II)(*R*,*S*-DACH)Cl_2 with purified double-stranded calf thymus DNA. Analysis, involving digestion of the platinated DNA with enzymes followed by separation of the products using HPLC, showed that, although both compounds are again produced, the isomer having the cyclohexane moiety on the same side of the coordination plane with O(6) of guanine, Adduct A, was found in greater abundance. This result suggests that, in the interaction with DNA, the cyclohexane group of the DACH complex prefers to be located in the major groove of DNA rather than over the sugar-phosphate backbone of the polymer.

ANTITUMOUR EFFECTS

Preclinical studies

Spectrum of activity. Following the initial report of the anti-tumour properties of a complex containing 1,2-diaminocyclohexane [66], a number of DACH compounds have been synthesized and tested. Unless otherwise noted, all of the complexes described contain the racemic form of the so-called *trans* isomer, i.e., equal molar amounts of the *R*,*R* and *S*,*S* optical isomers (*Figure 4.1*).

The most important complex in this series, (7), was found to be comparable with cisplatin against L-1210 cells *in vitro*, whereas the citrate and gluconate derivatives were somewhat weaker [34]. Cytotoxic effects of (7) against human colon tumour cell lines were 10-fold less than cisplatin [36]. Most of the DACH compounds have been shown to have antitumour effects equal to or better than cisplatin against L-1210 and P-388 leukaemias *in vivo* [60, 70]. In tests against other tumour systems, the sulphato aquo complex (8) was curative against Lewis Lung carcinoma; compounds (7), (8) and the oxalato derivative (9) were effective against B16 melanoma in the historic range for cisplatin [40].

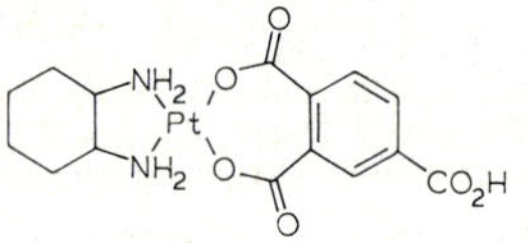

(7) 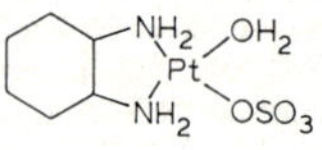 (8)

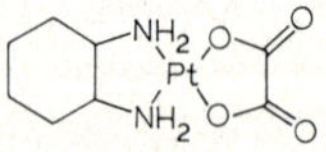

(9) 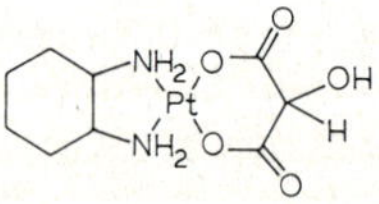 (10)

The DACH complexes have been the focus of special attention [73] since several were found fully inhibitory to a line of L-1210 leukaemia made totally resistant to cisplatin treatment *in vivo*. This property is not universal, however, since the 2-hydroxymalonato compound, (10), was found to have greatly reduced effectiveness against L-1210 resistant to cisplatin [74]. Two groups independently prepared L-1210 cells resistant to one or more DACH complexes and demonstrated that these cells retained sensitivity to cisplatin as well [75, 76].

As was originally shown by Kidani, Inagaki and Tsukagoshi [69], platinum(II) complexes containing the various isomeric forms of the DACH ligand possess different antitumour activities. Generally, the antitumour activity of compounds containing the *R*,*R* isomer was found to be greater than those containing either the *S*,*S* or *R*,*S* (*meso*) form of the ligand, irrespective of the leaving groups present. However, isomeric differentiation appeared to be dependent on tumour line, and little difference in activity between the various compounds was observed in P-388 [59, 70] and Sarcoma 180 [67]. In a test of *R*,*R*-(7) and *S*,*S*-(7) against L-1210 and L-1210-resistant cells, the two isomers were equally effective in prolonging survival of mice implanted with the parent line, but *R*,*R*-(7) was considerably more effective than *S*,*S*-(7) in inhibiting the cisplatin-resistant line (Rose and Bradner, unpublished observations).

The mechanism of L-1210 resistance to cisplatin and other platinum antitumour agents remains to be clearly defined. Both Waud [77] and others [78] suggest that reduced cisplatin uptake may be responsible in part for the resistance of the cell line. It has also been found that glutathione levels are high in human ovarian carcinoma cells made resistant to cisplatin *in vitro* [79]. However, reduction in the level did not alter the resistance to the drug [80]. Decreases in amino-acid transport and changes in amino-acid substrate specificities have also been suggested as the basis for the resistance [81], but definitive evidence has been difficult to obtain.

Potency. As might be expected, the DACH compounds with more labile leaving groups tend to be more potent as measured both by LD_{50} value and optimum dose level in tumour inhibition studies [44, 48, 82]. For example, the aquo sulphato and carboxyphthalato analogues of (7) and (8) require about 1.5- to 2-times the dose of cisplatin to achieve maximum antitumour effects on multidose schedules.

Toxicity. All of the DACH complexes studied have shown reduced kidney toxicity when compared with that of cisplatin [48, 83]. Likewise, (7) and (8) have been shown to be at least as leukopoenic as cisplatin when given at doses related to the LD_{50} value as a measure of potency [48]. These two compounds have also been shown to be emetic in dogs [48], but possibly to a lesser extent than cisplatin [83, 84].

Clinical studies

At least three DACH complexes have been subjected to clinical trial. The malonato derivative, (11), was reported to cause mild vomiting and thrombocy-

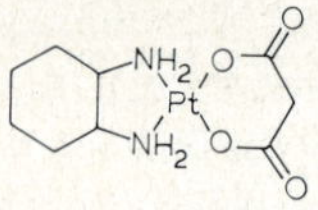

(11)

topenia but no renal toxicity in hydrated patients during Phase I trials [85]. Partial or minor tumour shrinkage was noted in three of seventeen evaluable patients with solid tumours. All responders had tumours clinically resistant to cisplatin therapy. It was also reported that trials of (11) were severely hampered by its poor solubility [85]. In Phase I trial of (7) [84], the drug was found less toxic than cisplatin in both vomiting and in renal effects. In this instance the dose-limiting toxicity was thrombocytopoenia. Objective antitumour effects were seen, some in patients who had received cisplatin, but whether or not they were refractory to or relapsing on the latter drug was not stated. Judging from the description of the drug preparation, stability of the formulation was clearly a problem. Furthermore, since (7) is a mixture of optical isomers and the resolved compounds, i.e., *R*,*R*-(7) and *S*,*S*-(7), show differing biological properties, one might question the feasibility of continuing development of racemic (7) for clinical treatment. Recently, the optically active oxalato compound, *R*,*R*-(9), was selected for further study. The potency and activity of *R*,*R*-(9) was found comparable with that of cisplatin in animal tumour tests [86]. In Phase I clinical trials, nausea and vomiting, similar to cisplatin, appeared to be dose-limiting. Renal and haematologic toxicity with *R*,*R*-(9) were not observed. Several objective responses for the compound were also reported [87].

CIS-DIAMMINE-1,1-CYCLOBUTANEDICARBOXYLATE-PLATINUM(II) (CARBOPLATIN) (12)

CHEMISTRY AND MECHANISM

An analogue of (1) containing a bidentate dicarboxylate as a leaving ligand is (12). X-ray structural analysis of the compound shows that the coordinated dicarboxylate ligand is not flat, but rather is puckered, having the cyclobutane moiety significantly displaced above the donor plane of the compound [88, 89]. Conductance measurement on aqueous solutions of (12) demonstrated, that unlike (1), the complex resists aquation [90]. Although this observation is probably due to the fact that the dicarboxylate is a bidentate ligand and thus is more difficult to displace from the platinum ion than a monodentate ligand, e.g., Cl-steric factors associated with the puckered dicarboxylate ring may also play a rôle in substitution reactions with the compound. Substitution processes on square planar Pt(II) compounds are known to occur *via* an associative mechanism involving approach of the substituting ligand on the axis of the complex [91]. Since the cyclobutane moiety of the dicarboxylate ligand is displaced significantly above the coordination plane, and, as is evident from NMR studies, is oscillating between both sides of the plane [88], its presence may partially block substituting ligands from approaching the Pt ion. This type of steric effect would explain the observed slow rates of binding of (12) and the related hydroxymalonato complex (13) to components present in human

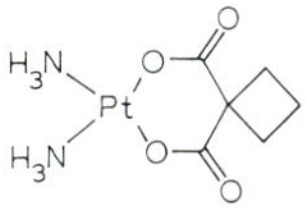

(12) 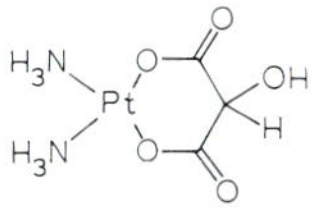 (13)

plasma [92]. If steric factors are important, a complex containing the unsubstituted malonate ligand, e.g., (11), would be expected to react more quickly with plasma components than either (12) or (13). A recent study has shown that this is in fact the case [92]. The reduced ability of (12) to undergo substitution reactions is also evident from studies with L-1210 cells [93], wherein it was shown that the peak levels of DNA-protein and DNA interstrand cross-links induced by (12) were found to occur 6 to 12 h after those produced by cisplatin.

The DNA binding properties of (12) have not been investigated to any great extent. However, it is known that the compound can bind to and unwind super-coiled PM2-DNA [94]. Comparative studies with (12), and cisplatin revealed that the amount of (12) required to produce a degree of DNA unwinding comparable with that of cisplatin was many times that necessary for the latter compound. Although further study is required, this difference in behaviour is probably related to different DNA platination rates (slower for (12)) of the two complexes.

ANTITUMOUR EFFECTS

Preclinical studies

Spectrum of activity. Using a 1 h drug exposure in a colony-forming assay of colon tumour cell lines, it was found that (12) was essentially non-cytotoxic. In a microtitre system with continuous exposure against several murine and human tumour cell lines, (12) was found on the average to be 22-times less cytotoxic than cisplatin [95] (Catino, J.J., unpublished observations). Initial *in vivo* tests suggested that carboplatin was less effective than cisplatin against L-1210 leukaemia [96, 97], but that it had comparable activity against the Adj/PC6A plasmacytoma which the Institute of Cancer Research group used for screening. Superior activity against a human bronchus tumour xenograft P246 was later identified as one of the key reasons for selecting (12) from among several analogues for further development [98]. Carboplatin has also been found to have at least comparable activity with that of cisplatin in almost all solid tumour systems involving treatment from a distal site [99]. Finally, in tests against a line of L-1210 leukaemia with acquired resistance to cisplatin, (12) was found to be totally inactive, suggesting complete cross-resistance on the part of these cells [74].

Potency. As noted above, (12) was considerably less potent than cisplatin in direct cytotoxicity *in vitro* against various tumour cells. This difference was also reflected *in vivo*, since optimum antitumour effects required 16-times as much (12) for single dose treatment of L-1210 leukaemia, 32-times for multidose treatment and at least 10-times for multidose treatment of B16 melanoma [44].

Toxicity. Studies in a variety of animal species have indicated that carboplatin is generally less nephrotoxic than cisplatin [99, 100]. Likewise, in the ferret, which has proved to be a useful animal model for studying emesis, carboplatin was found to cause considerably less vomiting than cisplatin [101]. The myelosuppressive effect of (12) was identified by CFU-C test of the bone marrow of mice, but not by peripheral blood counts [102]. Thrombocytopoenia, though observed in dogs, was not quantitatively compared with that produced by cisplatin [99]. No studies of thrombocytopoenia caused by platinum complexes have been reported using the ferret, an animal model which in tests of other types of antitumour agents may have some utility in predicting this type of toxicity [103].

Pharmacology. The pharmacology of (12) appears to be strongly influenced by the slower substitution kinetics of the compound when compared to cisplatin. Tests performed *in vitro* have shown that (12) is considerably more stable in plasma at 37 °C than cisplatin, with half-lives of 30 h and 1.5–3.6 h, respectively [104]. Pharmacokinetics in the rat indicated α and β phase half-lives of 2.2 and 25.2 min, respectively, for (12) [105]. The drug is mainly excreted through the kidney, and, unlike cisplatin, is recovered largely unchanged. In all animal species studied so far, carboplatin appears to be more rapidly and completely cleared from the tissue and organs than cisplatin after administration of doses of comparable potency.

Clinical studies

Carboplatin has demonstrated effectiveness as a single agent in the treatment of ovarian carcinoma, small cell lung cancer and head and neck cancer. It may also be active in treatment of cancer of the testis and the cervix [106]. Carboplatin does not appear to be active in cancers of the upper gastro-intestinal tract, in breast cancer [107] or non-small cell lung cancer [108]. It is of interest than in treatment of advanced ovarian cancer with (12), the response rate is markedly lower among patients who had received prior therapy with cisplatin compared with those with no prior cisplatin (26% vs. 57%). This suggests some degree of clinical cross-resistance between the two drugs. Pharmacokinetic studies in humans reveal a pattern similar to that of animals in that (12) is less rapidly protein bound than cisplatin and is excreted largely unchanged in the urine [104].

1,1-BIS(AMINOMETHYL)CYCLOHEXANEAQUOSULPHATO-PLATINUM(II) HYDRATE (SPIROPLATIN) (14)

CHEMISTRY AND MECHANISM

Compound (14) was first synthesized by the Institute of Organic Chemistry, TNO (Utrecht, The Netherlands) as a potential second-generation cisplatin analogue. The complex contains an unusual bidentate amine ligand which, when bound to a metal ion, produces a non-planar six-membered chelate ring. A recent X-ray structural analysis of the compound [109] revealed that both the chelate ring and the cyclohexane moiety adopt a chair-type conformation. Since ^{13}C-NMR studies on the diaquated form of the compound show the presence of symmetry-equivalent carbon atoms in the cyclohexane ring, the amine ligand must be rapidly interconverting between various conformational forms in solution.

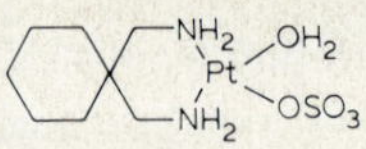

(14) 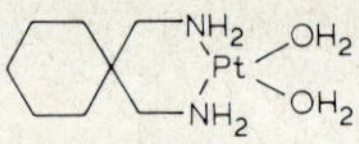 (15)

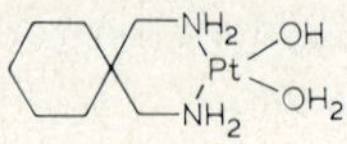

(16) 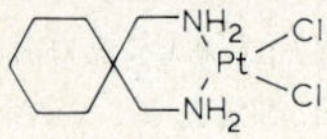 (18)

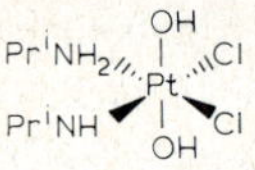

(17)

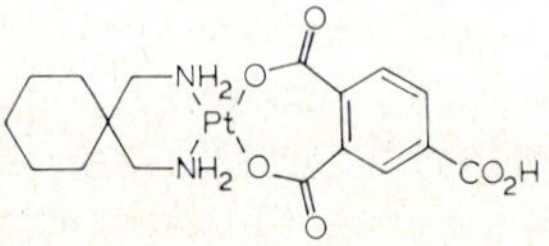

(19) 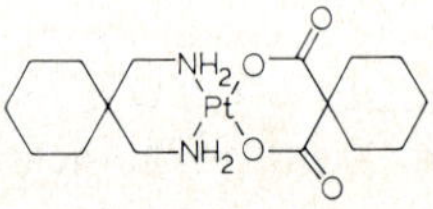 (20)

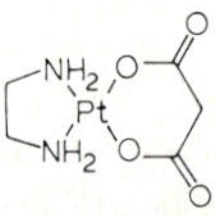

(21)

[195]Pt-NMR studies show that dissolution of (14) in water results in the partial displacement of sulphate ion and the production of the diaquo species (15). At low concentrations in aqueous media at pH $\approx$ 7, the compound probably exists primarily as the highly reactive mononuclear hydroxo-aquo species (16). However, at high concentrations, about 20 mM or more, the monomer reacts to form a dimer and trimer which are structurally analogous to those which form with (1) (*Scheme 4.2*). Although no studies with the compound pertaining to mechanism have been reported, it, like other Pt(II) compounds, almost certainly utilizes DNA as a biological target site.

ANTITUMOUR EFFECTS

Preclinical studies

Spectrum of activity. Among the series of 1,1-bis(aminomethyl)cyclohexane (DAMCH) complexes prepared, (14) has been most extensively studied. In tests against a line of Lewis Lung carcinoma cells *in vitro*, (14) was found to be as cytotoxic as cisplatin [110]. Against human colon tumours *in vitro* it was 2- to 5-fold less cytotoxic than cisplatin [36], and was on the average 8-fold less cytotoxic against a panel of murine and human tumours (Catino, J.J., unpublished data). Compound (14) has also been compared with the Pt(II) complexes cisplatin and (12) as well as the Pt(IV) complex (17) in a human tumour cloning assay of fresh tumour explants [11]. Qualitatively, the response incidence did not differ between (14) and cisplatin, since 6/11 tumours were inhibited by each drug. Several complexes containing the DAMCH ligand have been tested *in vivo*. Compounds (14), (18) and (19) had activity against L-1210 comparable with that of cisplatin, while the 1,1-cyclohexane dicarboxylate derivative, (20), was only weakly active [44, 74]. Compounds (14) and (18) were active against B16 melanoma, C26 colon carcinoma and L-1210 resistant to cisplatin. Compound (14) was as effective as cisplatin in inhibiting an immunocytoma of the LOU/M rat [112]. However, in tests against a xenograft of human ovarian carcinoma MR1-H-207, (14) was distinctly inferior to cisplatin, (12), (17) and (21) [113]. Likewise, in studies with M5076 murine sarcoma, cisplatin, (12), (14) and (17) were all effective against the tumour implanted i.p. with i.p. treatment, but (14) failed to inhibit the s.c. implanted tumour with i.v. treatment, whereas the other compounds were still active [45].

Potency. As with other series of compounds, the potency of the DAMCH complexes is influenced by the leaving group. Thus, (14) and (18) are close to cisplatin in potency, based on optimum dose for tumour therapy and LD_{50} value in non-tumour bearing animals [44, 48, 74]. Compounds (19) and (20), on the other hand, were 7- and 10-times less potent than cisplatin based on comparative LD_{50} values.

Toxicity. The renal toxicity of (14), (18) and (19) has been studied in mice and all were found less toxic than cisplatin [48, 74]. Spiroplatin (14) was also shown to be without renal toxicity in rats, but the compound caused severe kidney damage in dogs [112, 114].

Pharmacology. Spiroplatin is more rapidly protein-bound than cisplatin and also has 10-times the biliary excretion of the latter drug over a 6 h interval in rats [115]. Rapid protein binding appears to be a direct result of the fact that in solution the compound possesses excellent leaving ligands. Pharmacokinetics of both drugs were studied in dogs and the half-lives of distribution were 4.0–5.1 min for (14) and 9.7 min for (1), while elimination half-lives were 3.6–6.6 days and 5.9 days, respectively [114]. These results were rather well duplicated in humans [116].

Clinical studies

Initial clinical trials with spiroplatin suggested that it might produce less nausea and vomiting than cisplatin and could be given without hydration because of less renal toxicity [116]. However, as doses were raised, the limiting toxicity proved to be both renal and haematologic. As data accumulated from a number of studies, it became apparent that spiroplatin was devoid of meaningful antitumour effects in advanced ovarian cancer, a disease where some response might be expected [117]. This report also noted minimal activity in other solid tumours and the occurrence of severe renal toxicity in 8 patients out of 274 treated. It was in view of these findings, and the fact that other platinum complexes such as (12) and (17) were yielding positive antitumour effects in the clinic without renal complications, that further development of spiroplatin was stopped.

CIS-DICHLORO-*TRANS*-DIHYDROXOBIS(ISOPROPYLAMINE)-PLATINUM(IV) (IPROPLATIN), (17)

CHEMISTRY AND MECHANISM

The potential of Pt(IV) complexes as antitumor agents was first recognized in 1969 [1]. Most of the compounds of this type which have been studied to date have been synthesized by oxidation of active *cis*-diamineplatinum(II) compounds with chlorine, bromine or hydrogen peroxide [1, 38, 118–121]. In view of the nature of the oxidation, the resulting Pt(IV) compounds possess chloride, bromide or hydroxide ions above and below the plane defined by the donor groups present in the starting Pt(II) complex. Although many such complexes have been made, due to their generally high water solubility and good antitumour properties, Pt(IV) compounds having *trans* hydroxide ligands, i.e., H_2O_2 oxidation products of Pt(II) complexes, have been most extensively studied. Of these, the Pt(IV) complex, *cis,cis,trans*-$Pt(NH_2iPr_2)_2Cl_2(OH)_2$ (iproplatin) (17) is currently undergoing clinical trials in the United States as a potential second-generation cisplatin analogue.

A clue to the nature of the mechanism by which Pt(IV) compounds exert their antitumour effects was uncovered in clinical trials involving iproplatin [122].

Analysis of the urine and plasma of cancer patients receiving (17) identified significant amounts of the antitumour active reduction product of the compound, namely *cis*-dichlorobis(isopropylamine)platinum(II) (22). The presence of the divalent complex strongly suggested that (17) was deriving its antitumour effects *via in vivo* reduction to (22) or other biologically active Pt(II) compounds. This observation, reduction, was consistent with earlier views on the mechanism of action of Pt(IV) antitumour agents [90, 120, 123] and on the results of studies with (17) in tissue culture media [124]. *In vitro* studies with iproplatin showed that, although the compound does not bind to components in human plasma [92, 122], it probably utilizes DNA as a biological target site [17, 125].

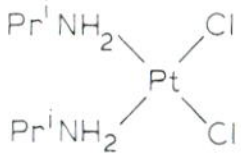

(22) 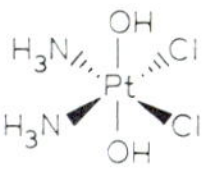 (23)

In attempting to uncover the chemical and biochemical events underlying the antitumour effects of (17) it was reported that the compound was capable of cleaving the closed circular form of PM2-DNA [94, 126, 127]. This was an interesting observation and it suggested that the Pt(IV) complex possessed a mechanistic profile similar to that of the natural product antitumour agents bleomycin and neocarzinostatin [128, 129]. However, subsequent X-ray structural analyses of iproplatin and the related complex *cis,cis,trans*-$Pt(IV)(NH_3)_2Cl_2(OH)_2$ (oxoplatin) (23) revealed that both compounds formed stable perhydrate complexes with hydrogen peroxide, the reagent used in the synthesis of the Pt(IV) complexes from their Pt(II) precursors [130, 131]. Since further study utilizing the non-perhydrate forms of (17) and (23) showed that neither complex was capable of DNA strand scission, lattice hydrogen peroxide appeared to be the agent responsible for strand breakage in the earlier DNA experiments involving (17). Inspection of the electrophoretic mobilities of DNA incubated with either (17) or (23) revealed that neither compound binds, in the manner analogous to (1), to DNA. In subsequent investigations pertaining to mechanism it was shown that biologically common reducing agents such as Fe(II) and ascorbate could reduce (17) and (23) to their divalent counterparts (22) and (1), respectively, and that, as expected, both reduction products can bind to and unwind supercoiled DNA [132, 133]. The prospect that DNA itself or its components may also be capable of carrying out the reduction to Pt(II) has recently been uncovered. Studies have shown that incubation of the Pt(IV) compound, (17), with the mononucleotide 5′GMP results in a Pt(II) product containing two co-ordinated mononucleotides [134]. It has also been reported that incubation of calf thymus DNA with (17) and (23) for extended periods of time, 12–14 days, results in platination of DNA [135]. Although binding without reduction may in fact be occurring, the oxidation state of platinum bound to DNA was not established in the latter study.

ANTITUMOUR EFFECTS

Preclinical studies

Spectrum of activity. Cytotoxicity tests *in vitro* have shown that iproplatin is virtually ineffective against colon tumour cell lines, but it was found more effective than cisplatin against L-1210 in a colony assay [34, 36]. In a plate assay of several murine and human solid tumours, (17) was, on the average, 31-times less potent than cisplatin (Catino, J.J., unpublished data). Compound (17) has also been tested against human ovarian carcinoma cells from 45 patients in the clonogenic assay [136], where it was concluded that the compound is as active as (12) and cisplatin. In addition, iproplatin may be effective against some tumours unresponsive to (12) and (1). Several platinum IV complexes have been tested against animal tumours *in vivo* and the results have been revealing regarding structural effects. As an example, (17) was highly active on L-1210 leukaemia, but the malonato analogue, (24), was totally inactive [97]. The Pt(IV) compounds (25) and (26) were highly effective against both L-1210 and the cisplatin-resistant L-1210 as well as B16 melanoma.

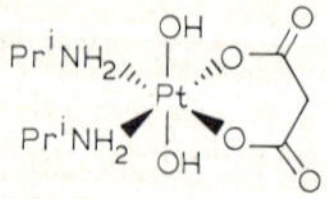

(24) 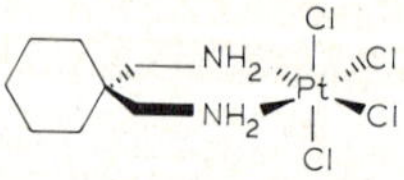 (25) 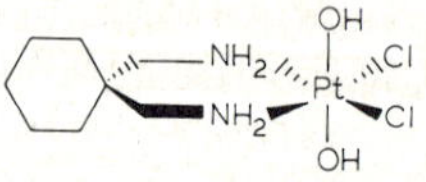 (26)

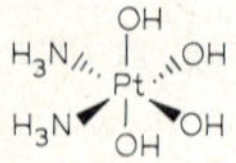

(27) 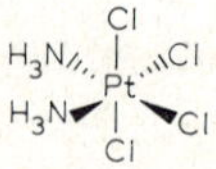 (28) 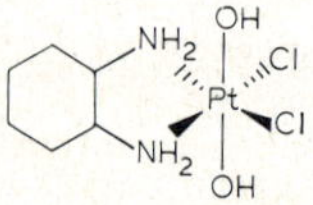 (29)

Compound (25) was active on Lewis Lung but not C26 colon tumour, whereas (26) had the reverse profile on these tumours [74]. Several simple platinum (IV) derivatives, including the tetrahydroxo (27) and tetrachloro (28) complexes were prepared and tested on L-1210, but none was as effective as cisplatin or iproplatin [118]. Oxoplatin, (23), has nevertheless been shown to be highly effective against a number of transplanted tumours [74, 137]. Compound (17) has also been tested and found effective against one of the two human ovarian carcinoma xenografts [113].

The results of comparative antitumour tests between (17) and its reduction product (22) are generally supportive of reduction as being the means of activation *in vivo*. Both compounds, (17) and (22), were equally inhibitory to L-1210 leukaemia, regardless of the treatment regimen used, and were equally effective against B16 melanoma when they were actually tested in the same experiment [44]. Against Lewis Lung carcinoma, (17) seemed superior, but it was not tested in the same experiment and the effectiveness of cisplatin seemed greater in the experiment involving the Pt(IV) than in the one where (22) was tested. Thus, without a head-to-head test of the compounds against Lewis Lung carcinoma, there is no firm evidence that (17) and (22) differ from each other in antitumour effectiveness.

A group of Pt(IV) complexes containing the various isomeric forms of the DACH ligand have recently been synthesized and evaluated for their antitumour properties [138]. While *R,R*-(29) was found to be considerably more active than its mirror image *S,S*-(29) against L-1210 leukaemia, the relative activities of the two compounds were reversed, i.e., *S,S*-(29) was more active than *R,R*-(29) against B16. The activity differences observed for the enantiomeric Pt(IV) compounds were, in general, greater than those observed for the analogous Pt(II) complexes.

Potency. Nearly all of the platinum(IV) complexes tested are less potent than cisplatin, with the single exception of (25) [74]. Iproplatin is 4–16-times less potent, with multidose schedules tending toward the higher number whether measured by antitumour O.D. or LD_{50} determinations. Oxoplatin, (23), is likewise about 10-times less potent than cisplatin.

Toxicity. Iproplatin has less renal toxicity than cisplatin in rats at levels eliciting comparable gastrointestinal and bone marrow toxicity [139]. Similar results were seen in mice, where (17) caused severe leukopoenia but no azotaemia at high doses [97]. Studies in dogs indicated that iproplatin caused vomiting to about the same extent as cisplatin [48].

Pharmacology. Pharmacokinetic experiments have shown that (17) is much more rapidly eliminated from plasma than cisplatin both in rats [139] and dogs [140]. The respective terminal half-lives for (17)/(1) were 14/69 h for rats and

39/120 h in dogs. It was also demonstrated that (17) did not bind at all to plasma protein *in vitro* over 50 h, whereas (1) was extensively bound within the first 5 h in a similar preparation [140]. More rapid renal clearance was also shown in both animal species and was believed associated with reduced renal toxicity relative to cisplatin. Tissue levels of radioactive ^{195m}Pt from labelled (17) were higher in the kidney, however, than those from cisplatin [141]. It has also been found that, unlike cisplatin, a substantial portion of iproplatin is excreted through the intestine.

Clinical studies

Phase I studies determined the starting dose of (17) to be 270 mg/m^2 in hematologically intact patients [142]. Thrombocytopoenia was dose-limiting and renal toxicity was not observed, even without hydration. In many ongoing Phase II studies with (17), the compound has been found to have activity in cancer of the ovary [143, 144], cancer of the cervix [145, 146] and small cell lung cancer in combination with etoposide [147]. Responses have been reported in head and neck cancer in two different studies [148, 149], but not if the patients had been heavily pretreated [150]. One objective response was observed in a breast cancer patient receiving (17) [107]. In some of these studies, (17) was directly compared with (12). The numbers are too small to draw firm conclusions at this time, though there were trends favouring iproplatin in cervix cancer and (12) in head and neck cancer. Evidence from several of the studies suggested cross-resistance between iproplatin and cisplatin as well as other alkylating agents.

μ-HYDROXO-BRIDGED DIMERS AND TRIMERS

CHEMISTRY AND MECHANISM

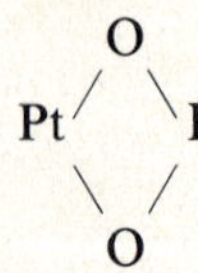

As was earlier outlined, a μ-hydroxo-bridged dimer (3) and trimer (4) can be synthesized via oligomerization of the aquo-hydroxo form of (1) (*Scheme 4.2*). Although a variety of complexes of this type have been reported [151–153], only those having *cis* ammonia ligands and the racemic form of 1,2-diaminocyclohexane (DACH) have been analyzed *via* X-ray structural analysis [13–16, 154, 155]. Structurally, the dimers are planar complexes containing a strained four membered Pt(O)(O)Pt ring (a Pt–O–Pt–O ring, drawn with one O above and one O below the two Pt atoms). The strain present in the ring has a pronounced effect on the ^{195}Pt-NMR resonance of the complex, causing the signal to appear about 500 ppm to lower field relative to those of the strain-free monomer and trimer [5, 109]. Although relatively little is known about the reactivity of the dimer (3), the complex reacts with the DNA nucleobases, 1-methylthymine and 1-methyluracil, to yield novel compounds containing two *cis*-$Pt(NH_3)_2^{2+}$ units bridged *via* two base moieties [156, 157]. Recently, it has also been reported [158], that Pt(II) dimers containing NH_3, $EtNH_2$, and Pr^iNH_2 can be oxidized with hydrogen peroxide to dinuclear Pt(IV) compounds. One of the complexes, (30), has been characterized by X-ray structural analysis.

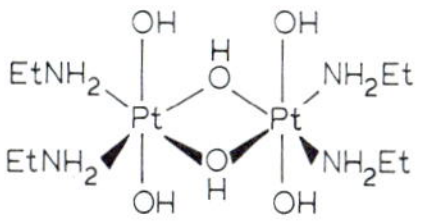

(30) 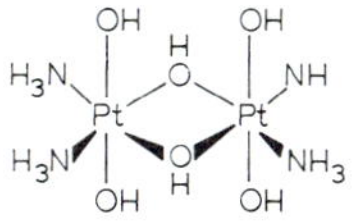 (31) 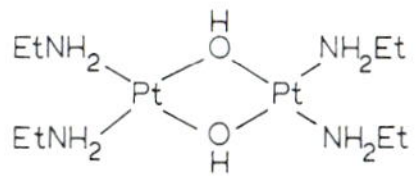 (32)

The dimer (3) has been reported to react with calf thymus DNA [159]. Extended X-ray absorption fine structure (EXAFS) analysis of the platinated DNA revealed that the two platinum ions remained close to one another in the initially formed adduct (1 day incubation), but that over an extended period of time (8 day incubation) they separate from each other. In addition, compound (3) as well as its $EtNH_2$ and Pr^iNH_2 analogues, but not the hydrogen peroxide oxidation products of the complexes, are known to bind to and unwind supercoiled PM2-DNA (Peritz and Dabrowiak, unpublished data).

Structural studies on the trimer (4) showed that the conformation of the six-membered Pt_3O_3 ring is dependent on the counter-ion present with the cation [14, 15]. On this basis it was suggested that the ring does not have a rigid conformation in solution. No studies with the trinuclear complexes and DNA have been reported.

ANTITUMOUR EFFECTS

Preclinical studies

Spectrum of activity and potency. *In vitro* cytotoxicity of Pt(II) polynuclear compounds has apparently not been previously reported. The trimer (4) required 1.65-times the dose of cisplatin on a weight basis to inhibit a panel of cell lines (Catino, unpublished data). Thus, the compound is actually more potent than cisplatin when calculated on a molar basis ((4)/cisplatin dose ratio = 0.54). The dimer (3) required 6.2-times the dose of cisplatin (3.02 by M.W.). The Pt(IV) dimer, compound (31), possessed substantially reduced potency, requiring 32-times the dose of cisplatin by weight. The divalent ethylamine dimer (32) likewise had considerable cytotoxicity, with a dose ratio to cisplatin of 1.67, whereas the platinum (IV) equivalent, (30), dropped sharply in potency, requiring 52-times the dose of (1).

To our knowledge, detailed *in vivo* antitumour test results with platinum oligomers have not been published. However, Rosenberg [160] reported in text that the ammine dimer and trimer, (3) and (4), were more toxic than the monomer and devoid of antitumour effects. We have confirmed this observation using L-1210 leukaemia and have shown that both compounds, as well as (31), had little or no antitumour effect in this system, while having toxicity (lethality) in the range of cisplatin (Schurig, J.E., unpublished observations). On the other hand, the dimer and trimer containing the racemic form of 1,2-diaminocyclohexane (DACH) have been reported to be active antitumour agents [161].

Toxicity. Organ toxicity in rats, of several platinum(II) dimers, has been studied [151]. All of the compounds tended to demonstrate more rapid lethality, greater nephrotoxicity and less stomach distention than their *cis* diaquo- and dichloroplatinum(II) counterparts.

Clinical studies

We have not seen any report of a clinical trial involving an oligomeric platinum complex. However, based on the observations presented here, the simpler complexes are clearly unlikely candidates for clinical use considering their toxicity and lack of experimental antitumor effects.

CONCLUSIONS

Biologically active platinum complexes have now been under investigation for nearly two decades. The large data base on structure-activity relationships has revealed a number of principles as well as raised new questions. Mechanistically, the aquation of the compounds and their ability to cause intrastrand cross-links in defined regions of DNA appear to be the chemical events most closely associated with antitumour activity. The reaction kinetics of the compounds in aqueous systems which may be influenced by chelate effects, steric hindrance of bulky ligands or metal oxidation state have been studied for some complexes and are amenable to reasonably precise investigation in the future. The pharmacological behaviour of the complexes in animals and in humans is, however, much less well defined. Although the pharmacokinetics of a number of compounds have been studied, conclusions regarding toxic effects have only been inferred and reasons for varying antitumour effects are even less well understood. With intense motivation both from the oncologic and commercial communities for clinical success, attention is heavily focused on the most active compounds in terms of antitumour effects. This underscores a possible missed opportunity for learning from some of the failures. For example, the inferior effects of (14) on solid tumours might be explained by its high reactivity, but the reason for its unusual nephrotoxicity relative to cisplatin remains unexplained. An even more fascinating question is why certain oligomers retain high potency for toxicity but lose antitumour effects.

Finally, in reviewing both the preclinical and clinical data on the first wave of so-called 'second generation' complexes, we suspect that those chosen for clinical trial were less than ideal, since many related complexes already known were superior. At this stage, only carboplatin (12) seems certain of becoming a commercial success and this is based almost exclusively on reduced toxicity compared with cisplatin. Clearly there is room for improvement, particularly in antitumour effectiveness and specificity. We believe this is achievable if the large body of knowledge on platinum complexes is diligently applied in the design of new compounds and in improvements in both biological evaluation methods and systems for selection of those compounds that will go to clinical trial.

ACKNOWLEDGEMENTS

One of us (J.C.D.) is grateful to the American Cancer Society (CH-296) and the Bristol Myers Company for supporting research on the platinum-based antitumour agents. One author (J.C.D.) is also thankful to the library staff of the Max Planck Institute for Biophysics, Göttingen, West Germany, for their help with literature searches concerning the review.

REFERENCES

1 B. Rosenberg, L. Van Camp, J.E. Trosko, and V.H. Mansour, Nature (London), 222 (1969) 385.
2 B. Rosenberg and L. Van Camp, Cancer Res, 30 (1970) 1799.
3 S.J. Lippard, Science (Washington, DC), 218 (1982) 1075.

4 A.T.M. Macelis and J. Reedijk, Recl. Trav. Chim. Pays-Bas, 103 (1983) 121.
5 B. Rosenberg, Biochemie, 60 (1978) 859.
6 A.W. Prestayko, S.T. Crooke and S.K. Carter, eds., Cisplatin: Current Status and New Developments (Academic Press, New York, 1980).
7 J.J. Roberts and M.P. Pera, Jr., in Molecular Aspects of Anticancer Drug Action, eds. S. Neidle and M.J. Waring (Macmillan Press, London, 1983) p. 183.
8 H. Sigel, ed., Metal Ions in Biological Systems, Vol. 11 (Marcel Dekker, New York, 1980).
9 B. Rosenberg, in Nucleic Acid Metal Ion Interactions, ed. T.G. Spiro (Wiley, New York, 1980) pp. 2–29.
10 K.W. Lee and D.S. Martin, Jr., Inorg. Chim. Acta, 17 (1976) 105.
11 M.C. Lim and R.B. Martin, J. Inorg. Nucl. Chem., 38 (1976) 1911.
12 T.G. Appelton, R.D. Berry, C.A. Davis, J.R. Hall and H.A. Kimlin, Inorg. Chem., 23 (1984) 3514.
13 R. Faggiani, B. Lippert, C.J.L. Lock and B. Rosenberg J. Am. Chem. Soc., 99 (1977) 777.
14 R. Faggiani, B. Lippert and B. Rosenberg, Inorg. Chem., 16 (1977) 1192.
15 R. Faggiani, B. Lippert, C.J.L. Lock and B. Rosenberg, Inorg. Chem., 17 (1978) 1941.
16 B. Lippert, C.J.L. Lock, B. Rosenberg and M. Zvagulis, Inorg. Chem., 17 (1978) 2971.
17 J.J. Roberts and H.N.A. Fraval, Biochemie, 60 (1978) 869.
18 R.B. Ciccarelli, M.J. Solomon, A. Varshavsky and S.J. Lippard, Biochemistry, 24 (1985) 7533.
19 A. Eastman, Biochemistry, 22 (1983) 3927.
20 A. Eastman, Biochemistry, 24 (1985) 5027.
21 J.C. Dewan, J. Am. Chem. Soc., 106 (1984) 7239.
22 A.M.J. Fichtinger-Schepman, P.H.M. Lohman and J. Reedijk, Nucleic Acids Res., 10 (1982) 5345.
23 A.M.J. Fichtinger-Schepman, J.L. van der Veer, J.H.J. den Hartog, P.H.M. Lohman and J. Reedijk, Biochemistry 24 (1985) 707.
24 N.P. Johnson, A.M. Mazard, J. Escalier and J.-P. Macquet, J. Am. Chem. Soc., 107 (1985) 6370.
25 A.L. Pinto and S.J. Lippard, Proc. Natl. Acad. Sci. USA, 82 (1985) 4616.
26 J.L. van der Veer, G.J. Ligtvoet, H. van den Elst and J. Reedijk, J. Am. Chem. Soc., 108 (1986) 3860.
27 S.E. Sherman, D. Gibson, A.H.-J. Wang and S.J. Lippard, Science (Washington DC), 230 (1985) 412.
28 J.-P. Macquet and J.-L. Butour, Biochimie, 60 (1978) 901.
29 R.C. Srivasto, J. Froehlich and G.L. Eichhorn, Biochimie, 60 (1978) 879.
30 G.L. Cohen, W.R. Bauer, J.K. Barton and S.J. Lippard, Science (Washington DC), 203 (1979) 1014.
31 W.M. Scovell and V.J. Capponi, Biochem. Biophys. Res. Commun., 107 (1982) 1138.
32 B. Van Hemelryck, E. Guittet, G. Chottard, J.-P. Girault, T. Huynh-Dinh, J.-Y. Lallemand, J. Igolen and J.-C. Chottard, J. Am. Chem. Soc., 106 (1984) 3037.
33 W.T. Bradner, in Proc. 10th Internat. Cong. Chemother. Zürich Switzerland, 1977, American Society for Microbiology, Washington, DC, 1 (1985) 114.
34 E. Balazova, M. Hrubisko and V. Ujhazy, Neoplasma, 31 (1984) 641.
35 B.F. Issell, J.J. Catino and E.C. Bradley, Recent Results Cancer Res. 94 (1984) 237.
36 B. Drewinko, L-Y. Yang and J.M. Trujillo, Invest. New Drugs, 3 (1985) 335.
37 B. Rosenberg, Cancer Treat. Rep. 63 (1979) 1433.
38 M.J. Cleare and J.D. Hoeschele, Bioinorg. Chem. 2 (1973) 187.

39 L.S. Hollis, A.R. Amundsen and E.W. Stern, J. Am. Chem. Soc., 107 (1985) 274.
40 M.K. Wolpert-DeFilippes, in Ref. 6, pp. 183–191.
41 W.T. Bradner and C.A. Claridge, in Antineoplastic Agents, ed. W.A. Remers (John Wiley, New York, 1984).
42 R.J. Speer, H. Ridgeway, L.M. Hall, D.P. Stewart, K.E. Howe, D.Z. Lieberman, A.D. Newman and J.M. Hill, Cancer Chemother Rep., 59 (1975) 629.
43 B. Rosenberg, in Ref. 6, pp. 9–20.
44 W.T. Bradner, W.C. Rose and J.B. Huftalen, in Ref. 6 pp. 171–182.
45 W.C. Rose and W.T. Bradner, Proc. Am. Assoc. Cancer Res. Abs., 1019, 27 (1985) 258.
46 U. Schaeppi, I.R. Heyman, R.W. Fleischman, H. Rosenkrantz, V. Ilievski, R. Phelen, D.A. Cooney and R.D. Davis, Toxicol. Appl. Pharmacol., 25 (1973) 230.
47 W.T. Bradner and J.E. Schurig, Cancer Treat. Rev., 8 (1981) 93.
48 J.E. Schurig, W.T. Bradner, J.B. Huftalen, G.J. Doyle and J.A. Cylys, in Ref. 6, pp. 227–236.
49 A.W. Prestayko, in Ref. 6, pp. 1–7.
50 A.J. Repta and D.A. Long, in Ref. 6, pp. 285–304.
51 F.M. Jaeger and L. Bijkerk, Z. Anorg. Chem., 233 (1937) 97/139, 109.
52 C.J. Hawkins, Absolute Configuration of Metal Complexes (Wiley-Interscience, New York, 1971).
53 F.D. Rochon, R. Melanson, J.-P. Macquet, F. Belanger-Garcepy and A.L. Beauchamp, Inorg. Chim. Acta, 108 (1985) 17.
54 M.A. Bruck, R. Bau, M. Noji, K. Inagaki and Y. Kidani, Inorg. Chim. Acta, 92 (1984) 279.
55 K.P. Larson and H. Toftlund, Acta Chem. Scand., A31 (1977) 182.
56 F. Maruma, Y. Utsumi and Y. Saito, Acta Cryst., B26 (1970) 1492.
57 J.-P. Macquet, S. Cros and A.L. Beauchamp, J. Inorg. Biochem., 25 (1985) 197.
58 C.J.L. Lock and P. Pilon, Acta Cryst., B37 (1981) 45.
59 M. Noji, K. Okamoto and Y. Kidani, J. Med. Chem., 24 (1981) 508.
60 P. Schwartz, S.J. Meischen, G.R. Gale, L.M. Atkins, A.B. Smith and E.M. Walker, Cancer Treat. Rep., 61 (1977) 1519.
61 G.R. Gale, E.M. Walker, L.M. Atkins, A.B. Smith and S.J. Meischen, Res. Commun. Chem. Pathol. Pharmacol., 7 (1974) 529.
62 H. Ridgway, L.M. Hall, R.J. Speer, D.P. Stewart, G.R. Edwards and J.M. Hill, Wadley Med. Bull., 6 (1976) 11.
63 R.J. Speer, H. Ridgway, L.M. Hall, A.D. Newman, K.E. Howe, D.P. Stewart, G.R. Edwards and J.M. Hill, Wadley Med. Bull., 5 (1975) 335.
64 S.J. Meischen, G.R. Gale, L.M. Lake, C.J. Frangakis, M.G. Rosenblum, E.M. Walker, Jr., L.M. Atkins and A.B. Smith, J. Natl. Cancer Inst., 57 (1976) 841.
65 R.J. Speer, H. Ridgway, D.P. Stewart, L.M. Hall, A. Zapata and J.M. Hill, J. Chim. Hematol. Oncol., 7 (1977) 210.
66 T.A. Connors, M. Jones, W.C.J. Ross, P.D. Braddock, A.R. Khokhar and M.L. Tobe, Chem. Biol. Interact., 5 (1972) 415.
67 Y. Kidani, K. Inagaki, M. Iigo, A. Hoshi and K. Kuretani, J. Med. Chem., 21 (1978) 1315.
68 Y. Kidani, M. Noji and T. Toshiro, Gann, 71 (1980) 637.
69 Y. Kidani, K. Inagaki and S. Tsukagoshi, Gann, 67 (1976) 923.
70 Y. Kidani, K. Inagaki, R. Saito and S. Tsukagoshi, J. Clin. Hematol. Oncol., 7 (1977) 197.
71 K. Inagaki, K. Kasuya and Y. Kidani, Chem. Lett., (1984) 171.
72 K. Inagaki and Y. Kidani, Inorg. Chem., 25 (1986) 1.
73 J.H. Burchenal, K. Kalaher, T. O'Toole and J. Chisholm, Cancer Res., 87 (1977) 3455.

74 W.C. Rose, J.E. Schurig, J.B. Huftalen and W.T. Bradner, Cancer Treat. Rep., 66 (1982) 135.
75 M. Hrubiski, E. Balazova and V. Ujhazy, Neoplasma, 31 (1984) 649.
76 A. Eastman and S. Illenye, Cancer Treat. Rep., 68 (1984) 1189.
77 W.R. Waud, Proc. Am. Assoc. Cancer Res. Abstr., 1038, 27 (1986) 262.
78 R.A. Hromas, J.A. North and C.P. Burns, Proc. Am. Soc. Cancer Res. Abstr., 1044, 24 (1986) 263.
79 P.A. Andrews, M.P. Murphy and S.B. Howell, Proc. Am. Soc. Cancer Res. Abstr., 1148 27 (1986) 289.
80 V.M. Richon, N.A. Coday and A. Eastman, Proc. Am. Assoc. Cancer Res. Abstr., 1155, 27 (1986) 291.
81 R.B. Gross and K.J. Scanlon, Chemoterapia, 5 (1986) 37.
82 R.J. Speer, L.M. Hall, D.P. Stewart, H.J. Ridgeway, J.M. Hill, Y. Kidani, K. Inagaki, M. Noji and S. Tsudagoshi, J. Clin. Hematol. Oncol., 8 (1978) 44.
83 A.M. Guarino, D.S. Miller, S.T. Arnold, M.A. Urbanek, M.K. Wolpert-DeFilippes and M.P. Hacker, in Ref. 6, pp. 237–248.
84 D.P. Kelson, H. Scher, N. Alcock, B. Leyland-Jones, A. Donner, L. Williams, G. Greene, J.H. Burchenal, C. Tan, F.S. Phillips and C.W. Young, Cancer Res., 42 (1982) 4831.
85 P. Ribaud, D.P. Kelson, N. Alcock, E. Garcia-Giralt, P. Dubouch, C. C. Young, F.M. Muggia, J.H. Burchenal and G. Mathe, Recent Results Cancer Res., 74 (1980) 156.
86 G. Mathe, Y. Kidani, R. Maral, C. Bourut, E. Chenu, I. Florentin and M. Timus, Proc. Am. Assoc. Cancer Res. Abstr., 1040, 26 (1985) 264.
87 G. Mathe, Y. Kidani, J.L. Misset, K. Triani, D. Machover and M. Musset, Proc. Am. Assoc. Cancer Res. Abstr., 752, 27 (1986) 190.
88 S. Neidle, I.M. Ismail and P.J. Sadler, J. Inorg. Biochem., 13 (1980) 205.
89 B. Beadley, D.W.J. Cruickshank, C.A. McAuliffe, R.G. Pritchard and A.M. Zaki, J. Mol. Struct., 180 (1985) 97.
90 M.J. Cleare, P.C. Hydes, D.R. Hepburn and B.W. Malerbi, in Ref. 6, p. 149.
91 K.F. Purcell and J.C. Kotz, Inorganic Chemistry (W.B. Saunders, Philadelphia, 1977).
92 R. Momburg, M. Bourdeaux, M. Sarrazin, F. Roux and C. Briand, Eur. J. Drug Metab. Pharmacokinet., 10 (1985) 77.
93 K.C. Miocetich, D. Barnes and L.C. Erickson, Cancer Res., 45 (1985) 4043.
94 S. Mong, C.H. Huang, A.W. Prestayko and S.T. Crooke, Cancer Res., 40 (1980) 3318.
95 J.J. Catino, D.M. Francher, K.J. Edinger and D.A. Stringfellow, Cancer Chemother. Pharmacol., 15 (1985) 240.
96 R. Wilkinson, P.J. Cox, M. Jones and K.R. Harrap, Biochemie, 60 (1978) 851.
97 A.W. Prestayko, W.T. Bradner, J.B. Huftalen, W.C. Rose, J.E. Schurig, M.J. Cleare, P.C. Hydes and S.T. Crooke, Cancer Treat. Rep., 63 (1979) 1503.
98 K.A. Harrap, Cancer Treat. Rev. (Suppl. A.), 12 (1985) 21.
99 W.C. Rose and J.E. Schurig, Cancer Treat. Rev., (Suppl. A), 12 (1985) pp. 1–19.
100 B.S. Levine, M.C. Henry, C.D. Port, W.R. Richter and M.A. Urbanek, J. Natl. Cancer Inst., 67 (1981) 201.
101 J.E. Schurig, A.P. Florczyk and W.T. Bradner, in Platinum Coordination Complexes in Chemotherapy, eds. M.P. Hacker, E.V. Douple and I.R. Krakoff (Martinus Nijhoff, Boston, 1984) pp. 187–199.
102 J.E. Schurig, A.P. Florczyk and W.T. Bradner, Cancer Chemother. Pharmacol., 16 (1986) 243.
103 W.T. Bradner, W.C. Rose, J.E. Schurig, A.P. Florczyk, J.B. Huftalen and J.J. Catino, Cancer Res., 45 (1985) 6475.

104 S.J. Harland, D.A. Newell, Z.H. Siddid, R. Chadwick, A.H. Calvert and K.R. Harrap, Cancer Res., 44 (1984) 1693.
105 Z.H. Siddik, D.R. Newell, M. Jones and F.E. Boxall, Proc. Am. Assoc. Cancer Res. Abstr., 659, 23 (1982) 168.
106 R. Canetta, M. Rozencweig and S.K. Carter, Cancer Treat. Rev. (Suppl. A), 12 (1985) 125.
107 J.B. Vermorken, S. Gunderson, J.F. Smyth, P. Dodion, P. Siegenthaler, J. Renard and H.M. Pinedo, Proc. Am. Soc. Clin. Oncol. Abstr., 284 5 (1986) 73.
108 I.N. Oliver, R.C. Donehower, D.A. van Echo, D.S. Ettinger and J. Aisner, Cancer Treat. Rep., 70 (1986) 421.
109 H.A. Meinema, F. Verbeek, J.W. Marsman, E.J. Bulten, J.C. Dabrowiak, B.S. Krishnan and A.L. Spek, Inorg. Chim. Acta, 114 (1986) 127.
110 A. Sacchi, G. Zupi, F. Calabresi, C. Greco and A. Caputo, Proc. 13th Internat. Congr. Chemother., 16 (1983) 257/62–65.
111 M. Rozencweig, Proc. 13th Internat. Congr. Chemother., 11 (1983) 224/49-52.
112 W.H. deJong, P.A. Steerenberg, J.G. Vos, E.J. Bulten, F. Verbeek, W. Kruizinga and E.J. Ruitenberg, Cancer Res., 43 (1983) 4927.
113 E. Boven, W.J.F. van der Vijgh, M.M. Nauta, H.M.M. Schluper and H.M. Pinedo, Cancer Res., 45 (1985) 86.
114 P. Lilieveld, W.J.F. van der Vijgh, R.W. Veldhuizen, D. van Velzen, L.M. van Putten, G. Atassi and A. Danguy, Eur. J. Cancer Clin. Oncol., 20 (1984) 1087.
115 W.J.F. van der Vijgh, P.C.M. Verbeek, I. Klein and H.M. Pinedo, Cancer Lett., 28 (1985) 103.
116 J.B. Vermorken, W.W.T. Bokkel Huinink, J.G. McVie, W.J.F. van der Vijgh and H.M. Pinedo, in Ref. 101, pp. 331–343.
117 N. Colombo, E. Sartori, F. Landoni, G. Favalli, L. Vassena, L. Zotti, E. Materman, S. Pecorelli and C. Mangioni, Cancer Treat. Rep., 70 (1986) 793.
118 R.J. Brandon and J.C. Dabrowiak, J. Med. Chem., 27 (1984) 861.
119 T.A. Connors, M. Jones, W.C.J. Ross, P.D. Braddock, A.R. Khokhar and M.L. Tobe, Chem.-Biol. Interact., 11 (1975) 145.
120 M.L. Tobe and A.R. Khokhar, J. Clin. Haematol. Oncol., 7 (1977) 114.
121 P.C. Hydes, in Ref. 101, p. 216.
122 L. Pendyala, L. Cowens and S. Madajewicz, in Ref. 101, p. 114.
123 E. Rotondo, V. Fimiani, A. Cavallaro and T. Anis, Tumori, 69 (1983) 32.
124 M. Laverick, A.H.W. Nias, P.J. Sadler and I.M. Ismail, Br. J. Cancer, 43 (1981) 732.
125 E. Bocian, M. Laverick and A.H.W. Nias, Br. J. Cancer, 47 (1983) 503.
126 S. Mong, P.C. Eubanks, A.W. Prestayko and S.T. Crook Biochemistry, 21 (1982) 3174.
127 S. Mong, J.E. Strong, J.A. Busch and S.T. Crooke, Antimicrob. Agents Chemother., 16 (1979) 398.
128 J.C. Dabrowiak, Adv. Inorg. Biochem., 4 (1982) 69.
129 J.H. Goldberg, T. Hatayama, L.S. Kappen, M. Napier and F.F. Povirk in Molecular Action on Targets for Cancer Chemotherapeutic Agents, eds. A.C. Sartorelli, J.R. Bertino, and S.S. Lajo (Academic Press, New York, 1981) p. 163.
130 J.F. Vollano, E.E. Blatter and J.C. Dabrowiak, J. Am. Chem. Soc., 106 (1984) 2732.
131 C.F.J. Barnard, P.C. Hydes, W.P. Griffiths and O.S. Mills, J. Chem. Res. (M) (1983) 2801.
132 E.E. Blatter, J.F. Vollano, B.S. Krishnan and J.C. Dabrowiak, Biochemistry, 23 (1984) 4817.
133 E.E. Blatter, J.F. Vollano, B.S. Krishnan and J.C. Dabrowiak, Prog. Clin. Biol. Res., Mol. Basis Cancer, Pt. B, 172 (1985) 185.

134 J.L. van der Veer, A.R. Peters and J. Reedijk, J. Inorg. Biochem., 26 (1986) 137.
135 O. Vriana, V. Brabec and V. Kleinwächter, Anticancer Drug Design, 1 (1986) 95.
136 D.S. Alberts, L. Young, and S.E. Salmon, Clin. Pharmacol. Therap., 39 (1986) 177.
137 M.A. Presnov, A.L. Konovalova, A.M. Kozlov, V.K. Brovtsyn, and L.F. Romanova, Neoplasma, 32 (1985) 73.
138 J.F. Vollano, S. Al-Baker, J.C. Dabrowiak and J.E. Schurig, J. Med. Chem., 30 (1987) 716.
139 M. Pfister, Z.P. Pavelic, G.A. Bullard, E. Mihich and P.J. Creaven, Biochemie, 60 (1978) 1057.
140 L. Pendyala, J.W. Cowens and P.J. Creaven, Cancer Treat. Rep., 66 (1982) 509.
141 J.D. Heoschele, L.A. Ferren, J.A. Roberts and L.R. Whitefield, in Ref. 101, p. 103.
142 P.J. Creaven, S. Madajewicz, L. Pendyala, A. Mittelman, E. Pontes, M. Spaulding, S. Arbuck and J. Solomon, Cancer Treat. Rep., 67 (1983) 795.
143 V.H.C. Bramwell, D. Crowther, S. O'Malley, R. Swindell, R. Johnson, E.H. Cooper, N. Thatcher and A. Howell, Cancer Treat. Rep., 69 (1985) 409.
144 C. Sessa, F. Cavelli, S. Kaye, A. Howell, W.T. Bokkel Huinink, J. Renard, J. Vermorken and H. Pinedo, Proc. Am. Soc. Clin. Oncol. Abstr., 479, 5 (1986) 123.
145 W. McGuire, J. Blessing, F. Stehman and K. Hatch, Am. Soc. Clin. Oncol. Abstr., C-472, 4 (1985) 121.
146 V.L. Puerto, A. Carugati, F. Muggia, N. Colombo, and S. Pavlovsky, Proc. Am. Soc. Clin. Oncol. Abstr., 464, 5 (1986) 119.
147 H. Takita, S. Madajewicz, J. Plager, N. Botsoglon, and P.J. Creaven Proc. Am. Soc. Clin. Oncol. Abstr., 755, 4 (1985) 193.
148 J. Kish, J. Ensley, M. Al-Sarraf, D. von Hoff, and D. Schuller, Proc. Am. Soc. Clin. Oncol. Abstr., C-504, 4 (1985) 130.
149 C.R. Franks, G. Nys, H.P. Schultz, H.J. Schmoll, R. Canetta and M. Rozencweig, Proc. Am. Soc. Clin. Oncol. Abstr., 501, 5 (1986) 129.
150 R. Abele, M. Clavel, A. Rossi, U. Bruntsch and H.M. Pinedo, Proc. Am. Soc. Clin. Oncol. Abstr., 501, 5 (1986) 147.
151 S.K. Aggarwal, J.A. Broomhead, D.P. Fairlie and M.W. Whitehouse, Cancer Chemother. Pharmacol., 4 (1980) 249.
152 L.A. Roos, D.P. Fairlie and M.W. Whitehouse, Chem. Biol. Interact., 35 (1981) 111.
153 J.A. Broomhead, D.P. Fairlie and M.W. Whitehouse, Chem. Biol. Interact., 31 (1980) 113.
154 J.A. Stanko, L.S. Hollis, J.A. Schriefels and J.D. Hoeschele, J. Clin. Hematol. Oncol., 7 (1977) 138.
155 J.-P. Macquet, S. Cros and A.L. Beauchamp, J. Inorg. Biochem., 25 (1985) 197.
156 C.J.L. Lock, H.J. Peresie, B. Rosenberg and G. Turner, J. Am. Chem. Soc., 100 (1978) 3371.
157 R. Faggiani, C.J.L. Lock, R.J. Pollock, B. Rosenberg and G. Turner, Inorg. Chem., 20 (1981) 804.
158 S. Al-Baker, J.F. Vollano and J.C. Dabrowiak, J. Am. Chem. Soc., 108 (1986) 5642.
159 A.P. Hitchcock, C.J.L. Lock, W.M.C. Pratt and B. Lippert, ACS Symposium Series, 209 (1983) 209.
160 B. Rosenberg, Cancer Treat. Rep., 63 (1979) 1433.
161 D.S. Gill and B. Rosenberg, J. Am. Chem. Soc., 104 (1982) 4598.

Progress in Medicinal Chemistry – Vol. 24, edited by G.P. Ellis and G.B. West

5 Towards cannabinoid drugs

R. MECHOULAM, M.Sc., Ph.D. and J.J. FEIGENBAUM, Ph.D.

Brettler Medical Research Center, Faculty of Medicine, Hebrew University of Jerusalem, Ein Kerem, Jerusalem 91 120, Israel

INTRODUCTION

Due to its psychotropic properties, *Cannabis sativa* was one of the first plants to be used by man, both in social-religious rites and in medicine [1–3]. In ancient times, the distinction between use in religious ceremonies and medical use was blurred. Later, in most countries, the ceremonial rites associated with Cannabis were forgotten or discarded, but the medical use continued for many centuries. In various parts of the world, Cannabis was used in a very wide range of medical conditions. This is not surprising, as the placebo effects of drugs with psychotropic action are particularly strong. However, many of the actions produced by Cannabis are undoubtedly distinct from the placebo effect and in recent years some have been confirmed in animals and in man. In the present survey, we shall try to compare past use with modern data, to present an overview of modern use and developments and to predict future pathways in medicinal chemistry research in this area.

In the past decade, several reviews on various aspects of these topics have appeared [3–8].

The following abbreviations are used in the present review: THC (tetrahydrocannabinol); CBD (cannabidiol); CBN (cannabinol); DMH (1,1-dimethylheptyl); SAR (structure-activity relationship); cisplatin (*cis*-diamminedichloroplatinum(II)); GABA (γ-aminobutyric acid); MES (maximal electroshock seizures); PTZ (pentylenetetrazol); AGS (audiogenic seizure); IOP (intraocular pressure).

THE PHARMACOHISTORY OF *CANNABIS SATIVA*

THE ANCIENT WORLD

In Assyria, *Cannabis sativa* was used both in rituals and as a medicine [9]. From the records available, these two uses are difficult to distinguish. Indeed it is probably impossible for man today to comprehend fully the attitudes of an ancient population towards drugs and their associations with beliefs, religion, superstitions and social life and interactions.

Thus, azallu (one of the names for Cannabis) is cited in connection with the term 'hand of ghost', which is apparently an as yet unidentified disease; to us, the term has a superstitious connotation. The use of azallu to 'annul witchcraft' may be somewhat difficult to demonstrate in animal models today. However, gan-zi-gun-nu (another name for or formulation of Cannabis) reputedly 'a drug which takes away the mind', seems more familiar. Some of the strictly medical uses were for binding temples, for a disease named 'arimtu' (which seems to mean some loss of control of the lower limbs), for depression, for impotence, and for kidney stones and together with some other plants it was found useful in 'a female ailment'. The use of Cannabis in female ailments has been reported in various cultures and it is surprising that no research has been done on this aspect in recent years. Cannabis fumes were a drug for the 'poison of all limbs' (presumably arthritis).

In ancient Egypt, Cannabis (the hieroglyph šmšmt in several papyri) was used in incense and as an oral medication for 'mothers and children' (probably for the prevention of haemorrhage in childbirth), in enemas, in eye medications and in ointments in bandages [3, 10].

Although the fun-loving Romans and Greeks were unaware of the hedonistic properties of Cannabis, it was apparently well established as a medicament [11]. Pliny the Elder (died *circa* 79 A.D.) described its use in detail: "Its seed is said to make the genitals impotent. The juice from it drives out of the ears the worms, it regulates the bowels of beasts of burden. The root boiled in water eases cramped joints, gout too and similar violent pains. It is applied raw to burns, but is often changed before it gets dry".

Both Dioscorides (died *circa* 99 A.D.) and Gallen (died *circa* 199 A.D.) found the juice from the seeds to be analgetic for pains caused by ear-obstruction. Gallen also found that, while hashish eased the muscles of the limbs, it also produced senseless talk [12, 13].

Hashish was (and still is) well known in the Near East [14]. A Syrian medical manuscript apparently compiled in the first centuries A.D. recommends that for painful teeth with cavities, extract of fresh Kunbare (Cannabis) should be injected into the nose of the patient; if the tooth had to be pulled out the extract was to be rubbed on the gums [15]. The medical treatises of the Greeks and Romans were also well known [11].

It is certainly of interest that Pliny, Dioscorides and Gallen indicate the roots and the seeds as active parts of the plant. Modern analyses have shown that free Δ^1-THC, the active cannabimimetic constituent, is not found there, and it may be worthwhile to re-examine these plant tissues for new analgetic components.

In ancient China, Cannabis was used together with other herbs for rheumatic pains, disorders of the female reproductive tract, absent-mindedness, malaria (probably for the headache caused by the disease) as well as for beri-beri [3, 16–18]. When however, the fruits of hemp were taken in excess, it was known that they could cause 'seeing devils'. Cannabis in wine was used as an anaesthetic in major operations [19]. As Cannabis is only a minor pain killer, one can assume that large stupefying doses were administered.

Cannabis oil and leaf juice were employed externally for various skin diseases, wounds and even in leprosy. The topical antibiotic properties of cannabinoids as known today justify the use in appropriate skin diseases. Cannabis was also used against vomiting. This use was widespread in India as well. It was rediscovered during the 19th century in Europe and again in the 1970's in the U.S. (*vide infra*). None of the modern articles on the subject refers to the Chinese, the Indian or even to the 19th century use and experience.

INDIA [3, 16, 20, 21]

Cannabis was part of the religious lore of the Aryans (*ca.* 2000 B.C.). It is mentioned in the sacred Vedas and is associated there with the god Siva. In one of the books of the Vedas, the Atharveveda (written in Sanscrit *circa* 1500–1200 B.C.), the plant is described as a sacred grass, and bhang, the mild drink prepared from Cannabis, is mentioned as an antianxiety herb. Due to its power to suppress the appetite, its virtue as a febrifuge and its 'thought-bracing qualities', the bhang leaf was considered to be the home of the great Yogi. Bhang was associated with rites required to clean away evil influences and offerings of bhang to the gods were made. Initially the medical properties of bhang were closely tied to religion. Fever was considered to be possession by the hot angry breath of the great gods Brahma, Vishnu and Shiva. If the fever-stricken performed certain religious rites with bhang, the god Shiva was pleased, his breath cooled and the fever ceased. Although the use of bhang retained its ties to religion, not all medical applications had mythological background. Cannabis was described in the famous Ayurveda medical treatise Susruta-samhita (7th century A.D.) as a remedy for catarrh and diarrhoea, and as a cure for biliary fever. Cannabis gradually became one of the major drugs in the Indian systems of medicine. Cannabis-derived drugs were used in the treatment of cramps, convulsions in children, headaches (migraine?), hysteria, neuralgia and tetanus. The emphasis on the nervous system is much closer to contemporary understanding of Cannabis activity. Early in the 19th century O'Shaugnessy showed that many of these claims were well founded [22].

Hemp drugs were also used in dysentery and cholera. As THC has been shown recently to reduce intestine motility, the use against these conditions makes therapeutic sense.

Cannabis was also used in hay fever, bronchitis, asthma and coughs. Contemporary work has shown that, at least as regards asthma, Indian tradition had a factual basis.

One of the most common uses was for the relief of pain and as a febrifuge, the drug being used either locally or given orally. Poultices were applied over inflamed, painful parts, and small fragments of charas were placed in a carious tooth to relieve toothache. Systemically, the drugs were used for labour pains, dysmenorrhoea, and even in minor operations (circumcision).

WESTERN EUROPE AND NORTH AMERICA

The medicinal properties of Cannabis were mentioned in the much copied herbals in Europe during the Middle Ages [23, 24]; however, it is doubtful whether is was widely used. It was apparently prescribed for 'distended stomachs', some cardiac conditions, pains in the anal region, and as a plaster for boils and carbuncles. Applied to wounds it relieved pain and a decoction of its roots and seeds mixed with white lead and oil of roses was used to treat erysipelas. The vapours were supposed to ease headaches. Some herbals present a picture of a panacea: from antiparasitic through coagulant and antiepileptic activity, and even "hempe seede, given to Hennes in the winter, when they lay fewest egges, will make them lay more plentifully".

The use of Cannabis as an antihelmintic is of some interest, as it is mentioned throughout many centuries in several countries. A modern antihelmintic, hexylresorcinol (3), has a chemical structure related to the structure of the cannabinoids. Experimental work in this area with cannabinoids has not been reported and may be worth pursuing.

The apparently limited medicinal use of Cannabis during the Middle Ages seems to have further diminished later on until the wide use of Cannabis in India was brought to the attention of British practitioners in the 19th century by O'Shaugnessy, who also conducted some animal and human experiments and applied his knowledge in the clinic [22, 25]. He administered tincture of Cannabis to patients with rheumatism, tetanus, rabies, infantile convulsions and cholera with considerable reported success, though in some cases the huge doses caused side-effects, such as total catalepsy or uncontrollable behaviour, which today would be considered quite unacceptable. Tolerance was noted in some cases. A further observation made by O'Shaugnessy was that Cannabis was a potent antiemetic.

The reports by O'Shaugnessy made Indian hemp an accepted drug in therapy, first in England and later, to a limited extent, in other European countries and in North America.

Christison [26] has reviewed the therapeutic uses of Cannabis in mid-19th century England. The analgetic power of hemp tincture was stressed, in particular in rheumatic and tooth pains. A marked mitigation of various types of spasm, the relief of asthmatic paroxysms and hypnotic effects were noted. A remarkable power of increasing the force of uterine contractions, together with a significant reduction of labour pain was observed. Cannabis acted within minutes and was faster than ergot, though the action of the latter lasted longer. However, Cannabis action was much more energetic than that of ergot. Christison also noted that Cannabis restrained uterine haemorrhage. These claims, substantiated by numerous other clinicians [27–29], have not been investigated recently.

By the end of the 19th century, enough experience had been accumulated to allow Reynolds [30] to state: "... Indian hemp when pure and administered carefully is one of the most valuable medicines we possess". He found Cannabis to be "absolutely successful for months, indeed years without any increase of dose" in cases of senile insomnia. In "almost all painful maladies". Indian hemp was found to be "by far the most useful of drugs". Neuralgias of several kinds were successfully treated for years. He emphasized the use in trigeminal neuralgia, but found it useless in "sciatica" and all pains which "occur only on movement". In cases of migraine, he found that "very many victims of this malady have for years kept their sufferings in abeyance by taking hemp at the moment of threatening or onset of the disease". Contrary to previous reports, he found Cannabis useless in epilepsy, tonic spasms, general chorea or tetanus, but very valuable in "nocturnal cramps of old and gouty people" and in cases of "simple spasmodic dysmenorrhoea". He also found that Cannabis in some cases relieved spasmodic asthma.

The reason for the relatively wide use of Cannabis in England by many of the British practitioners was probably due to the availability of Indian resin. In Europe, although known, its medicinal use was marginal. There are several reports from France on the successful use of Cannabis as oxytocic as well as on ocular pain and inflammation and in rheumatic pain [31–33]. Reports from Germany indicated its use as a hypnotic and in gastric disorders [34, 35]. It was found also to relieve vomiting and was said to be "a true sedative of the stomach" without causing any of the inconveniences experienced after the administration of opium, chloral or the bromides.

In North America, extracts of local hemp were used [36–40]. Most of the work reported was based on the British experience. Attacks of neuralgia, including trigeminal neuralgia, were successfully treated. It was considered a valuable remedy in the treatment of persistent headaches, in spasmodic asthma and as an oxytocic. The antivomiting effect was recorded again, as was the general hypnotic effect.

Around the turn of the century, the medical use of Cannabis in Europe and North America declined as reproducible clinical effects could not always be obtained. The active constituent in Cannabis had not been isolated in pure form; plant extracts were generally used and these were known to deteriorate with great rapidity. No simple test of the potency of the drug existed and the practitioner could not readily determine the required dose. Large doses caused the well-known psychotropic side-effects; hence, with the ready availability of standardized drugs, Cannabis was slowly neglected.

In the late 1930's and early 1940's important chemical work was reported by Adams [41] and by Todd [42]. Compounds with cannabimimetic activity were prepared. Extensive pharmacological work by Loewe [43] indicated that some of these synthetic compounds possessed therapeutic potential as analgetics and/or sedatives. Indeed one of these synthetic compounds, pyrahexyl (synhexyl) (4b), reached clinical stage of testing [44–47]. Unfortunately, most of the clinical work performed was on depression, which is a psychiatric condition known today not to be relieved by cannabinoids.

Pyrahexyl (4b) was found to be active in the treatment of alcohol and opiate withdrawal symptoms [44] and the homologue (5) was found to be antiepileptic [48].

Except for some work on the topical antibiotic properties of cannabinoid acids [49], medicinal chemical work in this field essentially stopped until 1964, when the major active constituent, Δ^1-THC, was isolated in a pure form and its structure was elucidated [50]. Since then, thousands of papers on the chemistry, pharmacology, metabolism and clinical effects of Δ^1-THC (1) have been published. On the basis of the knowledge accumulated, a considerable

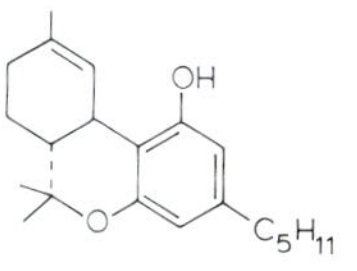

(1) Δ^1 –THC
(Δ^1 –tetrahydrocannabinol)
or
(Δ^9 –tetrahydrocannabinol
by another nomenclature)

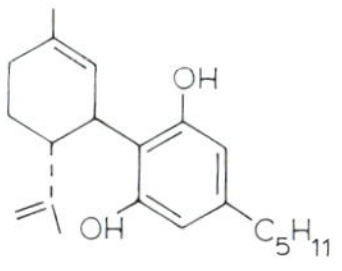

(2) CBD
(cannabidiol)

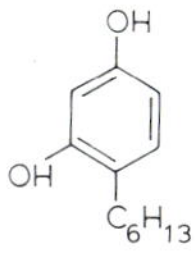

(3) hexylresorcinol

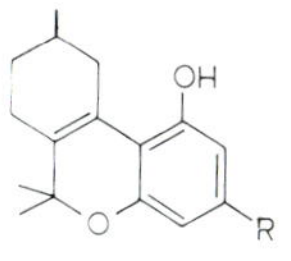

(4) a, $R = C_5H_{11}$, Δ^3–THC
b, $R = C_6H_{13}$
(5) R = 1,2–dimethyl-heptyl

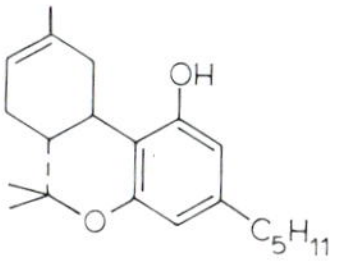

(6) Δ^6 – THC
(Δ^8 – THC by another nomenclature)

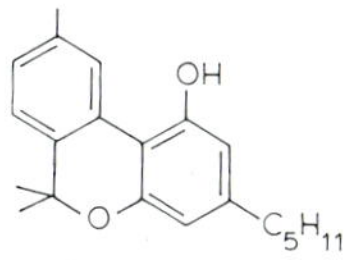

(7) CBN
(cannabinol)

amount of work on the medicinal chemical aspects of the problem has appeared and will be discussed below.

CANNABIMIMETIC ACTIVITY

INTRODUCTION

Cannabis and its active constituent Δ^1-THC produce a typical effect in man – the famous marihuana 'high'. The most appropriate and specific term for this action seems to be 'cannabimimetic effect' or 'cannabimimetic activity' [51].

The pharmacology of Cannabis, and, in the last 20 years, of Δ^1-THC, has been the object of thousands of publications and has been thoroughly reviewed. Most of the work has centered on the cannabimimetic effects in man as well as in various animal species. As the cannabimimetic effect has no medicinal value, we shall refrain from a detailed recapitulation of the well-reviewed material on this particular aspect. We would like to refer the reader to some recent reviews as well as to several older, critical, thoughtful and surprisingly topical reviews by Sir W.D.M. Paton and his colleagues from 1973 and 1975 [3–8, 52–61]. In the present review, we shall address ourselves to and summarize only four aspects related to cannabimimetic action: (a) pharmacological test methods; (b) structure-activity relationships (SAR); (c) stereochemical requirements and (d) recent approaches to the biochemical basis of cannabimimetic action.

PHARMACOLOGICAL TEST METHODS

The animal methods used for detecting and quantifying cannabimimetic activity have been surveyed [52, 56], hence here we shall only present a short list of methods currently used.

(a) Overt behaviour in dogs [62, 63]. This is a simple, very sensitive test which is semiquantitative in nature. Δ^1-THC at 0.2 mg/kg injected i.v. causes static ataxia and depressed activity and the tails are typically tucked.

(b) Overt behaviour in monkeys [64]. Rhesus monkeys (or baboons) when injected i.v. with active cannabinoids show sensitivity and pattern of behaviour similar to that of humans. This test is also semiquantitative in nature. At low doses (0.05 mg/kg), Δ^1-THC causes tranquility, drowsiness, decreased motor activity, occasional partial ptosis, occasional head drop; at higher doses (0.1–0.25 mg/kg) the drug causes stupor, ataxia, suppression of motor activity, full ptosis, typical crouched posture (thinker position) kept for up to 3 h (the

animal may, however, regain normal behaviour for short periods of time if external sensorial stimuli are applied); at doses above 0.5 mg/kg the drug causes severe stupor and ataxia, full ptosis, immobility, crouched posture lasting for more than 3 h, and absence of reaction to external stimuli. A compound is not considered to be cannabimimetic if at 5 mg/kg i.v. it fails to induce the above characteristic syndromes.

A schedule-controlled-behaviour test in squirrel monkeys has been described [62].

(c) Spontaneous activity in mice and rats. Two methods of measurement are generally employed. In one of them, mice injected i.v. with the drug are placed in a photocell activity chamber [62], wherein interuptions of the photocell beams are recorded to quantify the locomotor activity. In a standard variation, rats are used in an open field test with ambulation, rearing, defaecation, urination and grooming noted [65]. In the ring immobility assay [66], a mouse injected s.c. with the drug is placed on a horizontal ring and the percentage of time spent immobile during a 5 min exposure is determined. Unlike the former two tests, the latter is generally used only as a measure of psychotropic drug activity.

(d) Reduction of body temperature in mice [62]. This is a rather nonspecific effect caused also by numerous drugs of other types. Δ^1-THC causes a hypothermic response of *ca.* 3 C° at 4 mg/kg. Surprisingly, Δ^6-THC (6), which in other animal tests, as well as in humans, is less active than Δ^1-THC, in this test is more active than Δ^1-THC.

(e) Drug discrimination procedures [67–71]. These are probably the most specific tests available. Rats or pigeons are trained to emit one response when trained with Δ^1-THC and an alternative response when trained without drug. There is no generalization to Δ^1-THC from morphine, diazepam, pentobarbital, phencyclidine, 3-PPP and other drugs. However, some exceptions have been noted and it has been suggested that "the discriminative stimulus properties of THC may have some commonality with the effect of diazepam in a subpopulation of rats trained to discriminate THC" [71]. These cannabimimetic effects of diazepam are antagonized by specific benzodiazepine receptor antagonists.

STRUCTURE-ACTIVITY REQUIREMENTS FOR CANNABIMIMETIC ACTIVITY

Nearly 15 years ago [72, 73], we formulated some tentative rules for cannabimimetic structure-activity relationships (SAR). Most of these rules have withstood the erosion of time and of new data, although exceptions have been noted and certain refinements have to be made [74, 75, 75a]. The following generalizations can be made today.

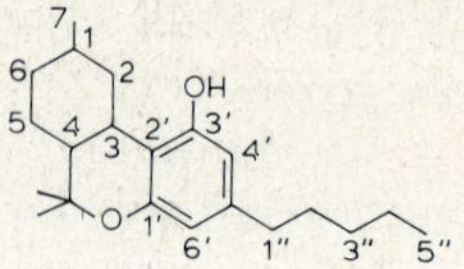

(8)

1. A dihydrobenzopyran-type structure (see (8) for numbering system) with a hydroxyl group at the 3′ aromatic position and an alkyl group on the 5′ aromatic position seems to be a requirement. Opening of the pyran ring leads to complete loss of activity.

However, three major exceptions have been found: (*a*) Compound (9) and some of its derivatives [76]. (*b*) Compound (10), numbered CP-47497. This is a major simplification of the cannabinoid molecule [76–78]. Compounds (9) and (10) are potent analgetics and will be further discussed in the appropriate section; (*c*) Compounds of type (11) and (12) [79, 80]. These compounds are not planar. Hence, planarity is not an absolute requirement.

Although these novel compounds seem to contravene the basic assumption that a dihydrobenzopyran is a central requirement for cannabimimetic activity, the fact remains that most cannabimimetic substances possess this moiety. Future work will probably throw additional light on these discrepancies.

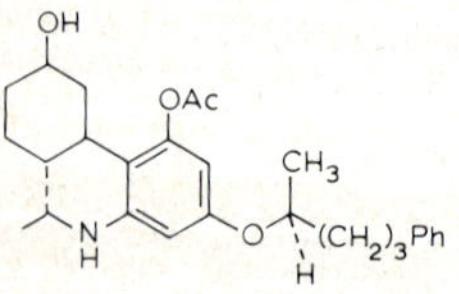

(9) Levonantradol
(Nantradol is the racemic form; no separation of side chain isomers)

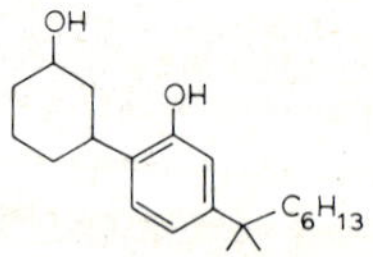

(10) CP – 47497

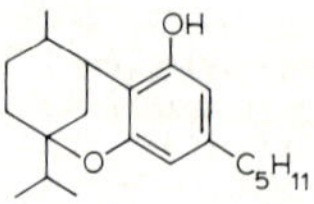

(11)

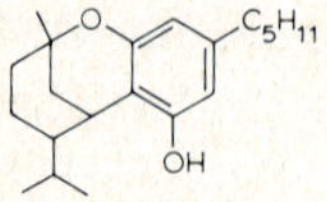

(12)

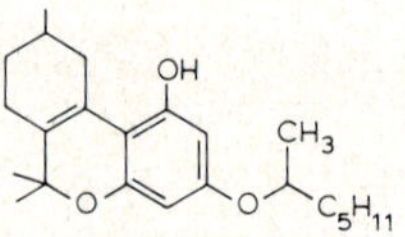

(13)

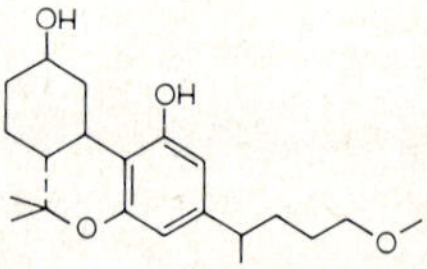

(14)

2. The aromatic hydroxyl group has to be free or esterified. Blocking of the hydroxyl group as an ether inactivates the molecule.

It is quite possible that the phenolic esters are also inactive as such but undergo hydrolysis *in vivo* to the free phenol. When tested *in vitro* in some biochemical reactions (in which the free phenols are active) the esters show complete lack of activity [81]. Replacement of the phenolic group by an amino group in some cannabinoids retains activity; replacement by a thiol group in all cases tested so far eliminates activity [82].

3. When alkyl groups are substituted on the phenolic ring at C-4′, activity is retained. Substitution at C-6′ eliminates activity. Electronegative groups such as carboxyl, methoxycarbonyl, acetyl at either C-4′ or C-6′ eliminate activity.

4. An all-carbon side-chain is not an absolute requirement. Several active cannabinoids have been synthesized in which the side-chain contains an etheric oxygen (see compounds (13), (14)) [76, 83, 84].

5. Hydroxylation at C-7, which is a major metabolic path, leads to the very potent compounds (15) and (16) [85, 86]. The activity of compounds which have very low cannabimimetic activity (for example, cannabinol) may be significantly increased by the introduction of a 7-OH group (75). Monohydroxylation on other positions of the terpene ring also usually leads to active derivatives [87, 88]. Dihydroxylation generally causes loss of activity.

Oxidation of the C-7 methyl group to a carboxyl, causes inactivation [89]. These types of acid are major metabolites.

6. Hydroxylation on the C-1″ of the side-chain abolishes activity. Can-

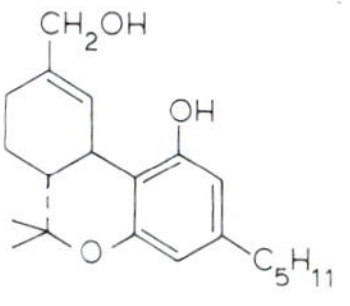

(15)

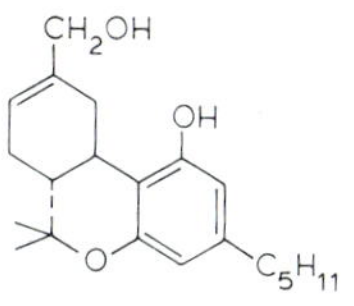

(16)

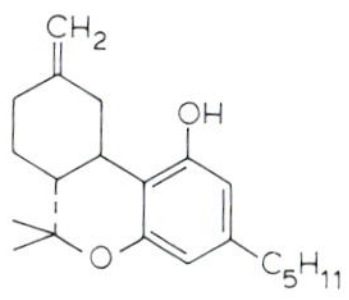

(17) Δ^7-THC

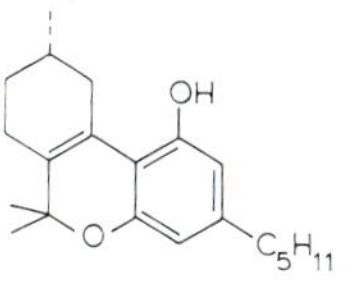

(18) (1S)-Δ^3-THC

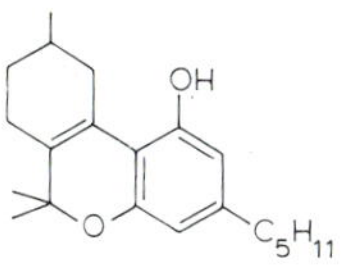

(19) (1R)-Δ^3-THC

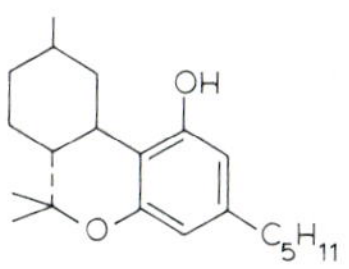

(20)

nabinoids hydroxylated at the other side-chain carbons are still active; hydroxylation at the C-3″ increases activity [90]. Some of these hydroxylated compounds have been detected as major metabolites.

7. The activity of the double-bond isomers of the THC in humans is as follows: Δ^1-THC > Δ^6-THC > (1*S*)-Δ^3-THC > (1*R*)-Δ^3-THC [91, 92]. The (1*R*)-Δ^3-THC enantiomer may in fact be inactive. Δ^5-THC and Δ^7-THC are inactive in animal tests, while Δ^2-THC and Δ^4-THC have not yet been tested.

8. The hexahydrocannabinol enantiomers ((20) and (21)) are both active, though the isomer in which the methyl group is equatorial (20) is much more active than that in which the methyl group is axial (21). The same relationship has been observed in 7-hydroxyhexahydrocannabinols [93]. Replacement of the hydroxy group in the 7-hydroxyhexahydrocannabinols with a methylamino or a dimethylamino group eliminates cannabimimetic activity, although the axial isomer causes yawning in monkeys [94].

9. Hexahydrocannabinols with additional substitution present a more complicated picture. Two groups of THC-type cannabinoids which differ only in that the chemical groupings placed at C-1, C-2 in one of them are situated at

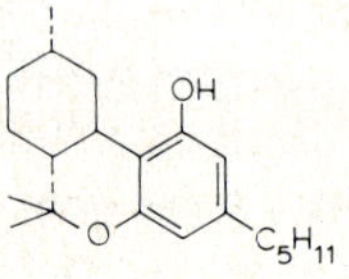

(21) 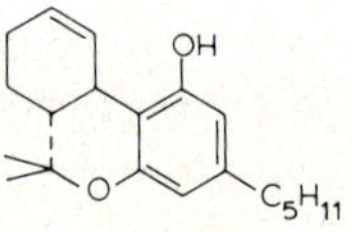 (22) 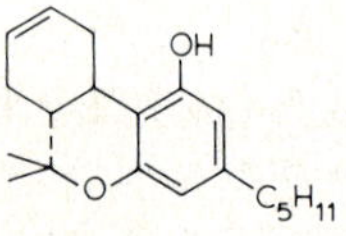 (23)

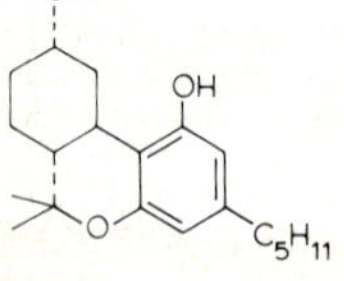

(24) 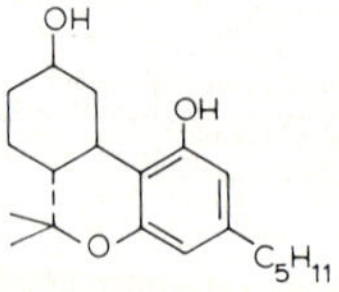 (25) 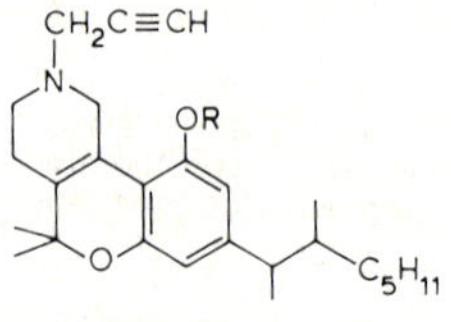 (26a) R=H

(26b) R=CO$(CH_2)_3$N$(CH_2)_5$

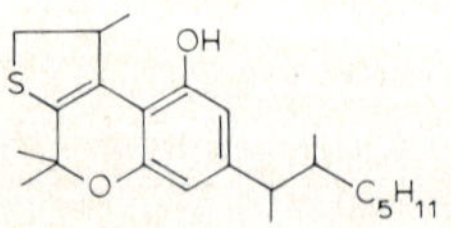

(27) 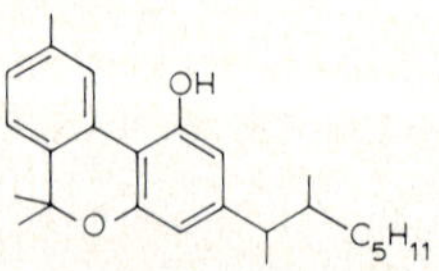 (28)

C-1, C-6 in the other (but retain their stereochemistry) have almost equivalent psychotropic activity [93].

10. The 7-methyl group is not an absolute requirement for activity. 7-Nor-Δ^1-THC (22) and 7-nor-Δ^6-THC (23) are active in dogs [95]; both 1-hydroxy-7-norhexahydrocannabinols (24) and (25) are cannabimimetic in several animal tests, the 1α-hydroxy isomer (24) being about 1/5 as potent as the 1*β*-hydroxy isomer (25) [96], but only the *β*-OH isomer (25) is active in rat discrimination tests [97].

11. The terpenoid ring may be exchanged by some heterocyclic systems (for example, compounds (26a) and (27) [75, 98]. The full scope of this modification has yet to be fully explored.

12. The 1,1- or 1,2-dimethylheptyl (DMH) side-chain occasionally confers cannabimimetic activity on compounds which lack or may have low activity in the *n*-pentyl series (cf compound (28)) [83].

STEREOCHEMICAL REQUIREMENT FOR CANNABIMIMETIC ACTIVITY

As mentioned above, the hexahydrocannabinol (21) in which the C-7 methyl group protrudes from the plane of the molecule is much less active than the isomer (20) in which the methyl group is in the plane of the molecule. The same relationships hold true for the related 7-hydroxyhexahydrocannabinols. It seems reasonable to expect that, at the active site, the approximate planarity of the molecule (or at least the planarity of the top part of the molecule) plays a rôle. This supposition can be tested only when the active site is identified. Several exceptions to this hypothesis exist, as mentioned above, for example, compounds of type (11) and (12). The pharmacology of these compounds is not well known and their activity may not be truly cannabimimetic.

Generally the 3,4-*cis*-THC's are much less active in all animal tests than the 3,4-*trans* compounds [100]; however, the *cis* compounds have not been studied over a wide range of tests.

Another aspect of the stereochemical requirements for cannabinoid activity is the need for enantiomeric purity. The natural Δ^1-THC and Δ^6-THC are (3*R*,4*R*) isomers [101]. The synthetic route which was developed by our laboratory for Δ^1-THC and Δ^6-THC nearly 20 years ago [102] makes possible also the synthesis of the (3*S*,4*S*) enantiomers: (+)-verbenol on condensation with olivetol leads to (29), which on ring opening leads to (+)-(3*S*,4*S*)-Δ^6-THC (30), which can be converted with ease into (+)-(3*S*,4*S*)-Δ^1-THC (31) (*Scheme 5.1*). In several tests for cannabimimetic activity, (+)-Δ^1-THC (31) was *ca.* 13–20-times less active than the (−)-isomer [72, 103]. These results indicate pharmacologic enantiomeric preference rather than absolute stereo-

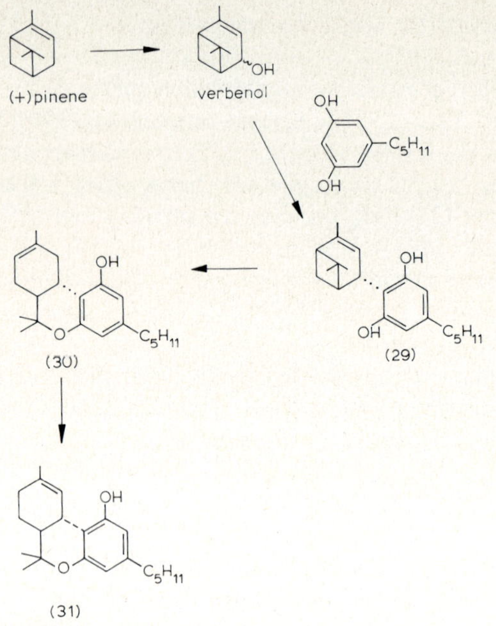

Scheme 5.1. Synthesis of (+)-(3S,4S)-Δ^1-THC (31).

selectivity. Indeed, Martin, in a recent review [8], states that, "while cannabinoid SAR supports the concept of a specific cannabinoid receptor, a disconcerting element is the apparent lack of greater stereospecificity (5–100-fold) in some animal models." However as the starting material, (+)-α-pinene, was not necessarily absolutely pure, this conclusion is tentative at best. Recently in our laboratories we were able to prepare a pair of *crystalline* enantiomeric cannabinoids (compounds (32) and (33)) which were recrystallized to absolute purity (*Scheme 5.2*) [104].

In generalization tests with rats and pigeons, Järbe found that (−)-7-OH-Δ^6-THC-DMH (32) was *ca.* 87-times more active than natural Δ^1-THC in the rat and *ca.* 73-times more active in pigeon. The (+)-enantiomer (33) was inactive at doses *ca.* 1000-times and *ca.* 4500-times (for rats and pigeons, respectively) higher than those of the ED_{50} value of the (−)-enantiomer [105].

The same type of results were observed in the rotarod neurotoxicity test in rats [106]. The (−)-enantiomer (32) was *ca.* 260-times more potent than natural (−)-Δ^6-THC; the (+)-enantiomer (33) was inactive at doses *ca.* 2000-times higher than those of the ED_{50} value of the (−)-enantiomer. Qualitatively,

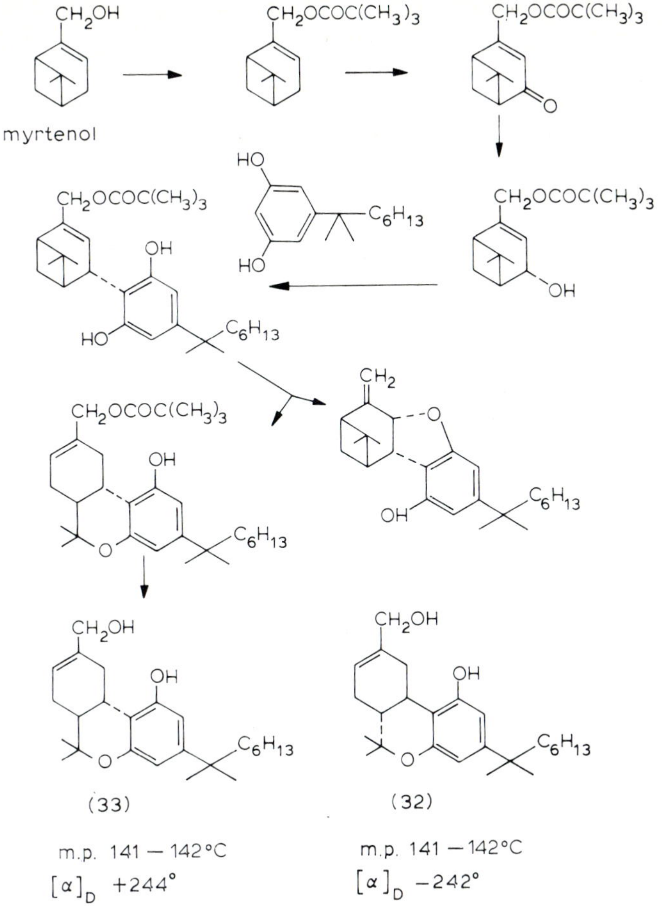

(33) m.p. 141 – 142 °C $[\alpha]_D$ +244°

(32) m.p. 141 – 142 °C $[\alpha]_D$ −242°

Scheme 5.2. Synthesis of 7-OH-Δ⁶-THC-DMH.

the same type of results were obtained in the mouse ring test: the (+)-enantiomer was inactive in all doses tested; the (−)-enantiomer was *ca.* 100-times more active than natural (−)-Δ^6-THC [107]. Compound (32) is thus one of the most active cannabimimetic substances prepared so far. The above results indicate that in the four tests used, cannabimimetic activity resides exclusively in the (−)-(3*R*,4*R*)-enantiomer.

The results of animal tests of the types described above with Δ^1-THC, Δ^6-THC and other cannabinoids have been shown to parallel activity in man

[91]. Hence we assume that the results with (32) and (33) likewise indicate parallel activity in man. If this is correct, cannabimimetic activity has a strict stereochemical requirement, which indicates a probable interaction with a chiral biological system (enzyme, receptor site, etc.) and not just an unspecific action due to the high lipid solubility of the cannabinoids.

RECENT APPROACHES TO THE BIOCHEMICAL BASIS OF CANNABIMIMETIC ACTION

There seems to be an inverse relationship between the quantity of publications on cannabinoids and our understanding of the biochemical and molecular basis of cannabimimetic action.

A simplistic comparison of THC with other drugs does not lead very far. As Paton pointed out nearly 15 years ago: “One does not readily find another substance so ‘contradictory’, capable of taming yet producing aggressiveness, of both enhancing and depressing spontaneous activity, of being an anti-convulsant yet generating epileptiform cortical discharges” [108].

All cannabinoids are highly lipophilic substances and it has been suggested that THC is probably related in its properties to the anaesthetics. As related cannabinoids, such as CBD, CBN and the (+)-enantiomers are inactive (see above), one can assume an enantiospecific structural fit into a hydrophobic environment in some cell membrane. Indeed, Lawrence and Gill, using spin-labelled liposomes, demonstrated lipid interactions between several cannabinoids and lecithin-cholesterol bilayers [109]. They showed differences between (+)- and (−)-Δ^1-THC, suggesting stereospecificity; CBD and CBN gave a response different from that observed with (−)-Δ^1-THC; 7-OH-THC (15) was more active than THC. These observations parallel the known SAR of cannabinoids and support the lipid interaction theory of THC action.

THC and CBD differ in their action on membrane electrical properties. It has also been found that THC causes large shifts in the transition temperature of membrane lipids [110, 111]. A possible relationship between this effect and microsomal demethylase activity was observed, and it was suggested that THC could influence enzymatic action by membrane effects.

A recent study on the interaction between cannabinoids and membranes by NMR has been published [112]. THC and CBD were found to induce changes in the size of the vesicles, the cation binding capacity of the membrane surface and the motional state of the hydrocarbon chains of the phospholipids. At high concentrations, both drugs induce fusion of the vesicles, tighten the packing of the paraffinic chains of the phospholipid molecules and reduce the binding of Pr^{3+} ion to the surface. Δ^1-THC is more effective than CBD in causing these changes. At low concentrations, neither drug has a detectable effect when incorporated into vesicles of pure egg yolk lecithin (EYL). The introduction of cholesterol into the vesicles causes significant changes in the effect of cannabinoids: THC increases the mobility of the hydrocarbon chains of EYL in the vesicles, while CBD has the opposite influence.

In a recent study, a dose-response relationship was found between THC and the transition temperature shift for lecithins [113]. The cholesterol content was again found to influence this effect.

While all these observations strongly support the reasonable hypothesis that the highly liposoluble cannabinoids act through membrane lipids (apparently with structural and enantiomeric specificity), they have not brought us any closer to the identification of the biochemical system responsible for cannabimimetic action.

The alterations produced by THC and other cannabinoids in biogenic amine levels as well as on uptake, release and synthesis of neurotransmitters and effects on enzymes have been the subject of numerous investigations (for reviews see [8, 52, 55, 114, 115]). It is beyond the scope of the present summary to try to analyse and put into a proper perspective the wealth of data published so far. It is our subjective view that the mode of action of cannabimimetic compounds is somehow directly associated with prostaglandin metabolism (see, in particular, the series of papers by Burstein [115, 116]), and/or reduction of hippocampal acetylcholine turnover observed in rats [117, 118]. The latter effect is enantiospecific and follows the known SAR of the cannabinoids. This *in vivo* selectivity of action suggests that the THC may activate specific transmitter receptors which indirectly modulate the activity of the cholinergic neurons in the septalhippocampal pathway.

Recent work has shown that Δ^1-THC increases homovanilic acid (a dopamine metabolite) in rat prefrontal cortex but not in the caudate, indicating specific dopaminergic action [118a]. On the other hand, GABA involvement is suggested by the increased diazepam and flunitrazepam binding to benzodiazepine receptors caused by THC and synthetic cannabinoids [118b, c]. Other recent work indicates brain histamine involvement in THC tolerance [118d].

Several groups have reported work on a THC-receptor or binding site [8, 119, 120]. Until recently, studies of [^{3}H]cannabinoid binding to brain tissue failed to reveal stereospecific and saturable binding sites with high affinities. A possible reason for this lack of success was again the high lipophilicity of cannabinoids which may have interfered with specific receptor binding due to extensive partitioning into the lipid of brain membranes. Recently, a synthetic THC derivative with a hydrophilic side-chain, 5″-trimethylammonium-Δ^6-

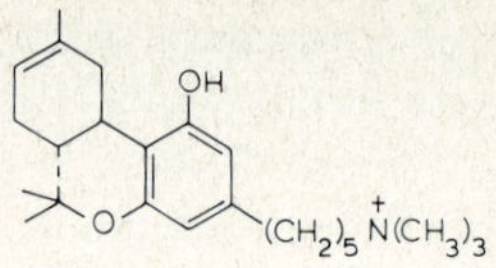

(34) 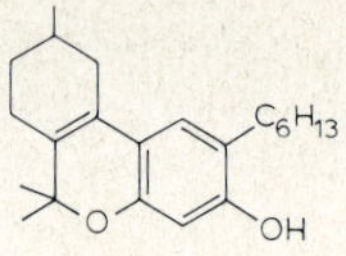 (35)

THC (34), was prepared and its binding to rat neuronal membranes was studied [120]. This cannabimimetic compound was found to bind saturably and reversibly to brain membranes with high affinity to apparently one class of site. Δ^1-THC potently, stereoselectively and competitively inhibited [^{3}H]-(34) binding. However, for numerous cannabinoids (CBD for example), potency in behavioural and physiological tests did not parallel their affinity for the binding site of (34). It seems that this binding site is not a specific one for THC and other cannabimimetics. It is possible, however, that the binding site is relevant to cannabimimetic activity, but represents only part of a more complicated picture.

In summary, we can sadly state again that at the biochemical level the mechanism of action of the psychoactive cannabinoids remains elusive.

ANALGESIA

As described in some detail above, Cannabis was used as an analgetic from ancient times up to the end of the 19th century. An early report by Christison is typical: he took orally 4 grains of extract for toothache; within an hour the pain ceased; he noted a pleasant numbness of the limbs, giddiness, sleepiness and a fast pulse [121]. Numerous other 19th century case reports are available. However, the action was not dependable, not only because of the variability of Cannabis preparations but also due to personal idiosyncracy. Indeed, more recently Ames has reported that, whereas one subject experienced no pain on venepuncture under Cannabis, another found it agonizing [122].

In 1949 a British group [123] tested some of the synthetic cannabinoids prepared a few years before by Todd [42]. They recorded substantial analgetic activity with racemic Δ^3-THC (4a) in a rat tail test and also unexpectedly found that the C_6H_{13} homologue of the positional isomer of Δ^3-THC (35) had considerable analgetic activity. Few other members of the latter series have been tested since then [123a], although isomers of this type are known to have negligible cannabimimetic activity, i.e., separation of analgesia from cannabimimetic activity may have been achieved.

A considerable amount of work on analgesia has been reported with Δ^1-THC since it was isolated in 1964 [50]. Our group reported in 1968 that both Δ^1-THC and Δ^6-THC when administered i.p. were about 50% as effective as morphine (administered s.c.) in the standard writhing, tail flick and hot-plate tests in mice. Although limited in scope, this report indicated for the first time that natural THC, a non-opioid, non-alkaloidal substance, was a potent analgetic [124]. As Cannabis was known to have low (if any) physical dependence liability as well as low toxicity, it seemed reasonable to expect that novel cannabinoids would soon be developed as clinical analgetics. This expectation was further strengthened when cannabinoids were found to induce no respiratory depression or other major side-effects at low doses. Yet today, 20 years later, no synthetic cannabinoid is used as an analgetic. The main reason is that, in spite of a considerable amount of work in this field, until recently no separation could be achieved between the cannabimimetic and the analgetic actions.

In the late 1960's and early 1970's, a heated debate took place as to whether Δ^1-THC was analgetic and, if it were, what the level of its potency was. Depending on the stringency of the end-points used by the various groups of investigators, major differences in the results were obtained. Buxbaum [125] found Δ^1-THC to be equiactive with morphine in the hot-plate and tail flick tests in rats, but to be less active than morphine in mice. Another report indicated that Δ^1-THC was not active in the tail flick, hot-plate or writhing tests in mice below doses that produced severe behavioural and psychomotor impairment [126]. A flat dose-response curve was obtained between 20 and 80 mg/kg, with a maximum activity of 65%. However, a Brazilian group [127] confirmed the effectiveness of Δ^1-THC in mice at 5 and 10 mg/kg when tested by the hot-plate method, and an industrial group also reported an ED_{50} of *ca.* 10 mg/kg Δ^1-THC in the tail pinch, hot-plate and writhing tests in mice [128, 129]. Today it is generally accepted that Δ^1-THC and Δ^6-THC are indeed analgetic in many animal tests; however, this activity is seen only at doses which are close to those producing cannabimimetic activity. (For a recent review see [129].)

The above discrepancies seem to be due mainly to the stringencies of the end-points in each test. However, one can also doubt the applicability of these tests for cannabinoids. These tests were developed for opiates and may not necessarily parallel and be fully relevant to THC activity in man. Indeed, recent work on electrodes implanted into several distinct brain areas of the rat led to the conclusion that the "apparent analgesia produced by the cannabinoids is more related to their disruption of discharge in response to some synthetic impact rather than to a depression of pain pathways" [130]. The question whether cannabinoids cause analgesia by interference with pain reception or with pain perception is still unresolved.

Recently, Martin [131] found that when Δ^1-THC was administered i.v. to mice, its activity was much higher than when administered s.c.

The uncertainty as regards the level of analgetic activity of Δ^1-THC in animals was not resolved on testing in humans. In a small group of volunteers, one group found some analgesia which the authors interpreted as both a direct action as well as a disruption of a cognitive effect [132]. Two studies by another group [133, 134] showed good analgetic activity. In the first one, the subjects reported positive effects in headaches, migraine, menstrual cramps and post-surgical pain. This preliminary report with crude Cannabis was followed by a relatively well-controlled study with Δ^1-THC using patients with chronic pain from metastatic carcinoma. The activity found with a dose of 10 mg was equivalent to that observed with codeine, and no major side-effects were observed. However, at a higher dose (20 mg), the expected cannabimimetic effects such as sedation and drowsiness were noted. However, other clinical groups have failed to observe significant analgesia. A study of surgical and experimental pain with 15–30 mg of THC administered i.v. failed to establish a consistent significant analgesic effect. However at the low dose of THC, 3 of 10 patients reported good pain relief [135].

In summary, these results seem to indicate that pure THC is not a valuable general analgetic suitable for modern use. It is possible, however, that if it is administered together with cannabidiol, which is known to eliminate some of the cannabimimetic effects of THC [136], it may prove to be of some value possibly in specific pain conditions.

The first industrial group to devote appreciable time and effort in developing a new synthetic cannabinoid analgetic was that of Sisa in Cambridge, MA. This group has reviewed its work [98]. The most active compounds prepared were those in which the terpene ring of the natural product was replaced with a heterocyclic one.

One of the Sisa compounds, named nabitane (26b) has been tested clinically. Nabitane was found to be about 6-times more potent than codeine in the relief of chronic pain. Cannabimimetic side-effects were, however, recorded which obviously make this derivative unsuitable for clinical use [137, 138].

7-Hydroxy-Δ^1-THC (15) and 7-hydroxy-Δ^6-THC (16) are potent analgetics [139] (see *Table 5.1*). This activity parallels their cannabimimetic activity. But 7-nor-Δ^1-THC (22) and 7-nor-Δ^6-THC (23), which are cannabimimetic, are not analgetic. For the first time, a possible separation of these pharmacological effects was recorded: but unfortunately in the therapeutically undesirable direction. Additional results of the same type were obtained with 7-nor-1-hydroxyhexahydrocannabinols (24) and (25). Both isomers were cannabimimetic but only the equatorial isomer (25) was analgetic, with a potency

Table 5.1. EFFECTS OF SOME NATURAL CANNABINOIDS, THEIR METABOLITES AND SYNTHETIC CANNABINOIDS ON CANNABIMIMETIC ACTIVITY (DOGS) AND ANALGESIA (MICE) [96, 139]
n.a., not active.

	Overt behaviour in dogs effective dose (mg/kg)	*Analgesia (hot plate) (s.c.) ED_{50} (mg/kg)^{50}*
Δ^6-THC (6)	0.4	8.8
Δ^1-THC (1)	0.2	9.6
7-OH-Δ^6-THC (16)		1.9
7-OH-Δ^1-THC (15)	0.01	1.9
7-Nor-Δ^6-THC (23)	0.2	n.a.
7-Nor-Δ^1-THC (22)	0.2	n.a.
7-Nor-1β-OH-hexahydroCBN (25)	0.05	1.6
7-Nor-1α-OH-hexahydroCBN (24)	0.50	n.a.
Morphine		1.2

2–3-times higher than that of morphine [96]. These results indicated that the two effects – the cannabimimetic and the analgetic – are separable and that if such a separation were obtained the analgetic activity could be in the opiate range.

The above observations were the basis of an extensive project initiated at Pfizer [76–78, 140, 142]. By a systematic investigation and modification of the various structural moieties within the THC molecule, the Pfizer group found that the oxygen atom of the dihydropyran moiety could be replaced by nitrogen with retention of activity. Derivatives of this type had been prepared previously; their cannabimimetic activity had been reported as either non-existent or very low. The Pfizer group then made use of other molecular modifications (known to increase cannabimimetic and/or analgetic activity) such as (a) substitution of the pentyl side-chain with either a dimethylheptyl moiety or a long-chain ether and (b) modification of the terpene ring as described above (in accordance with the SAR findings [96]). After some additional structural manipulation, compound (9), named nantradol, was chosen for detailed biological studies. Nantradol is a mixture of four isomers (two diastereoisomers). Nantradol was found to be 20–100-times more potent than Δ^1-THC as an analgetic, and between 2- and 7-times more potent than morphine across a battery of analgetic tests (see *Table 5.2*). Nantradol does not bind to the opiate receptor and its analgetic activity is not blocked by naloxone. Another group has, however, found partial blocking of nantradol action by naloxone [143]. Moreover, surprisingly, nantradol, like the opiates, was found to have antidiarrhoeal and antitussive properties.

Nantradol was separated into two diastereoisomers and the more potent of the two was then resolved into its two optically active enantiomers. The laevorotatory enantiomer, named levonantradol (9), was found to be at least 100-times more potent than the dextrorotatory enantiomer.

Further work by the same group based on theoretical grounds led to the observation that compound (10) (CP-47497) is a potent analgetic. This was interpreted to mean that the dihydropyran moiety present in THC was unnecessary for either cannabimimetic or analgetic activity and apparently it "serves no other purpose than to firmly anchor the phenolic ring and cyclohexyl ring in an active conformation".

On the basis of calculations of minimum energy conformations, the same group suggested that CP-47497 readily assumes a conformation quite similar to the rigid THC molecule and is recognized by a biological system as being equivalent to a tricyclic cannabinoid. They also postulated that CP-47497 (and presumably also other cannabinoids) interact at a receptor site by a three-point contact, the three binding sites being the equatorial alcohol, the phenol and the C-5 side-chain. The analgetic activity of (10) is presented in *Table 6.2*. It is equipotent with morphine but unfortunately it retains many of the cannabimimetic properties of THC.

The observation that the dihydropyran ring is not an absolute requirement for analgetic activity led the Pfizer group to synthesize numerous analogues such as (36), and (37). These compounds are extremely potent analgetics, compound (37) being up to 100-times more active than morphine (*Table 5.2*).

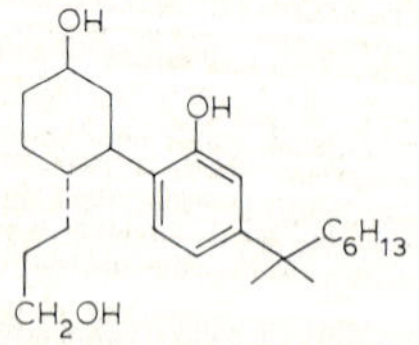

OH OH C_6H_{13} CH_2OH

(36)

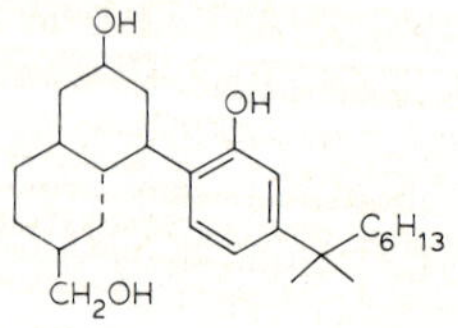

OH OH C_6H_{13} CH_2OH

(37)

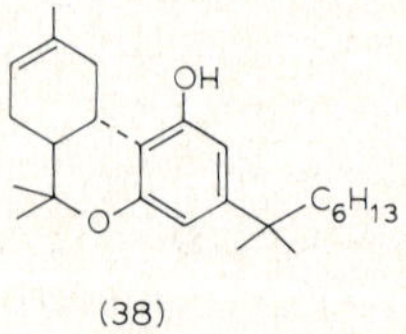

OH C_6H_{13} O

(38)

Table 5.2. ANALGESIC ACTIVITY OF SOME CANNABINOIDS [76]

	Tests (MPE_{50}[a] (mg/kg, s.c.))		
	PBQ writhing (mouse)	*Tail flick (mouse)*	*Tail clamp (rat)*
Compound (9) (nantradol)	0.4	0.7	1.0
Compound (9) (levonantradol)	0.07	0.2	0.2
Compound (10) (CP-47497)	1.0	4.4	4.7
Compound (37)	0.02	0.03	0.06
Compound (36)	0.07	0.04	0.13
Morphine	1.8	5.7	4.8
Δ^1-THC (1)	5.9	55	29.1
Compound (25) (7-nor-1β-OH-hexahydroCBN)	0.63	9.1	7.0

[a] The dose to produce 50% of the maximum possible effect.

Table 5.3. (+)-7-HYDROXY-Δ^6-THC-DMH (33) AND MORPHINE: COMPARATIVE DOSES PRODUCING 50% OF THE MAXIMUM POSSIBLE ANALGESIC EFFECT (MPE_{50})

	MPE_{50} (mg/kg s.c.) at time of estimated peak activity of compound[a]			
	Mouse (sabra strain)		*Rat (sabra strain)*	
Compound	*Acetic acid writhing*	*Hot plate*	*Tail clamp*	*Hot plate*
(+)-7-OH-Δ^6-THC-DMH (33)[b]	0.2	1.1	4.9	1.1
Morphine-HCl	1.2	3.5	5.5	4.6

[a] All morphine-injected animals observed 1 h post injection; all (+)-7-OH-Δ^6-THC-DMH-injected animals observed 2 h post injection; refer to test methods given in the original article.

[b] Dissolved in 5% EtOH, 5% emulphor 620 and 90% ddH_2O, with 5×10^{-4} M $CuCl_2$ added subsequently.

Although data on the cannabimimetic activity of these compounds have not yet been published, it seems reasonable to expect that, in view of their similarity to CP-47497, they will show the same profile.

(+)-Δ^1-THC is not analgetic; hence, until recently very little work on analgesia was done with the unnatural (+)-(3*S*,4*S*) cannabinoids. In a lecture published a few years ago we reported some analgetic activity for the dimethylheptyl homologue of (+)-Δ^6-THC (38); however, as this compound was not optically pure it was difficult to draw firm conclusions [144]. As mentioned above, recently we synthesized both enantiomers of the dimethylheptyl homologue of 7-hydroxy-Δ^6-THC (32, 33) (*Scheme 5.2*). These compounds are crystalline and can be obtained with absolute optical purity. The (−)-isomer (32) is a highly potent cannabimimetic. The (+)-compound (33), which is not cannabimimetic, was found to have considerable analgetic activity in several mice and rat tests [145] (*Table 5.3.*). In all tests undertaken, compound (33), when administered with 5×10^{-4} M $CuCl_2$, showed activity above the potency level of morphine. However, contrary to morphine, the dose-response curve was not sharp, but flat. The activity was retained over 3–4 days. Although cannabinoids are not considered to act through an opioid mechanism [145a], unexpectedly, the analgetic activity of (33) was blocked by naloxone. This seems to be the first case of complete separation of analgetic from cannabimimetic activity in the cannabinoid series and if this separation is valid in humans, compound (33) may prove to be a valuable addition to our armamentarium of new analgetic drugs as cannabinoids generally lack many of the side-effects of opiates such as high addiction liability and respiratory depression.

Of potential interest and relevance to cannabinoid analgesia is the recent observation that some novel prenylflavonoids isolated from *Cannabis sativa* exhibit anti-inflammatory activity [145b].

ANTIEMETIC ACTION OF CANNABINOIDS

Levitt [146] has recently commented, somewhat facetiously, that, "The use of cannabinoids as cancer chemotherapy antiemetics represents, in essence, using a drug with a relatively undefined mechanism of action to treat the side-effects of other drugs, also with relatively undefined mechanisms of action which are being used to treat cancer, a disease or series of diseases whose precise nature remains enigmatic". This is a situation not unknown in other areas of medicinal research, but apparently not to the extent noted in this field.

As noted above, Cannabis was used as an antiemetic drug in antiquity; it was used in India as such for hundreds of years (and its use probably still continues in popular medicine) and was well known to physicians in England during the 19th century. Yet, the discovery of the antiemetic effects of Δ^1-THC in 1975 was serendipitous and followed 'street reports' that marihuana helped to overcome the side-effects of anticancer chemotherapy [147] . The original report was followed by numerous other clinical trials (for recent reviews see [146, 148].). In 1986 THC was approved for use by the U.S. Food and Drug Administration and is being marketed under the generic name Marinol. It thus represents one of the very few constituents of higher plant origin to have become a new pharmacopoeal drug in recent years. However in spite of the very wide publicity given to THC as an antiemetic drug, little progress has been made in defining the structure-activity relationships in this area, in developing new animal models for testing antiemetic and antinausea activity, in clarifying the mode of action of THC as an antiemetic agent and even in determining the scope of antiemetic activity outside the cancer chemotherapy field.

Much of our understanding of the physiology of emesis and nausea is due to Borison and McCarthy, and so is also the little we know on the mode of action of cannabinoids as antiemetics [149–152].

Most drugs introduced in recent years have a very respectable background of SAR investigations. By contrast, the cannabinoid antiemetics have hardly any published, although it is plausible that some SAR data have accumulated in pharmaceutical firms. The reason for this paucity of data may be due to the apparent lack of a facile animal model. The cat, an animal extensively used by Borison and McCarthy, is expensive and difficult to handle. Few laboratories are willing (or can afford) to employ cats for single, lethal experiments with highly toxic emetogenic compounds such as cisplatin. Rodents cannot be used as they do not vomit. Recently, conditioned taste aversions in mice produced by emetic drugs were shown to be attenuated by some antiemetic drugs [153]. Prochlorperazine (1.0 mg/kg), Δ^1-THC (0.3 and 1 mg/kg) and nabilone (39) (0.01 and 0.03 mg/kg) significantly attenuated the taste aversions induced by cyclophosphamide. On this basis, it was suggested that this reaction "warrants investigation as a model for evaluating potential antiemetics". However levonantradol, a good antiemetic, did not react in the above model. Stark, in 1982, reported that nabilone has been successfully tested in a pigeon model for antiemetic activity [154].

In the cat model, with cisplatin as emetogenic drug, Δ^1-THC, administered

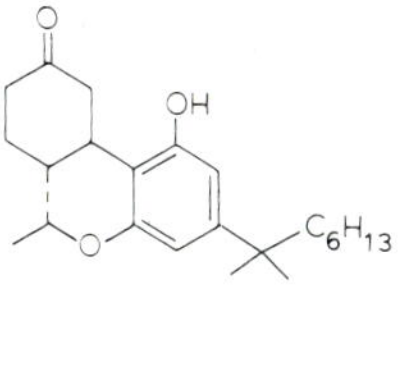

O OH C_6H_{13}

(39)

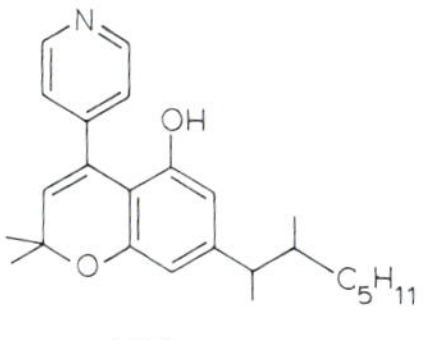

N OH C_5H_{11}

(40)

$OCO(CH_2)_3\overset{+}{N}Et_2$ C_7H_{15}

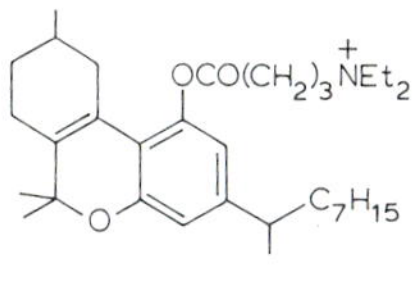

(41)

orally or i.m. at 4 mg/kg prevented vomiting in almost all animals [152]. In the one case (out of seven) in which vomiting occurred, the latency was prolonged and the emetic episodes were significantly lower. By comparison, 7-hydroxy-Δ^1-THC, which is more cannabimimetic than Δ^1-THC (in all animal models tested, including cats), causes less attenuation of vomiting in cats than Δ^1-THC. This observation may indicate that the cannabimimetic and the antiemetic actions do not have the same SAR and hence by molecular modification a non-cannabimimetic, antiemetic cannabinoid may be obtained. This possibility has been strengthened by the observation that (+)-7-hydroxy-Δ^6-THC-DMH (33), which shows no discrimination to THC in pigeons [156], significantly diminishes cisplatin-induced vomiting in this species [155].

Very few other cannabinoids have been tested (or at least reported) in animals. *N*-Methyllevonantradol (*cf.* 9) and nabilone (39) have been compared in the cat model [151]. Both cannabinoids showed dose-dependent antiemetic activity at the 20–100 mg/kg dose levels, *N*-methyllevonantradol being *ca.* 5-times more active than nabilone. The latter drug had previously been shown to suppress in the cat emesis induced by apomorphine, deslanoside and the anticancer drugs 1,3-bis(2-chlorethyl)-1-nitrosourea (BCNU) and cisplatin, but not nicotine. At the doses tested (25–100 μg/kg), nabilone produced behavioural disturbances, from mild ataxia and display of pleasure at 25 μg/kg to severe locomotor disturbance, catatonic behaviour and vocalization at 100 μg/kg.

The number of clinical studies with Δ^1-THC is large and the field has recently been reviewed [146, 148]; hence, only a few typical reports are mentioned here. In a 1980 study [157], 46 patients (9–70 years old) in a double-blind, crossover study were administered either Δ^1-THC (10 mg/m^2 orally, three times daily) or prochlorperazine (10 mg, orally, three times daily). Δ^1-THC was more active than prochlorperazine, regardless of the emetogenicity of the chemotherapeutic agent, and young patients were more likely to have complete response than older patients. In a 1982 study [158], 27 adults in a double-blind study with Δ^1-THC (10 mg/m^2, orally five times daily) and metoclopramide (2 mg/kg, i.v. five times daily) were compared. Metoclopramide was more potent than THC both as regards number and volume of emetic episodes.

From the above and other clinical studies, several basic facts emerge:

(*a*) Δ^1-THC is a potent antiemetic agent. It is very efficient in its action on nausea and vomiting caused by radiation therapy, as well as by many, but not all, anticancer drugs. Thus, Δ^1-THC is only marginally active in patients administered potent emetogens such as cisplatin. Δ^1-THC is active in adults in doses of *ca.* 15 mg oral total dose. This dose causes side-effects in many individuals. Indeed, most patients report somnolence and sedation.

Psychological 'high', anxiety, increased heart rate and orthostatic hypotension are frequently encountered. In some cases, mental confusion, blurred vision and even hallucinations are reported.
(*b*) Some of the early reports [159, 160], correlated antiemetic activity to blood levels of THC and cannabimimetic activity. However, it has been shown that "absolute correlations between the degree of intoxication, as measured by the subjective 'high' and log concentrations of plasma THC are not particularly strong" [161] and hence the correlation between blood levels of THC with antiemetic action does not seem to be a straightforward one.
(*c*) There are indications that patients less than 20 years old experience less side-effects and react with better results to Δ^1-THC than older patients. We do not consider these indications to constitute statistically significant results. In view of the clinical significance of such observations, clinical trials to verify these results may be of importance.

In our view, Δ^1-THC will not be used for long as an antiemetic drug in cancer chemotherapy. However, the experience gained with it will be of importance in the development of new drugs, one of which (nabilone) is already on the market in several countries, including the U.K. and Switzerland. A detailed review on the chemical, pharmacological and clinical aspects of nabilone development and clinical use has been published recently [162].

The synthesis of nabilone is a relatively simple one leading to a racemate [163]. It is of considerable practical interest that nabilone polymorphs are formed which are found to be biounavailable in animals when administered orally, and special codispersion formulations had to be prepared for the final drug product [164].

As with the natural cannabinoids, the metabolism of nabilone appears to be extensive and complex. Of particular interest are the primary metabolites – the reduction products of the 1-oxo moiety. While in rats both enantiomers present in the racemate were reduced *in vitro* and *in vivo* to the (1*S*)-alcohol, in dogs one of the enantiomers gave both the (1*S*)-ol and the (1*R*)-ol. The pharmacokinetic picture also differed. While the total carbinol metabolites accounted for 90% of metabolites in dog plasma after 24 h, in humans even after 8 h only 10% were carbinols. However, the doses administered were not fully comparable. These differences in metabolism (or possibly related ones) seem to account for a difference in toxicity. Rats, monkeys and apparently humans differ in their metabolism (and toxicological profile) from the dog, in which some deaths were observed upon prolonged administration [162].

Whatever the reason for the chronic toxicity in dogs, it does not appear to be relevant to clinical use, which is limited today to short courses of therapy, possibly repeated after 3–4 weeks. Hence, no chronic accumulation of metabolites or chronic toxicity would be observed.

As expected, nabilone has a pharmacological profile rather similar to that of THC: it causes dog ataxia; it discriminates for THC in rats trained on THC; in the THC-seizure-susceptible rabbit it causes convulsions; in rhesus monkeys it reduces blood pressure. All these effects are generally obtained at doses 10-times lower than those of Δ^1-THC needed to cause the same effects. In human volunteers, 1 mg nabilone produced relaxant and sedative effects. At higher doses, notably at 5 mg, postural hypotension, euphoria and dry mouth were observed. However, tolerance to these effects was noted after 7 days [162, 165].

Nabilone has a low abuse potential, as expected from the known low abuse potential of *oral* THC [166]. There is a considerable divergence in view as regards the euphoriant effect level of nabilone on comparison with that of THC. While one group has concluded that nabilone is 7-times more potent than THC, others have found it less so [162]. It is of more than casual interest that the antiemetic effect ratio of the two drugs is in the same range: 2 mg dose for nabilone, 15–20 mg for Δ^1-THC.

The antiemetic effect of nabilone in the clinic is well established in numerous studies. In an early study with 113 patients undergoing cancer chemotherapy with a wide variety of anticancer drugs, 80% experienced full or partial improvement of nausea and emesis while only 36% responded to prochlorperazine treatment. Drowsiness and dizziness were reported in many cases. Euphoria (16%), dry mouth and blurred vision (4.5%), orthostatic hypotension (1%) and visual hallucinations (1%) were the more serious side-effects [167]. This general picture has been repeated over and over again.

Nabilone is only partially effective in vomiting due to cisplatin [168] although it may be better than Δ^1-THC. Recently, nabilone has been administered together with metoclopramide, and with dexamethasone [162] as well as with prochlorperazine [162a], and the results appear to be encouraging.

Several other synthetic cannabinoids have reached the clinical stage. Levonantradol (9), which was discussed above, is, in addition to being an analgetic, also a potent antiemetic [169–171]. It is the only cannabinoid available for parenteral use. This is of considerable clinical importance, as oral administration is certainly not efficient in patients who are already vomiting. It appears to be slightly more effective than THC; however, because of the pronounced side-effects, it will probably not be introduced as a drug on the market.

A very well controlled and executed recent clinical trial is described as an example [170]. Levonantradol (1 mg every 4 h) was administered by i.m. injection and was compared with Δ^1-THC (15 mg every 4 h) administered orally in a double-blind, crossover study. The patients received severely emetogenic chemotherapy. With either drug *ca.* 25% of the patients had no nausea; the median number of emetic episodes was 2.0 (for levonantradol) and 3.0 (for Δ^1-THC). Almost all patients experienced some side-effects, with drowsiness and dizziness affecting *ca.* 50% of the patients, dysphoria (depression, anxiety, confusion) *ca.* 30%, dry mouth *ca.* 25% and hypotension *ca.* 20%. The side-effects appear to be somewhat lower with THC than with levonantradol.

Two additional compounds, (40) and (41), have received limited clinical trials, but although the results were encouraging it seems that their further development has been stopped [7, 172].

Although in the past Cannabis was used as a general antiemetic drug, today the two commercial antiemetics, Δ^1-THC and nabilone, are prescribed only in cases of nausea and vomiting caused by anticancer treatments. This is probably due to the somewhat exaggerated apprehensiveness of side-effects. In view of the lower cannabimimetic activity of Δ^6-THC (as compared with Δ^1-THC), we decided to try this THC isomer on patients with various conditions. We administered Δ^6-THC to two adult patients with intractable nausea and vomiting. One of the patients was a 60-year-old male physician with terminal primary amyloidosis; the second, a 51-year-old female teacher with terminal intestinal carcinoma; she was not under anticancer treatment any more. Both cases received 10 mg Δ^6-THC once or twice daily orally, in gelatine capsules with olive oil for up to 30 days. No further vomiting occurred and the nausea was slight [173].

In collaboration with Professor A. Avramov of the Pediatric Department of the Bikur Holim Hospital in Jerusalem, we administered Δ^6-THC, 3.5–5 mg total oral dose to 20 children, 0.5–10 years old, who had been admitted for various conditions (mostly G.I. infections) and whose vomiting could have led to dehydration. In 19 cases the vomiting ceased within *ca.* 30 min. In a few cases vomiting restarted after 6 h when a second dose was administered. No side-effects were observed [174].

It is doubtful whether additional cannabinoids will be introduced as antiemetic drugs in cancer chemotherapy until a better separation between the side-effects and the antiemetic properties is achieved or a cannabinoid with a potent action against emetogenic agents such as cisplatin is discovered. However, it is quite possible that cannabinoids – natural or synthetic – will be introduced as antinausea and antiemetic drugs in general medical practice outside the field of oncology. The dose levels needed for such activity may prove to be much below the doses causing cannabimimetic effects.

THE NEUROLOGICAL IMPLICATIONS OF CANNABINOIDS

Several cannabinoids have been found to be clinically useful in treating a number of neurological entities, including seizure and movement disorders. Their use in this regard is reviewed below.

SEIZURE DISORDERS

A seizure is a paroxysmal, transient disorder of the brain's normal electrical discharge, resulting in an alteration of motor, sensory and autonomic function. Seizures are dichotomized into partial seizures (largely restricted to one region or focus of cerebral tissue) and generalized seizures (which comprise petit and grand mal subtypes, and involve diffuse bilateral brain regions). Both partial and generalized (grand mal) seizures are due *inter alia* to disordered neurotransmission in a number of central pathways, especially those involving GABA mediated inhibition. Petit mal or absence seizures are typically slight, rarely involve autonomic responses, and are quite transient. Grand mal seizures are far more severe, usually involving a loss of consciousness and prolonged muscle spasms varying from those seen in minimal generalized seizures (i.e., tonic spasms or involuntary alternating muscular contractions and relaxations) to those of maximal generalized seizures (i.e., tonic spasms or continuous prolonged muscular contractions).

The primary drugs of choice currently used to treat seizure disorders include phenytoin, phenobarbital, carbamazepine, ethosuximide and valproic acid. However, all have some limitation associated with their use, including (a) an inability to adequately control the seizure, even in combination with other drugs; (b) untoward side-effects involving arousal, motor performance and mood; (c) some drug toxicity; and (d) an inability to provide effective long-term, control. The need for new highly effective and specific anti-convulsants with low psychotropic and sedative lability is evident, and has led to the investigation of cannabinoids for this purpose, especially cannabidiol (CBD) and Δ^1-THC.

The preclinical investigation of a newly proposed antiepileptic in experimental animals typically involves the use of electrical stimulation (electroshock) to induce maximal generalized seizures (MES), and the systemic administration of convulsant drugs (e.g., pentylenetetrazole) to elicit minimal seizures. The use of animals genetically susceptible to seizures caused by light or specific noise is also prevalent [175]. The efficacy of the antiepileptic is then tested against the seizures induced. (For leading references see [175].)

In MES models, both Δ^1-THC and CBD quite consistently block maximal (tonic) seizures and also elevate the electroshock seizure threshold after acute administration, but have only variable effects on minimal (clonic) seizures. In mice, both drugs elevated the electroshock seizure threshold at 6 Hz, but not at 60 Hz. The acute anticonvulsant profile of CBD was most like that of the standard antiepileptic phenytoin [176, 177].

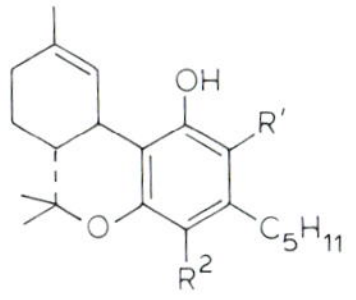

(42) a, R^1 = COOH, R^2 = H
b, R^1 = H, R^2 = COOH

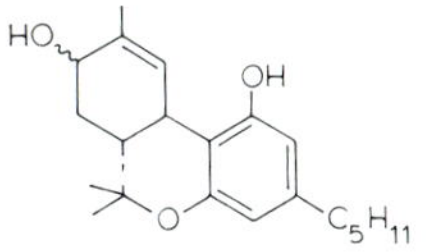

(43) a, 6—OH, α
b, 6—OH, β

MES activity was also attenuated by other cannabinoids [178]: Δ^6-THC and 7-nor-Δ^6-THC (23) were about as active as Δ^1-THC; 7-OH-Δ^1-THC (15) and the 1,2-dimethylheptyl homologues of Δ^3-THC (5) were 7–14-times more potent than Δ^1-THC. Surprisingly, the natural Δ^1-THC acids A and B (42a and 42b, respectively), which have no psychotropic activity, were also active in the MES test, though 2–4-times less so than Δ^1-THC. A certain stereospecificity was also observed. While 6α-OH-Δ^1-THC (43a) was inactive, 6β-OH-Δ^1-THC (43b) was as active as Δ^1-THC. Contrary to results in analgesia (*vide supra*) *both* C-1 isomers of 7-nor-1-hydroxyhexahydrocannabinol ((24) and (25)) were equiactive with Δ^1-THC in the MES test. The obvious conclusion is that the SAR of THC derivatives in the MES anticonvulsant test in mice does not parallel that of other biological tests with these derivatives.

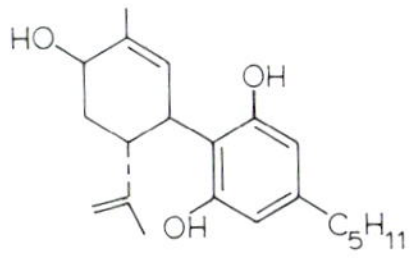

(44)

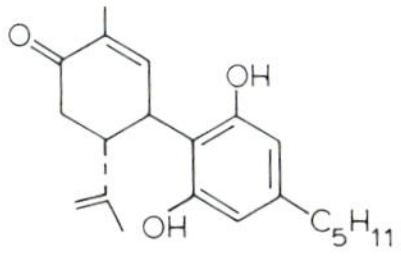

(45)

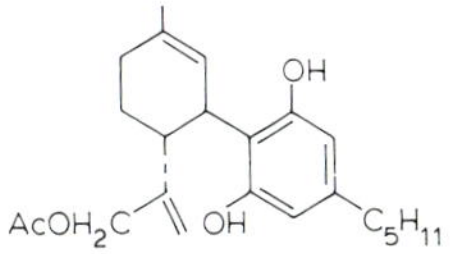

(46)

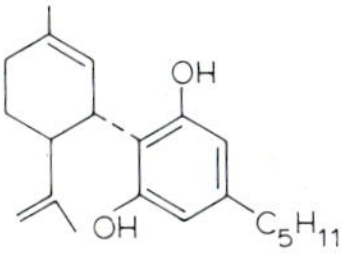

(47)

The anticonvulsant SAR of CBD-type compounds bears no relationship to either the cannabimimetic or the anticonvulsant activity of THC-type compounds. As mentioned above, CBD itself, which is non-cannabimimetic in animal tests as well as in humans, exhibits significant activity in the MES test [178–180]. The CBD derivatives (44), (45) and (46) were equiactive with CBD [179]. Unexpectedly, the unnatural (+)-enantiomer of CBD (47) was also shown to be at least as active as natural CBD in the MES test [180]. The DMH homologues of both (−)-CBD and (+)-CBD were several times more active than CBD [180]. These and numerous other side-chain homologues of (+)-and (−)-CBD were also shown to be active in audiogenic seizure susceptible (AGS) rats [181]. These observations are particularly striking since the unnatural (+)-enantiomer of THC as well as its DMH homologue (38) are much less active than natural (−)-Δ^1-THC in such rats. The highly purified, crystalline (+)-7-hydroxy-Δ^6-THC-DMH (33) is completely inactive in the AGS rats, while the (−)-enantiomer (32) is extremely potent [106].

There are various measures of antiepileptic drug effect in current use, including the anticonvulsant ED_{50} value; the neurotoxic TD_{50} value; and the protective index PI (ratio of ED_{50} value to TD_{50} value, i.e., the margin of safety for clinical or preclinical use). The ED_{50} value of cannabinoids against MES and the TD_{50} value of these drugs may vary considerably with regard to both the species used and the route of administration. However, CBD consistently has a PI > 1, irrespective of route or species, indicating a greater selectivity of anticonvulsant action relative to neurotoxicity. Furthermore, the PI of natural (−)-CBD has been consistently found to be comparable with that of widely used antiepileptics against both partial and grand mal seizures, such as phenobarbital, phenytoin and carbamazepine [176, 177, 182]. The PI of (+)-CBD is higher than that of (−)-CBD [181].

Chronic cannabinoid administration may cause either tolerance or reverse tolerance. After 3 days of treatment, tolerance developed to the MES effects of Δ^1-THC and CBD in mice [183]. After 22 days of Δ^1-THC, tolerance developed to its anti-convulsant effects at 6 Hz, while with CBD an anti-convulsant reverse-tolerance occurred at 60 Hz. Apparently, CBD is superior to Δ^1-THC not only in being relatively free of psychotropic effects but also in inducing less anticonvulsant tolerance after prolonged use [184].

In some seizure models using convulsant drugs, acute CBD usually blocks the tonic convulsion and lethality induced by a number of GABA antagonists [185]. However, the seizure endpoint of clonic convulsion produced by these drugs in mice and rats is not prevented by acute CBD and is neither effected nor even potentiated by Δ^1-THC. Moreover, Δ^1-THC offers no protection against the clonus and/or tonus and lethality of lidocaine, picrotoxin, nicotine and strychnine [186]. However, Δ^1-THC, Δ^6-THC and the THC metabolites 7-OH- or 7-oxo-Δ^1-THC delayed the latencies to pentylenetetrazole-(PTZ)-induced tonic and clonic seizures and reduced the incidences of PTZ-evoked tonic seizures and lethality, being more potent for all of these effects than Δ^1-THC and Δ^6-THC, thus underscoring the importance of metabolism in THC anticonvulsant activity [187].

Focal (i.e., partial) seizure activity may be elicited by several means, including the topical application of metals to the cerebral cortex to induce cortical focal seizures and the repetitive electrical stimulation (kindling) of limbic structures.

Acute Δ^1-THC has been found to inhibit (iron model) [188] or enhance (cobalt and aluminum) [189] focal potentials. Short-term Δ^1-THC in the cobalt model also produced anticonvulsant effects, probably by reducing the spread of seizures from cortical foci. In contrast, CBD did not affect the spontaneously firing foci in cobalt- or iron-induced epileptic rats [190]. On the other hand, CBD abolished spontaneous clonic convulsions in the cobalt model, probably by depressing seizure generation or spread. CBD also abolished convulsions in the iron-induced epileptic rat, and in this respect was even superior to phenytoin.

In contrast to its prophylactic effect in cats [191], Δ^1-THC in amygdaloid-kindled rats failed to suppress seizures in non-toxic doses, exerted no prophylactic effect on seizure development, and also induced tolerance after repeated administration [192]. In hippocampal-kindled cats, Δ^1-THC elevated seizure thresholds, but only following weak stimulus intensities [193].

The effects of Δ^1-THC, CBD and phenytoin were examined on kindled limbic convulsions in rats [194, 195]. Δ^1-THC augmented the threshold but had no effect on amplitude. It also bilaterally increased the duration of after-discharges at a number of central sites. In contrast, CBD diminished both the amplitude and duration of focal after-discharges and did not affect after-discharge durations at distant sites. By comparison, the electrophysiological effects of phenytoin were generally similar to those of Δ^1-THC on all parameters. Thus, while CBD tends to produce only a selective depressant effect on kindled seizures generally, Δ^1-THC and the standard anti-epileptic drug, phenytoin, are generally associated with stimulatory responses (either alone or in combination with depressant responses). Thus, CBD has more selective antiepileptic activity than either Δ^1-THC or standard anti-epileptic drugs.

In 1980, a controlled study was published concerning the effects of CBD in 15 epileptic patients having inadequately maintained complex partial seizures with secondary generalizations. Patients received either oral CBD (200–300 mg) or placebo daily in a double-blind fashion for up to 4.5 months, and were also maintained on standard antiepileptic drugs. Of the 8 patients receiving CBD, 4 were free of convulsions, 3 had partial improvement, and only 1 was unchanged. The clinical condition of 6 of the 7 placebo patients remained unchanged, with only 1 improved. No signs of toxicity or serious side-effects were found. While CBD was administered as a supplemental treatment, it is significant that improvement was seen in these patients who were previously inadequately controlled by standard antiepileptics without CBD. This is potentially quite important, as complex partial seizures with secondary generalization are difficult to treat with currently used drugs [196].

MOVEMENT DISORDERS

Movement disorders refer to those clinical syndromes involving (a) a deficit in non- or extrapyramidal function and (b) at least one non-epileptic abnormal movement. Typical symptoms include akinesia or bradykinesia, ataxia, catalepsy, chorea, spasm, tremor and dystonia (slow involuntary muscle contractions producing abnormal posture or position), either focal or generalized. Many movement disorders are poorly controlled with current therapies, especially those with dystonia, spasticity and iatrogenic dyskinesia, and better drug therapies are clearly indicated.

Marihuana, Δ^1-THC and THC analogues actually induce a variety of neurological effects in normal animals, usually manifested as alterations of motor function and altered locomotor activity, impaired co-ordination and balance, decreased muscle tone and reflexes and, at very high doses, tremors, myoclonic jerks, and convulsions. CBD is very considerably less potent than Δ^1-THC in producing these abnormalities [53–56].

Both central and peripheral sites of action have been implicated in these alterations of motor function. In the isolated sciatic nerve–sartorius muscle preparation of the frog, Δ^1-THC blocked neuromuscular transmission by depressing sartorius presynaptic acetylcholine release [197]. However, the precise mechanisms of cannabinoid action at cellular and neurobehavioural levels are as yet unknown.

Ataxia, tremors of the body and hands, and muscle weakness occur rather frequently after Δ^1-THC and may be considered side-effects. Occasionally, myoclonus is also seen. However, such effects have *not* been reported for CBD in humans [196].

In 1981, the results of a double-blind experiment with a crossover design were reported, with 9 multiple sclerosis (MS) patients given Δ^1-THC (5 or 10 mg) or placebo [198]. Relative to placebo, 10 mg of the drug significantly reduced spasticity as quantified by clinical and electromyographic measurements, and side-effects were negligible. The authors also found clinical benefit in 3 additional patients with tonic spasms who were administered Δ^1-THC. Finally, preliminary data from an uncontrolled study in progress on the efficacy of chronic CBD in the treatment of dystonia suggest that this cannabinoid may improve symptoms of both tonic (dystonic posture and pain) and phasic (dystonic spasms and tremor) components of this disorder. Reported side-effects of CBD were mild and did not include disturbances of autonomic function and mood associated with Δ^1-THC [175, 199].

In summary, THC-like cannabinoids exhibit antiepileptic activity in several animal models. The SAR observed differ considerably from those noted for cannabimimetic or analgetic activity in these series. The unnatural (+)-(3*S*,4*S*)-THCs exhibit either very low, or no activity in MES or AGS tests.

CBD-like cannabinoids also exhibit antiepileptic activity in several animal models, in contrast to lack of activity in cannabimimetic or analgetic models, in which these compounds are inactive. Surprisingly, the unnatural (+)-CBD's are at least as active as the (−)-enantiomers in all the animal models for antiepileptic activity in which they have been tested.

Hence, of all the cannabinoids, CBD has by far the greatest therapeutic potential in the treatment of neurological disorders. Unlike Δ^1-THC, acute CBD is free of psychotropic, cardiac and other major untoward effects in humans. In animals, CBD exhibits more selective anticonvulsant activity relative to neurotoxicity than Δ^1-THC and in this regard is similar to several well-established antiepileptic drugs such as phenytoin and phenobarbital. Tolerance does not readily develop to CBD as it does to Δ^1-THC, and CBD does not cause central stimulation or convulsions in seizure models as Δ^1-THC does. In kindled limbic epilepsy, CBD has more complete anti-seizure activity at the limbic focus and is superior to both Δ^1-THC and phenytoin and ethosuximide in producing no excitatory effects on seizure generation or spread. Moreover, CBD has selective depressant effects on neurotransmission in the CNS which, together with its anticonvulsant effects in animal models of epilepsy, suggests an inhibition of seizure spread as an additional mode of action.

Preclinical data suggest that CBD would be effective for partial seizures and generalized grand mal seizures, but not for petit mal, where it may even reduce the efficacy of currently used antiepileptic drugs. In patients, antiepileptic activity is observed with CBD at high doses. Hence, future research should aim at obtaining more active compounds with a CBD profile of activity.

Cannabinoids may also prove beneficial in the treatment of movement disorders, with CBD causing only selective depression of activity at central and peripheral sites associated with the induction of abnormal movements and muscle tone, in contrast to Δ^1-THC, which elicits both stimulatory and depressant effects.

Muscle spasms, particularly those associated with multiple sclerosis or dystonia, may improve significantly with cannabinoid therapy. However, while CBD has been found effective in combination with standard therapy (which was insufficient to adequately control dystonic motor symptoms without CBD), both Δ^1-THC and marijuana were found to have untoward side-effects at the doses required to reduce muscle tone. Additionally, CBD may prove beneficial for other hyperkinetic movement disorders such as drug-induced dystonia and possibly tardive dyskinesia.

It appears, then, that CBD may prove of significant benefit in the treatment of a variety of neurological dysfunctions, principally those involving seizure and movement disorders, and in this regard is superior to both marijuana and Δ^1-THC in both its selectivity of action and its absence of adverse side-effects.

Finally, in light of (a) the extremely different clinical profiles of CBD and Δ^1-THC in treating neurological entities, (b) the independence of CBD on stereospecificity for its therapeutic efficacy as compared with Δ^1-THC, (c) the relative abundance of psychotropic and neurological side-effects produced by Δ^1-THC but not by CBD and (d) the difference between these two drugs in generating tolerance to their therapeutic and side-effects, it is quite apparent that these two cannabinoids act by different mechanisms. Moreover, while the stereospecificity of Δ^1-THC for its (−)-isomer suggests that it acts at a central putative cannabinoid receptor, the similar biological activity found for both the natural (+)- and natural (−)-isomers of CBD means that it is unlikely to act directly through a receptor-linked mechanism.

GLAUCOMA

Glaucoma is one of the primary causes of blindness in the U.S.A., with over two million people debilitated in that country alone. Glaucoma generally involves a progressive increase in intraocular pressure (IOP), producing a continuous deterioration of vision that results, in some instances, in complete blindness. Primary in the development of glaucoma is a disruption in the equilibrium between the formation of aqueous humour and its outflow from the anterior chamber. Increased IOP thus results either from an increase in aqueous humour secretion and/or from some interference with its drainage or outflow facility.

In consonance with this, the drugs most widely used to treat glaucoma (for example, pilocarpine) are parasympathetomimetics, which induce miosis and augment the drainage of aqueous humour (by enlarging the canal in the anterior chamber). Other drugs include: (a) miotics other than pilocarpine (either alone or with the β-blocker, timolol, as a concomitant treatment); (b) carbonic anhydrase inhibitors, for use in chronic and acute angle-closure glaucoma, respectively; (c) potent AChE inhibitors such as isoflurophate for aphakic eyes (with congenital absence of the lens); and (d) epinephrine (adrenaline) (often used with pilocarpine). However, all of these drugs produce tolerance and untoward side-effects (powerful miotics may cause cataracts or retinal detachment and pilocarpine may precipitate acute bronchial asthma). Thus, a need for more effective drugs with minimal side-effects and a broad spectrum of activity is clearly indicated. The cannabinoid class of compounds seems to be of interest in this area (for reviews see [200–202]).

In 1971, Hepler and Frank [203] found by serendipity that smoking marijuana decreased lacrimation and intraocular pressure (IOP) of the eye in normal subjects. This effect has been confirmed (for marijuana and THC) by numerous reports on normal subjects and in glaucoma patients by different routes of systemic administration (smoking, i.v. and orally) with all groups realizing diminution in IOP of about 20–40% which lasted several hours [204, 205]. This duration in lowering IOP is comparable with that of pilocarpine, but is much longer than that of epinephrine. Thus, the cannabinoids act as long as or longer than the most commonly used antiglaucomic agents. However, undesirable side-effects have been noted at the doses administered in virtually all studies employing Δ^1-THC, including cardiovascular effects (tachycardia, hypotension), emotional distress (acute panic and/or paranoia) and a number of ocular effects (photophobia, hyperemia of conjunctiva, decreased lacrimation, corneal ulceration, conjunctivitis, keratitis and small change in pupil size) [202]. In addition, corneal opacification has been reported in dogs and cats (but not in rats, rabbits or humans) [206, 207].

The prevalence of these side-effects is a serious limiting factor in the clinical applicability of cannabinoids by systemic administration.

The order of potency in man for the IOP action on systemic administration was found to be as follows. Δ^6-THC (6) $\geqslant$ 7-OH-Δ^1-THC (15) $\geqslant$ Δ^1-THC (1) > 6β-OH-Δ^1-THC (43b) $\geqslant$ CBN (7) [208]. The higher (or even equal) potency of Δ^6-THC as compared with Δ^1-THC is unusual, since (as mentioned before) the reverse situation holds true for cannabimimetic action. If this result is reproducible, it might indicate that the known cannabimimetic SAR does not apply to the reduction of IOP.

An effort to diminish the side-effects attending systemic administration was by direct topical administration of the cannabinoids opthalmically. As Δ^1-THC is insoluble in water, initial attempts at topical administration consisted of suspending the Δ^1-THC in mineral oil as eye drops. However, no significant effect was found on the IOP of visually normal subjects [209, 210]; and while a drop in the IOP of glaucomic patients was noted, there was evidence of systemic absorption as a similar decrease in IOP was seen in the contralateral non-drug treated eye [211]. Later attempts at topical administration focussed on developing water-soluble derivatives of Δ^1-THC. Although some compounds were topically active in rabbits, no topically active cannabinoid was found to act reproducibly on man. Thus, naboctate (41), a water-soluble cannabinoid ester, is active systemically but apparently not topically [212].

A considerable amount of research has been done on the reduction of IOP by cannabinoids in animals, much of it led by Green [200]. These studies have demonstrated a considerable similarity to the action of THC in humans regarding its efficacy, latency and duration of action in lowering IOP. However, while topical application in rabbits causes reduction of IOP, this is (as mentioned above) not the case with humans.

On i.v. administration to rabbits, the following order of activity was found: 7-OH-Δ^1-THC (15) > Δ^1-THC (1) > 6,7-dihydroxy-Δ^1-THC > 6β-OH-Δ^1-THC (43b). CBD was inactive [213].

On topical application in mineral oil, a somewhat different order of activity was recorded: Δ^6-THC ⩾ Δ^1-THC ⩾ 6α-OH-Δ^1-THC > CBD ⩾ 7-OH-Δ^1-THC > 6,7-dihydroxy-Δ^1-THC > 7-OH-Δ^6-THC > 6β-OH-Δ^1-THC [214]. These results do not follow the well-established cannabimimetic SAR. This conclusion is strengthened by the recent observation that the naturally occurring cannabigerol (48), which is not cannabimimetic, reduces IOP in the cat and moreover showed reverse tolerance (in being more effective after chronic administration than in acute one) [215].

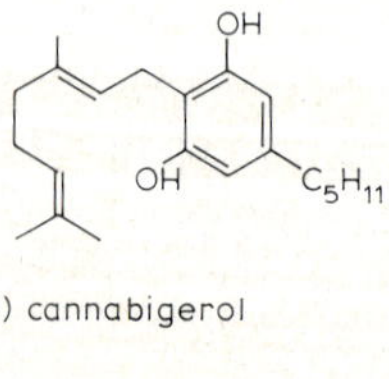

OH
OH
C_5H_{11}

(48) cannabigerol

A water-soluble non-cannabinoid fraction from *Cannabis sativa* has been found to reduce IOP [216–219]. It is apparently a glycoprotein. It seems strange that a plant should contain both small molecules (cannabinoids) and biopolymers with the same kind of activity.

The site of action of the cannabinoids is uncertain. Clear evidence is extant that some (or perhaps most) of the activity of Δ^1-THC derivatives is confined locally to the eye itself. However, there is also evidence of a central mode of action [201–203]. Thus, the study cited above of a crossover effect on IOP of THC applied to one eye and effecting IOP in the untreated contralateral eye strongly suggests that systemic absorption occurred, with a concomitant action exerted in the CNS [211]. On the other hand, cannabigerol reduces IOP, despite its having only minimal CNS effects [215]. Thus, the issue is far from settled.

The mode of action of cannabinoids in reducing IOP has also been examined, and apparently involves a diminution in the formation of aqueous humour (this applies equally to virtually all cannabinoids and marijuana-derived mixtures effecting IOP) and possibly also an increased drainage from the anterior chamber [220], though this is controversial, probably because of the use of different animal species in the conflicting studies [200]. One study suggested that the diminution in fluid production is due to an effect on β-adrenoceptors, and enhanced total outflow is the result of an action exerted on α-adrenoceptors [200, 221]. However, there are a number of problems associated with this rather over-simplified adrenoceptor attribution [201]. Thus, the mechanism underlying these effects remains largely unknown and awaits further studies for clarification.

Studies of tolerance have been quite contradictory and scarce in both rabbits and humans [202], and in two reports from the laboratory of Colasanti a reverse tolerance was seen in cats [206, 215]. It is therefore difficult to draw any conclusions at this point regarding the development of tolerance to the lowering of IOP by cannabinoids until more systematic, well-controlled studies become available.

In summary, the cannabinoids tested so far appear to be of limited use in the treatment of glaucoma. They appear to act only against a primary (but not sole) symptom of the disease (i.e., ocular hypertension), rather than against the underlying disease process, which remains uncertain. The side-effects of those cannabinoids particularly effective in lowering IOP restrict their clinical usefulness (with some exceptions such as cannabigerol). Cannabinoids administered intraocularly to humans cause no IOP reduction. The future development of new cannabinoids should focus on the topical application of very long-acting THC derivatives having low tolerance potential, minimal side-effects and water solubility to circumvent their inherent inability to penetrate the eye.

BRONCHODILATORY EFFECTS OF CANNABINOIDS

Δ^1-THC has been recognized for some years as a bronchodilator, with most of the relevant clinical studies having been conducted over the past 14 years (for a recent review see [222]). While presently undergoing re-evaluation and development, it has not inspired great confidence in this regard, as evidenced by such articles as 'THC as a bronchodilator. Why bother?' [223]. Δ^1-THC has equally good bronchodilatory effect when administered intravenously or by inhalation in aerosol form [224, 225], and both routes are superior to oral administration, which is unsatisfactory for this purpose [226]. While inhalation of Δ^1-THC in aerosol form has produced good results, there are drawbacks to smoke, which may adversely affect lung physiology in normal subjects when administered chronically [227] and may arouse coughing or constriction initially in asthmatics [228].

In comparison with β-stimulants, 2% marijuana administered by inhalation proved superior to isoprenaline, increasing conductance for more than 2 h and exhibiting a dilator effect slower in onset but lasting longer than isoprenaline [229]. However, oral THC proved less reliable than oral salbutamol [230]. In a double-blind study on stable asthmatics, 200 μg of Δ^1-THC was compared with 100 μg of salbutamol delivered in a fixed volume of 63 μl freon per inhalation. The onset of salbutamol was found to be faster, but the drugs were equiactive after 1 h and the THC effect was more prolonged [225]. Side-effects were minimal, though some initial irritation and coughing were noted. It was subsequently found that the irritation could be reduced by administering two successive 50 μg inhalations of Δ^1-THC [231].

There have been only a few comparative studies of the relative bronchodilatory efficacy of different cannabinoids [232]. One of the reasons may be the absence of suitable, generally accepted *in vivo* animal models, although *in vitro* models have been used [232a, b]. One of the clinical studies found that Δ^6-THC in doses of 50 and 75 mg orally produced a dose-related bronchodilation comparable to that of 20 mg of Δ^1-THC orally but with considerably less psychotropic and cardiovascular side-effects. On the other hand, cannabinol and cannabidiol were inactive, even at doses as high as 1200 mg. The study was conducted in healthy males who used marihuana recreationally.

The side-effects of Δ^1-THC were found in another study [233] to be reduced by combining a low dose of Δ^1-THC (5 mg) with cannabidiol (400 mg). However, no interaction was seen on bronchodilation. Apparently, nabilone (39) produces considerably less bronchodilation than Δ^1-THC. In a recent study comparing the bronchial effects of nabilone and the β-stimulant, terbutaline, on normal and asthmatic subjects, 2 mg of nabilone given orally produced only a slight bronchodilation in normal subjects and no effect in asthmatics. By comparison, terbutaline produced twice as much conductance (dilation) compared to controls [234].

Side-effects regularly encountered after high doses of Δ^1-THC specifically involving respiration appear to involve some irritation and coughing when first inhaled (*vide supra*), but this is generally of only brief duration and is seen only initially.

There does not appear to be a single mode of action responsible for the bronchodilatory effect of the cannabinoids. Though Δ^1-THC is not an acetylcholine antagonist on smooth-muscle receptors, it does reduce its release from autonomic fibres in the smooth muscle of guinea-pig ileum [235]. If this generalizes to the bronchioles, Δ^1-THC should produce dilatation by reducing constriction (which it does). The evidence that bronchial muscle relaxation is mediated by adrenergic receptors is contradictory at best, with various studies showing no effect [223, 236] or inhibition [237] induced by β-adrenoceptor antagonists on THC-elicited bronchial muscle relaxation.

In summary, relative to other therapeutic areas, studies of the use of cannabinoids as bronchodilators are inexplicably sparse. There are virtually no double-blind studies comparing their efficacy with that of β-stimulants and only very few comparative studies using naturally occurring cannabinoids other than Δ^1-THC. At present, the most effective cannabinoid by far, natural or synthetic, appears to be Δ^1-THC, especially when given by inhalation in an aerosol with low-dose multiple inhalations to reduce its side-effects. The few comparative studies to appear in the literature indicate that, while its onset of action may be slower than that of some β-stimulants, Δ^1-THC has at least as much activity and its duration of action may be considerably longer. Moreover, while the cannabinoids elicit relatively mild side-effects when properly administered, those of the β-adrenoreceptor stimulants are far more severe, including heart palpitations, exacerbations of angina, thyroid toxicosis and diabetes, hypertension and cardiac arrhythmia. However, as no reliable animal tests for bronchodilation caused by cannabinoids are in general use, this field of research may continue to lag behind those described in the previous sections.

CONCLUSION

Several of the traditional medicinal uses of cannabis have been substantiated by intense research in recent years. As a result two cannabinoids, a natural and a synthetic one, Δ^1-THC and nabilone, have become official drugs. We strongly believe that in the future we shall see cannabinoid analgetics, anticonvulsants, antiglaucoma agents and additional antiemetics on the market. However, it is quite possible that such drugs will be used for specific complaints (migraine, for example) rather than for all types of pain. These developments will depend on the separation by molecular modification of the multiplicity of pharmacological actions of active cannabinoids. Several avenues to such separations have been indicated throughout this chapter.

When the biochemical basis of cannabinoid action(s) is elucidated, we may be able to advance the field by an approach more sophisticated and intellectual than those used hitherto. But such a breakthrough seems to be far away.

REFERENCES

1 R.E. Schultes and A. Hofmann, The Botany and Chemistry of Hallucinogens (Charles C. Thomas, Springfield, IL) 2nd Edn. (1980) pp. 82–116.
2 E.L. Abel, Marihuana, The First Twelve Thousand Years (Plenum Press, New York, 1980).
3 R. Mechoulam, ed., Cannabinoids as Therapeutic Agents (CRC Press, Boca Raton, FL, 1986).
4 S. Cohen and R.C. Stilman, eds., The Therapeutic Potential of Marihuana (Raven Press, New York, 1976).
5 R. Mechoulam and E.A. Carlini, Naturwissenschaften, 65 (1978) 174.
6 Marihuana and Health – Report of a Study by a Committee of the Institute of Medicine (National Academy, Washington, 1982).
7 R.K. Razdan, Med. Res. Rev., 3 (1983) 119.
8 B.R. Martin, Pharmacol. Rev., 38 (1986) 45.
9 R. Campbell Thompson, A Dictionary of Assyrian Botany (British Academy, London, 1949) pp. 220–223.
10 H. von Deines and H. Grapow, Grundriss der Medicine der Alten Agypter VI. Worterbuch der Agyptischen Drogennamen (Academie Verlag, Berlin, 1959) p. 493.
11 T.F. Brunner, Bull. Hist. Med., 47 (1973) 344.
12 Dioscorides, The Greek Herbal, ed. R.T. Gunter (Oxford University Press, Oxford, 1934) pp. 390–392.
13 Al-Kindi, The Medical Formulary or Aqra-ba-dhin, translated with a study by M. Levey (University of Wisconsin Press, Madison, WI, 1966) p. 194.
14 F. Rosenthal, The Herb Hashish versus Medieval Muslim Society, (E.J. Brill, Leiden, 1971).
15 E.A.W. Budge, ed., The Book of Medicines (Oxford University Press, Oxford) Vol. 2 (1913) p. 189.
16 M. Touw, J. Psychoact. Drugs, 13 (1981) 23.
17 H.L. Li, in Cannabis and Culture, ed., V. Rubin (Mouton, The Hague, 1975) pp. 51–62.
18 F. Hubotter, Chinesisch-Tibetisch Pharmacologie und Receptur (Karl Hang Verlag, Ulm, 1957).
19 S. Julien, C.R. Acad. Sci., 28 (1849) 195.
20 I.C. Chopra and R.N. Chopra, Bull. Nar., 9 (1957) 4.
21 Report of the Indian Hemp Drugs Commission, 1893–1894 (Thos. Jefferson., Silver Spring, MD, 1969) pp. 174–183, 490.

22 W.B. O'Shaugnessy, in The Bengal Dispensatory and Pharmacopoeia (Bishop's College Press, Calcutta, 1941) pp. 579–604.
23 J. Parkinson, The Theater of Plantes, An Universall and Compleate Herbal (T. Cotes, London, 1640).
24 N. Culpeper, Complete Herbal (Richard Evans, London) reprinted from an earlier version (1813).
25 W.B. O'Shaugnessy, Pharmac. J. Trans., 2 (1843) 594.
26 A. Christison, Monthly J. Med. Sci., (1851) 13, 26 and 177.
27 J. Clendinning, Med. Chirurg, Trans., 26 (1843) 189.
28 J. Grigor, Monthly J. Med. Sci., 15 (1852) 124.
29 J.R. Reynolds, Arch. Med., 2 (1859) 154.
30 J.R. Reynolds, Lancet, i (1890) 637.
31 Michel, Montpellier Med., 45 (1880) 103.
32 F. Binard, Ann. Ocul., 23 (1850) 99.
33 Grumault, Bull. Gen. Ther. Paris, (1862) 400.
34 Fronmuller, J. Materia Med., 2 (1860) 474, quoted by C.L. Winek, Clin. Toxic., 10 (1977) 243.
35 G. See, Deut. Med. Wochenschr., 60 (1890) 679 and 771.
36 H.C. Wood, Proc. Am. Philosoph. Soc., 11 (1871) 226.
37 R. Dunglison, New Remedies (Blanchard & Lea, Philadelphia, 1856) pp. 175–184.
38 J. Aulde, Ther. Gazette, (1890) 523.
39 S. Mackenzie, J. Am. Med. Assoc., 9 (1887) 732.
40 T.L. Wright, Cincinnati Lancet and Observer, 6 (1863) 73.
41 R. Adams, Harvey Lectures, 37 (1941–1942) 168.
42 A.R. Todd, Experientia, 2 (1946) 55.
43 S. Loewe, Arch. Exp. Pathol. Pharmakol., 211 (1950) 175.
44 L.J. Thompson and R.C. Proctor, N.C. Med. J., 14 (1953) 520.
45 G.T. Stockings, Br. Med. J., 1 (1947) 918.
46 D.A. Pond, J. Neurol. Neurosurg. Psychiatry, 11 (1948) 271.
47 C.S. Parker and F. Wrigley, J. Ment. Sci., 96 (1950) 276.
48 J.P. Davis and H.H. Ramsey, Fed. Proc., 8 (1949) 284.
49 J. Kabelik, Z. Krejči and F. Šantavy, Bull. Nar., 12 (1960) (No. 3) 5.
50 Y. Gaoni and R. Mechoulam, J. Am. Chem. Soc., 86 (1964) 1646.
51 A. Weissman, J. Clin. Pharmacol., 21 (1981) 159 S.
52 W.D.M. Paton and R.G. Pertwee in Marijuana, Chemistry, Pharmacology, Metabolism and Clinical Effects, ed., R. Mechoulam (Academic Press, New York, 1973) pp. 191–333.
53 W.D.M. Paton, R.G. Pertwee and E. Tylden in Ref. 52, pp. 335–365.
54 W.D.M. Paton, Annu. Rev. Pharmacol., 15 (1975) 191.
55 M.C. Braude and S. Szara, eds., Pharmacology of Marihuana (Raven Press, New York, 1976).
56 L.S. Harris, W.L. Dewey and R.K. Razdan, Handbook Exp. Pharmacol. 45/II (1977) 371.
57 L. Lemberger, Annu. Rev. Pharmacol., 20 (1980) 151.
58 R.T. Jones, Annu. Rev. Med., 34 (1983) 247.
59 G.G. Nahas, Marihuana in Science and Medicine (Raven Press, New York, 1984) pp. 109–246.
60 W.L. Dewey, Pharmacol. Rev., 38 (1986) 151.
61 S. Agurell, M. Halldin, J.E. Lindgren, A. Ohlsson, M. Widman, H. Gillespie and L. Hollister, Pharmacol. Rev., 38 (1986) 21.

62 B.R. Martin, R.L. Balster, R.K. Razdan, L.S. Harris and W.L. Dewey, Life Sci., 29 (1981) 565.
63 R.P. Walton, L.F. Martin and J.H. Keller, J. Pharmacol. Exp. Ther., 62 (1938) 239.
64 Y. Grunfeld and H. Edery, Psychopharmacology, 14 (1969) 200.
65 T.U.C. Järbe and A.J. Hiltunen, Neuropharmacology, (in press).
66 R.G. Pertwee, Brt. J. Pharmacol., 46 (1972) 753.
67 T.U.C. Järbe, J.O. Johansson and B.G. Henriksson, Psychopharmacology, 48 (1976) 181.
68 T.U.C. Järbe, M.D.B. Swedberg and R. Mechoulam, Psychopharmacol., 75 (1981) 152.
69 A.J. Hiltunen and T.U.C. Järbe, Neuropharmacology, 25 (1986) 133.
70 A. Weissman, in Stimulus Properties of Drugs, eds. F.C. Colpaert and J.A. Rosencrans (Elsevier/North-Holland Biomedical Press, Amsterdam, 1978) pp. 99–122.
71 D.J. Mokler, D.B. Nelson, L.S. Harris and J.A. Rosencrans, Life Sci., 38 (1986) 1581.
72 H. Edery, Y. Grunfeld, Z. Ben-Zvi and R. Mechoulam, Ann. N.Y. Acad. Sci., 191 (1971) 40.
73 R. Mechoulam and H. Edery, in Ref. 52, pp. 101–136.
74 R. Mechoulam, N.K. McCallum and S. Burstein, Chem. Rev., 76 (1976) 75.
75 B.R. Martin, L.S. Harris and W.L. Dewey, in The Cannabinoids: Chemical, Pharmacologic and Therapeutic Aspects, eds. S. Agurell, W.L. Dewey and R.E. Willette (Academic Press, Orlando, FL, 1984) pp. 523–544.
75a R.K. Razdan, Pharmacol. Rev., 38 (1986) 75.
76 M.R. Johnson and L.S. Melvin, in Ref. 3, pp. 121–145.
77 A. Weissman, G.M. Milne and L.S. Melvin, J. Pharmacol. Exp. Ther., 223 (1982) 516.
78 L.S. Melvin, M.R. Johnson, C.A. Harbert, G.N. Milne and A. Weissman, J. Med. Chem., 27 (1984) 67.
79 S. Houry, R. Mechoulam, P.J. Fowler, E. Macko and B. Loev, J. Med. Chem., 17 (1974) 287.
80 S. Houry, R. Mechoulam and B. Loev, J. Med. Chem., 18 (1975) 951.
81 S.P. Banerjee, S.H. Snyder and R. Mechoulam, J. Pharmacol. Exp. Ther., 194 (1975) 74.
82 K. Matsumoto, P. Stark and R.G. Meister, J. Med. Chem., 20 (1977) 17.
83 B. Loev, P.E. Bender, F. Dowalo, E. Macko and P.J. Fowler, J. Med. Chem., 16 (1973) 1200.
84 D. Korbinits, J. Szejtli, A. Szoke, S. Antus, A. Guttsegen, M. Nogrady, Z. Furst and J. Knoll, Eur. J. Med. Chem. Chim. Ther., 20 (1985) 492.
85 Z. Ben-Zvi, R. Mechoulam and S. Burstein, J. Am. Chem. Soc., 92 (1970) 3468.
86 D.J. Harvey, Rev. Biochem. Toxicol., 6 (1984) 221.
87 Z. Ben-Zvi, R. Mechoulam, H. Edery and G. Porath, Science, 174 (1971) 951.
88 R. Mechoulam, H. Varcony, Z. Ben-Zvi, H. Edery and Y. Grunfeld, J. Am. Chem. Soc., 94 (1972) 7930.
89 R. Mechoulam, Z. Ben-Zvi, S. Agurell, I.M. Nilsson, J.L.C. Nilsson, H. Edery and Y. Grunfeld, Experientia 29 (1973) 1193.
90 A. Ohlsson, S. Agurell, K. Leander, J. Dahmen, H. Edery, G. Porath, S. Levy and R. Mechoulam, Acta Pharm. Suec., 16 (1979) 21.
91 L.E. Hollister, Pharmacology, 11 (1974) 3.
92 L.E. Hollister, H.K. Gillespie, R. Mechoulam and M. Srebnik (in press).
93 R. Mechoulam, N. Lander, T.H. Varkony, I. Kimmel, O. Becker, Z. Ben-Zvi, H. Edery and G. Porath, J. Med. Chem., 23 (1980) 1068.
94 H. Edery, G. Porath, R. Mechoulam, N. Lander, M. Srebnik and N. Lewis, J. Med. Chem., 27 (1984) 1370.

95 R.S. Wilson and E.L. May, J. Med. Chem., 17 (1974) 475.
96 R.S. Wilson, E.L. May, B.R. Martin and W.L. Dewey, J. Med. Chem., 19 (1976) 1165.
97 R.D. Ford, R.L. Balster, W.L. Dewey, J.A. Rosencrans and L.S. Harris, in ref. 75, pp. 545–561.
98 H.G. Pars, R.K. Razdan and J.F. Howes, Adv. Drug Res., 11 (1977) 97.
99 K. Matsumoto, P. Stark and R.G. Meister, J. Med. Chem., 20 (1977) 27.
100 W.L. Dewey, B.R. Martin and E.L. May, in CRC Handbook of Stereoisomers: Drugs in Psychopharmacology, ed. D.F. Smith (CRC Press, Boca Raton, FL, 1984) pp. 317–326.
101 R. Mechoulam and Y. Gaoni, Tetrahedron Lett., (1967) 1109.
102 R. Mechoulam, P. Braun and Y. Gaoni, J. Am. Chem. Soc., 89, (1967) 4552.
103 G. Jones, R.G. Pertwee, E.W. Gill, W.D. Paton, I.M. Nilsson, M. Widman and S. Agurell, Biochem. Pharmacol., 23 (1974) 439.
104 R. Mechoulam, N. Lander, M. Srebnik, A. Breuer and J. Zahalka, (unpublished observations).
105 T.U.C. Järbe (unpublished observations).
106 P. Consroe and R. Mechoulam, U.S. National Institute on Drug Abuse Monograph, Washington, DC (in press). Lecture presented on Oct. 31, 1986, Washington, DC.
107 M. Segal and R. Mechoulam (unpublished observations).
108 W.D.M. Paton and R.G. Pertwee in Ref. 52, p. 259.
109 D.K. Lawrence and E.W. Gill, Mol. Pharmacol., 11 (1975) 595.
110 D. Bach, I. Bursuker and R. Goldman, Biochim. Biophys. Acta, 469 (1977) 171.
111 D. Bach, A. Raz and R. Goldman, Biochem. Pharmacol., 25 (1976) 1241.
112 I. Tamir, D. Lichtenberg and R. Mechoulam, in NMR Spectroscopy in Molecular Biology, ed. B. Pullman (D. Reidel, Dordrecht, 1978) pp. 405–422.
113 E.P. Bruggemann and D.L. Melchior, J. Biol. Chem., 258 (1983) 8298.
114 S.H. Burstein and S.A. Hunter, Rev. Pure Appl. Pharmacol. Sci., 2 (1981) 155.
115 R. Mechoulam, M. Srebnik and S. Burstein, in Marihuana '84, ed. D.J. Harvey (IRL Press, Oxford, 1985) pp. 1–12.
116 S. Burstein, S.A. Hunter and L. Renzulli, J. Pharmacol. Exp. Ther., 235 (1985) 87.
117 A.V. Revuelta, D.L. Cheney and E. Costa, Life Sci., 30 (1982) 1841.
118 A.V. Revuelta, D.L. Cheney, E. Costa, N. Lander and R. Mechoulam, Brain Res., 195 (1980) 445.
118a M.B. Bowers and F.J. Hoffman, Brain Res., 366 (1986) 405.
118b B.K. Koe, G.M. Milne, A. Weisman, M.R. Johnson and L.S. Melvin, Eur. J. Pharmacol., 109 (1985) 201.
118c B.B. Sethi, J.K. Trivedi, P. Kumar, A. Gulati, A.K. Agarwal and N. Sethi, Biol. Psychiatry, 21 (1986) 3.
118d A.J. Verberne, M.R. Fennessy, S.J. Lewis and D.A. Taylor, Pharmacol. Biochem. Behav., 23 (1985) 153.
119 L.S. Harris, R.A. Carchman and B.R. Martin, Life Sci., 22, (1978) 1131.
120 J.S. Nye, H.H. Seltzman, C.G. Pitt and S.N. Snyder, J. Pharmacol. Exp. Ther., 234 (1985) 784.
121 R. Christison, A Dispensatory or a Commentary on the Pharmacopoeias of Great Britain (A. & C. Black, Edinburgh, 1848) p. 117.
122 F. Ames, J. Ment. Sci., 104 (1958) 972.
123 A.W.D. Avison, A.L. Morrison and M.W. Parkes, J. Chem. Soc., (1949) 952.
123a E.C. Taylor, K. Lenard and B. Loev, Tetrahedron, 23 (1967) 77.
124 H.I. Bicher and R. Mechoulam, Arch. Int. Pharmacodyn. Ther., 141 (1968) 24.

125 D.M. Buxbaum, Psychopharmacology, 25 (1972) 275.
126 W.L. Dewey, L.H. Harris and J.S. Kenedy, Arch. Int. Pharmacodyn. Ther., 196 (1972) 133.
127 I.G. Karniol and E.A. Carlini, Psychopharmacology, 33 (1973) 53.
128 R.D. Sofia, H.B. Vassar and L.C. Knobloch, Psychopharmacology, 40 (1975) 285.
129 M. Segal in Ref. 3, pp. 105–120.
130 J.M. Parker and T.C. Dubas, Int. J. Clin. Pharmacol., 7 (1973) 75.
131 B.R. Martin, Life Sci., 36 (1985) 1523.
132 P. Zeidenberg, W.C. Clark, J. Jaffe, S.W. Anderson, S. Chen and S. Malitz, Compr. Psychiatry, 14 (1973) 549.
133 R.N. Noyes Jr., S.F. Brunk, D.H. Avery and A. Canter, Clin. Pharmacol. Ther., 18 (1975) 84.
134 R. Noyes Jr., S.F. Brunk, D.A. Baram and A. Canter, J. Clin. Pharmacol., 15 (1975) 139.
135 D. Raft, J. Gregg, J. Ghia and L.H. Harris, Clin. Pharmacol. Ther., 21 (1977) 26.
136 A.W. Zuardi, I. Shirakawa, E. Finkelfarb and I.G. Karniol, Psychopharmacology, 76 (1982) 245.
137 L.S. Harris, in Mechanism of Pain and Analgetic Compounds, eds. R.F. Beers, Jr. and E.G. Basset (Raven Press, New York, 1979) pp. 467–473.
138 M. Stagnet, C. Gantt and D. Marchin, Clin. Pharmacol. Ther., 23 (1978) 397.
139 R.S. Wilson and E.L. May, J. Med. Chem., 18 (1975) 700.
140 M.R. Johnson, L.S. Melvin, T.H. Althius, J.S. Bindra, C.A. Harbert, G.M. Milne and A. Weissman, J. Clin. Pharmacol., 21 (1981) 217 S.
141 M.R. Johnson, L.S. Melvin and G.M. Milne, Life Sci., 31 (1982) 1703.
142 G.M. Milne, M.R. Johnson, B.K. Koe, A. Weissman and L.S. Melvin, Actual. Chim. Ther., 10 (1983) 227.
143 J.J. Jacob, K. Ramabadran and M. Campos-Medeiros, J. Clin. Pharmacol., 21 (1981) 327 S.
144 R. Mechoulam, N. Lander, M. Srebnik, I. Zamir, A. Breuer, B. Shalita, S. Dikstein, E.A. Carlini, J.R. Leite, H. Edery and G. Porath, in Ref. 75, pp. 777–793.
145 R. Mechoulam, N. Lander, M. Srebnik, A. Breuer, M. Segal, J.J. Feigenbaum, T.U.C. Järbe, A.J. Hiltunen and P. Consroe, U.S. National Institute of Drug Abuse Monograph, Washington, DC. (in press). Lecture presented on Oct. 31, 1986, Washington, DC.
145a W.A. Devane, J.W. Spain, C.J. Coscia and A.C. Howlett, J. Neurochem., 46 (1986) 1929.
145b M.L. Barrett, A.M. Scutt and F.J. Evans, Experientia, 42 (1986) 452.
146 M. Levitt in Ref. 3, pp. 71–83.
147 S.E. Sallan, N.E. Zinberg and E. Frei, N. Engl. J. Med., 293 (1975) 795.
148 B.J. Vincent, D.J. McQuiston, L.H. Einhorn, C.M. Nagy and M.J. Brames, Drugs (Suppl. 1) 25 (1983) 52.
149 H.L. Borison, R. Borison and L.E. McCarthy, J. Clin. Pharmacol., 21 (1981) 23 S.
150 S.W. London, L.E. McCarthy and H.L. Borison, Proc. Soc. Exp. Biol. Med., 160 (1979) 437.
151 L.E. McCarthy and H.L. Borison, J. Clin. Pharmacol., 21 (1981) 30 S.
152 L.E. McCarthy, K.P. Flora and B.R. Vishnuvajjala in Ref. 75, pp. 859–870.
153 M.R. Landauer, R.L. Balster and L.S. Harris, Pharmacol. Biochem. Behav., 23 (1985) 259.
154 P. Stark, Cancer Treat. Rev., 9 (Suppl. B) (1982) 11.
155 J.J. Feigenbaum and R. Mechoulam (unpublished observations).
156 T.U.C. Järbe (unpublished observations).
157 S.E. Salan, C. Cronin, M. Zelen and N.E. Zinberg, N. Engl. J. Med., 302 (1980) 135.
158 R.J. Gralla, L.B. Tyson, R.A. Clark, L.A. Bordin, D.P. Kelsen and L.B. Kalman, Proc. Soc. Am. Soc. Clin. Oncol., 1 (1982) 58.

159 A.E. Chang, D.J. Shiling, R.C. Stillman, N.M. Goldberg, C.A. Seipp, I. Barofsky, R.M. Simon and S.A. Rosenberg, Ann. Intern. Med., 91 (1979) 819.
160 V.S. Lucas and J. Laszlo, J. Am. Med. Assoc., 243 (1980) 1241.
161 A. Ohlson, J.E. Lindgren, A. Wahlen, S. Agurell, L.E. Hollister and H.K. Gillespie, Clin. Pharmacol. Ther., 28 (1980) 409.
162 R.A. Archer, P. Stark and L. Lemberger in Ref. 3, pp. 85–103.
162a D. Cunningham, G.J. Forrest, M. Soukop, N.L. Gilchrist and I.T. Calder, Br. Med. J., 291 (1985) 864.
163 R.A. Archer, W.B. Blanchard, W.A. Day, D.W. Johnson, E. Lavagnino, C.W. Ryan and J.E. Baldwin, J. Org. Chem., 42 (1977) 2277.
164 A.L. Thakkar, C.A. Hirsch and J.G. Page, J. Pharm. Pharmacol., 29 (1977) 783.
165 L. Lemberger and H. Rowe, Clin. Pharmacol. Ther., 18 (1975) 720.
166 M.D. Aceto, L.S. Harris and E.L. May, NIDA Res. Monogr. 41 (1982) 338.
167 T.S. Herman, L.H. Einhorn, S.E. Jones, C. Nagy, A.B. Chester, J.C. Dean, B. Furnas, S.D. Williams, S.A. Leigh, R.T. Dorr and T.E. Moon, N. Engl. J. Med., 300 (1979) 1295.
168 N. Steele, R.J. Gralla, D.W. Braun and C.W. Young, Cancer Treat. Rep., 64 (1980) 219.
169 J. Laszlo, V.S. Lucas, D.C. Hanson, C.M. Cronin and S.E. Sallan, J. Clin. Pharmacol., 21 (1981) 51 S.
170 M.L. Citron, T.S. Herman, F. Vreeland, S.H. Krasnow, B.E. Fossieck Jr., S. Harwood, R. Franklin and M.H. Cohen, Cancer Treat. Rep., 69 (1985) 109.
171 C.M. Cronin, S.E. Salan, R. Gelber, V.S. Lucas and J. Laszlo, J. Clin. Pharmacol., 21 (1981) 43 S.
172 M. Staquet, D. Bron, M. Rosencweig and Y. Kenis, J. Clin. Pharmacol., 21 (1981) 60 S.
173 Unpublished results from our laboratory.
174 A. Avramov and R. Mechoulam (unpublished results).
175 P. Consroe and S.R. Snider in Ref. 3, pp. 21–49.
176 R. Karler and S.A. Turkanis, J. Clin. Pharmacol., 21 (1981) 437 S.
177 P. Consroe, A. Martin and V. Singh, J. Clin. Pharmacol., 21 (1981) 428 S.
178 R. Karler and S.A. Turkanis, in Marihuana, Biological Effects, eds. G.G. Nahas and W.D.M. Paton (Pergamon Press, Oxford, 1979) pp. 619–640.
179 E.A. Carlini, R. Mechoulam and N. Lander, Res. Commun. Chem. Pathol. Pharmacol., 12 (1975) 1.
180 J.R. Leite, E.A. Carlini, N. Lander and R. Mechoulam, Pharmacology, 24 (1982) 141.
181 P. Consroe, A. Martin and R. Mechoulam in Ref. 115, pp. 705–712.
182 P. Consroe and A. Wolkin, J. Pharmacol. Exp. Ther., 26 (1977) 201.
183 R. Karler and S.A. Turkanis, Br. J. Pharmacol., 68 (1980) 479.
184 R. Karler, H.K. Borys and S.A. Turkanis, Naunyn-Schmiedebergs Arch. Pharmakol., 320 (1982) 105.
185 P. Consroe, M.A.C. Benedito, J.R. Leite, E.A. Carlini and R. Mechoulam, Eur. J. Pharmacol., 83 (1982) 293.
186 R.D. Sofia, T.A. Solomon and H. Barry, Eur. J. Pharmacol., 35 (1976) 7.
187 I. Yamamoto, K. Watanabe, K. Oguri and H. Yoshimura, Res. Commun. Subst. Abuse, 1 (1980) 287.
188 S.A. Turkanis and R. Karler, Neuropharmacology, 21 (1982) 7.
189 B.K. Colasanti, C. Lindamood and C.R. Craig, Pharmacol. Biochem. Behav. 16 (1982) 573.
190 P. Chiu, D.M. Olsen, H.K. Borys, R. Karler and S.A. Turkanis, Epilepsia, 20 (1979) 365.
191 J.A. Wada, A. Wake, M. Sato and M.E. Corcoran, Epilepsia, 16 (1975) 503.
192 M.E. Corcoran, J.A. McCaughran and J.A. Wada, Epilepsia, 19 (1978) 47.

193 D.M. Feeney in Ref. 178, pp. 643–657.
194 S.A. Turkanis, K.A. Smiley, H.K. Borys, D.M. Olsen and R. Karler, Epilepsia, 20 (1979) 351.
195 S.A. Turkanis and R. Karler, in Ref. 75, pp. 845–855.
196 J.M. Cunha, E.A. Carlini, E.E. Pereira, O.L. Ramos, C. Pimentel, R. Gagliardi, W.L. Sanvito, N. Lander and R. Mechoulam, Pharmacology, 21 (1980) 175.
197 N.M. Kumbaraci and W.L. Nastuk, Mol. Pharmacol., 17 (1980) 344.
198 D.J. Petro and C. Ellenberger, J. Clin. Pharmacol., 21 (1981) 413 S.
199 S.R. Snider and P. Consroe, Neurology, 34 (Suppl. 1) (1984) 147.
200 K. Green, Curr. Topics Eye Res., 1 (1979) 175.
201 A.D. Korczyn, Gen. Pharmacol., 11 (1980) 419.
202 M.W. Adler and E.B. Geller, in Ref. 3, pp. 51–70.
203 R.S. Hepler and I.R. Frank, J. Am. Med. Assoc., 217 (1971) 1392.
204 T.J. Zimmerman, B. Leader and H.E. Kaufman, Annu. Rev. Pharmacol. Toxicol., 20 (1980) 415.
205 J.C. Merritt, W.J. Crawford, P.C. Alexander, A.L. Anduze and S.S. Gelbart, Ophthalmology (Rochester, Minn.) 87 (1980) 222.
206 B.K. Colasanti, R.E. Brown and C.R. Craig, Gen. Pharmacol., 15 (1984) 479.
207 G.R. Thompson and C.L. Yang, in Ref. 4, pp. 457–473.
208 M. Perez-Reyes, D. Wagner, M.E. Wall and K.H. Davis, in Ref. 55, pp. 829–832.
209 J.C. Merritt, D.D. Perry, D.N. Russell and B.F. Jones, J. Clin. Pharmacol., 21 (1981) 467 S.
210 W.M. Jay and K. Green, Arch. Ophthalmol (Chicago) 101 (1983) 591.
211 J.. Merritt, J.L. Olsen, J.R. Armstrong and S.M. McKinnon, J. Pharm. Pharmacol., 33 (1981) 40.
212 Ref. 7, pp. 130–131.
213 K. Green and K. Bowman, in Ref. 55, pp. 803–813.
214 K. Green, H. Wynn and K.A. Bowman, Exp. Eye Res. 27 (1978) 239.
215 B.K. Colasanti, C.R. Craig and R.D. Allara, Exp. Eye Res., 39 (1984) 251.
216 H.M. Deutsch, K. Green and L.H. Zalkow, J. Clin. Pharmacol., 21 (1981) 479 S.
217 K. Green, K. Cheeks, K.A. Bowman, L.C. Hodges, H.M. Deutsch and L.H. Zalkow, Current Eye Res., 3 (1984) 751.
218 K. Green, K. Cheeks, T. Mittag, M.V. Riley, C.M. Symonds, H.M. Deutsch, L.C. Hodges and L.H. Zalkow, Curr. Eye Res., 4 (1985) 631.
219 K. Green, L.H. Zalkow, H.M. Deutsch, M.E. Yablonski, N. Oliver, C.M. Symonds and R.D. Elijah, Current Eye Res., 1 (1981) 65.
220 K. Green and J.E. Pederson, Exp. Eye Res., 15 (1973) 499.
221 K. Green and K. Kim, Invest. Ophthalmol., 15 (1976) 102.
222 J.P.D. Graham in Ref. 3, pp. 147–158.
223 B.J. Shapiro, D.P. Tashkin and L. Vachon, Chest, 71 (1977) 558.
224 L.A. Malit, R.E. Johnstone, D.I. Bourke, R.A. Kulp, V. Klein and T.C. Smith, Anaesthesiology, 42 (1975) 666.
225 S.J. Williams, J.P.R. Hartley and J.D.P. Graham, Thorax 31 (1976) 720.
226 N. Perez-Reyes, M.A. Lipton, M.C. Timmons, M.E. Wall, D.R. Brine and K.H. Davis, J. Clin. Pharmacol. Ther. 14 (1973) 48.
227 D.P. Tashkin, B.M. Calvarese, S. Simmons and B.J. Shapiro, Chest 78 (1980) 699.
228 L. Vachon, N. Engl. J. Med., 294 (1976) 160.
229 D.P. Tashkin, B.J. Shapiro and I.M. Frank, Am. Rev. Resp. Dis., 109 (1974) 420.
230 J.D.P. Graham, B.H. Davies, A. Seaton and R.M. Weatherstone, in Ref. 55, pp. 269.

231 J.P.R. Hartley, S.G. Nogrady, A. Seaton and J.D.P. Graham, Br. J. Clin. Pharmacol., 5 (1978) 523.

232 H. Gong, D.P. Tashkin, M.S. Simmons, B. Calvarese and B.J. Shapiro, Clin. Pharmacol. Ther., 35 (1984) 26.

232a L. Vachon, A.A. Mathe and B. Weisman, Res. Commun. Chem. Pathol. Pharmacol., 13 (1976) 345.

232b R.M. Orzelek-O'Neil, F.R. Goodman and R.B. Forney, Toxicol. Appl. Pharmacol., 54 (1980) 493.

233 I.G. Karniol, I. Shirakawa, N. Kasinski, A. Pfeferman and E.A. Carlini, Eur. J. Pharmacol., 28 (1974) 172.

234 H. Gong, D.P. Tashkin and B. Calvarese J. Clin. Pharmacol., 23 (1983) 127.

235 S. Roth, Can., J. Physiol. Pharmacol. 56 (1978) 968.

236 B.J. Shapiro and D.P. Tashkin in Ref. 55, pp. 277–278.

237 R.M. Orzelek-O'Neil, F.R. Goodman and R.B. Forney, Arch. Int. Pharmacodyn. Ther., 246 (1980) 71.

Progress in Medicinal Chemistry – Vol. 24, edited by G.P. Ellis and G.B. West

6 Hypoglycaemic Agents Which Do Not Release Insulin

KURT E. STEINER, Ph.D. and ERIC L. LIEN, Ph.D.

Wyeth Laboratories, Inc., P.O. Box 8299, Philadelphia, PA 19101, U.S.A.

INTRODUCTION

Diabetes mellitus afflicts a significant portion of the population in many countries. For example, an estimated 5.5 million individuals in the United States have some form of diabetes mellitus [1]. Of these, 500,000 are insulin-dependent diabetics (IDDM's), and require exogenous insulin therapy. The remainder of the diabetic population is categorized as non-insulin dependent diabetics (NIDDM's), most often characterized by obesity and insulin resistance. These individuals may be treated by diet, exercise and/or oral hypoglycaemic agents. The most commonly employed agents are sulphonylureas, which act primarily through the stimulation of insulin release [2]. In some cases, these therapies prove ineffective either because of difficulties with patient compliance or a lack of the proper physiological responses (e.g., either primary or secondary treatment failures with oral agents). The objective of this review is to summarize the status of a number of compounds which cause hypoglycaemia by non-insulin-releasing mechanisms. While none of these compounds has yet won approval for general patient use, some or all of these approaches may prove clinically efficacious or at least provide direction for future research. We have attempted to survey the literature in each of these areas in a comprehensive manner in order to provide a current profile of information on each agent. Two major areas, glucosidase inhibitors (e.g., acarbose) and biguanides, have not been included in this chapter, as they have recently been reviewed elsewhere [3, 4].

SOMATOSTATIN AND ANALOGUES

Somatostatin (1), a tetradecapeptide capable of inhibiting growth hormone release, was originally isolated from bovine hypothalamic tissue [5]. Inhibition of pancreatic hormone release (both insulin and glucagon) was subsequently observed in experimental animals [6, 7] and humans [8, 9]. These observations

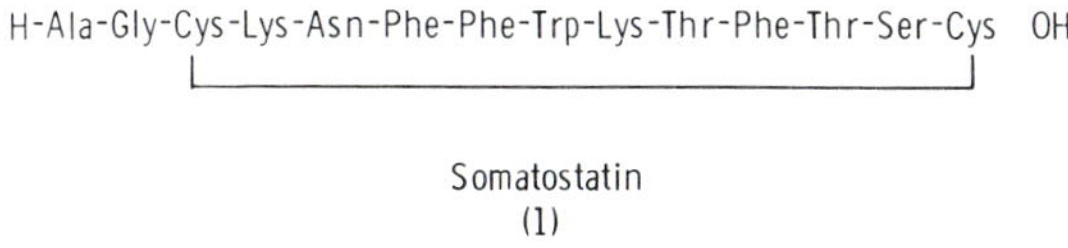

Somatostatin
(1)

suggested that somatostatin might improve glycaemic control in IDDM by suppressing release of the counterregulatory hormones glucagon and growth hormone [10, 11]. The rôle these counterregulatory hormones play in poor glycaemic control was clearly demonstrated in patients with IDDM by infusing combinations of somatostatin (to suppress endogenous hormone release) and either glucagon or growth hormone [12]. The unopposed action of glucagon or growth hormone produced a marked deterioration in glycaemic control; in addition, plasma levels of non-esterified fatty acids (NEFA) were elevated, resulting in increased rates of fatty acid oxidation and elevated ketone body levels. When insulin was initially infused to achieve euglycaemia, addition of glucagon caused only a moderate deterioration of glycaemic control. Addition of growth hormone in a similar study produced no change in glucose levels.

The efficacy of somatostatin in improving glycaemic control in IDDM has been demonstrated in a number of clinical studies. Patients with IDDM receiving infusions of somatostatin ranging in length from 1 to 3 days had consistently lowered blood glucose levels and reduced postprandial glucose excursions [13–15]. Plasma glucagon and growth hormone levels were also suppressed and glycosuria was abolished during the somatostatin infusions. In one of the studies [13], significant improvements in glucose levels in response to somatostatin were observed, even when insulin doses were lowered by 50%

(average diurnal glucose levels were 114 ± 9 mg/100 ml with somatostatin infusion versus 131 ± 9 mg/100 ml in the control); a rapid deterioration occurred when the somatostatin infusion was terminated and the dose of insulin held constant due to counterregulatory hormone rebound and relative insulin deficiency. In addition to suppressing the release of counterregulatory hormones, somatostatin may improve glycaemic control by slowing nutrient absorption from the gut and inhibiting exocrine pancreatic function [16, 17].

The usefulness of somatostatin therapy is limited, however, by the short half-life of the peptide (2.3 ± 0.3 min) [18], requiring continuous infusion, and the non-selective inhibition of hormone release (somatostatin suppresses insulin release as well as the release of glucagon and growth hormone [5, 8, 9]). This second problem is clearly demonstrated in NIDDM patients in whom somatostatin infusions suppressed basal insulin as well as glucagon release, resulting in sustained hyperglycaemia [19]. In these patients, somatostatin administration also resulted in a deterioration in oral glucose tolerance and total suppression of glucose-stimulated insulin release [20]. The non-selective inhibition of endocrine pancreatic function by somatostatin may also be undesirable in IDDM, since many patients with IDDM have residual β-cell function (detectable circulating C-peptide levels) [21].

The synthesis of hundreds of somatostatin analogues has been undertaken in hopes of finding an analogue that either selectively inhibits glucagon and growth hormone (but not insulin) release, and/or has a prolonged duration of action. A detailed consideration of all of these analogues is beyond the scope of the current chapter; some of the work in this area has been previously reviewed [22–24].

The DTrp^8 substitution routinely incorporated in most analogues increases somatostatin's potency 8–10-times, but does not alter selectivity [25, 26]. Surprisingly, the duration of action of this analogue is similar to that of somatostatin, despite the fact that *in vitro* incubation of [DTrp^8]-somatostatin in plasma produces a dramatic reduction in proteolytic degradation compared with the parent molecule. Other analogues may be divided into two broad classes: (1) analogues that are similar in size to the native molecule, but with elevated glucagon/insulin selectivity and little or modest increase in duration of action, and (2) shorter, more potent, long-acting analogues with little alteration in selectivity.

ANALOGUES WITH GLUCAGON AND GROWTH HORMONE SELECTIVITY

Several analogues, similar in size to the native hormone, suppress glucagon and growth hormone but not insulin release. For example, [DCys^{14}]- and

[DTrp^{8},DCys^{14}]-somatostatin potently inhibit arginine-induced glucagon, but not insulin, release in rodents [27–30]. In fasted alloxan diabetic dogs receiving a subcutaneous injection of 1 mg of the more potent of these analogues, [DTrp^{8},DCys^{14}]-somatostatin, a prolonged reduction of circulating glucagon and glucose levels has been observed [31]. The limited clinical experience with this analogue in IDDM has been disappointing. Infusions of the peptide did not alter plasma glucagon levels, although modest reductions in growth hormone levels and postprandial hyperglycaemia were observed [32, 33]. The relatively low doses of the peptide employed in these studies (25–40 μg/h) or the route of administration (intravenous infusion versus subcutaneous injection) may be responsible for the poor efficacy in clinical studies as opposed to the diabetic dog studies cited above. The fact that alterations at the Cys-14 residue may play an important rôle in conferring specificity is supported by results with two other analogues containing alterations in the 14 position. Both des-Ala^{1},Gly^{2}[DTrp^{8},DAsn^{3,14}]-somatostatin and des-Ala^{1},Gly^{2}[DTrp^{8},D-Asu^{3,14}]-somatostatin, an analogue with a covalent bond between residues 3 and 14, increase suppression of glucagon release relative to insulin release [34, 35]. (Des preceding amino acids designates deletion of these residues and Asu is aminosuberic acid.)

In addition to position-14 alterations, a somatostatin analogue with substitutions at positions 4 and 5, des-Ala^{1},Gly^{2}[His^{4,5}DTrp^{8}]-somatostatin (Wy-41, 747), selectively suppressed glucagon and growth hormone (in preference to insulin) release 15 min following subcutaneous administration in arginine-treated rats [36, 37]. This analogue also suppressed growth hormone but not glucagon or insulin release for 2 h in rats [36]. Subcutaneous administration of this analogue to fasted streptozotocin diabetic dogs resulted in reductions of both basal plasma glucagon (60%) and glucose (30%) for 5 h following a dose of 0.5 mg/kg [38]. Substantial improvements in postprandial hyperglycaemia and hyperglucagonaemia were also observed in these dogs. The postprandial improvements in glycaemic control are most likely due to a combination of suppression of glucagon levels (as suggested in the fasting experiments) and delayed nutrient absorption (3-*O*-[^{14}C]methylglucose absorption was delayed by this drug). The acute effects of Wy-41, 747 have been evaluated in meal studies in individuals with IDDM [39]. Postprandial hyperglycaemia was completely prevented by subcutaneous administration of 10–50 μg/kg of the analogue even when insulin doses prior to the meal were reduced almost 50%. Postprandial glucagon levels were suppressed for 2 h and only rose modestly above basal levels at time periods as long as 5 h. This is in marked contrast to the postprandial hyperglucagonaemia observed in these patients when the analogue was not administered. As in diabetic dogs, the improvements in glycaemic control appeared to be mediated through both hormonal (decreased glucagon levels) and absorptive mechanisms (slow appearance of circulating xylose following ingestion with the meal). The duration of action of des-Ala^1,Gly^2[His^{4,5}DTrp^8]-somatostatin in humans following subcutaneous administration is sufficient to provide an improvement in glycaemic control following one meal, but not following a second meal administered 4 h later [40]. Preparation of a zinc phosphate suspension provided a dramatic increase in duration of action in both normal rats (reduction in growth hormone for up to 8 h) and diabetic dogs [41, 42]. This improvement is most likely due to a complex between the zinc and adjacent histidines at positions 4 and 5.

CORE ANALOGUES

The analogues discussed above have relative specificity for suppression of glucagon and growth hormone as opposed to insulin release. This type of specificity apparently requires most of the cyclic skeleton of somatostatin (residues 3–14). A central core of somatostatin, including Phe-7, Trp-8, Lys-9 and Thr-10, appears to be essential for specific somatostatin binding to its receptors [43]. Deletion studies with analogues such as des-AA^{1,2,4,5,12,13}-somatostatin demonstrate retention of growth hormone, insulin and glucagon suppressing activity in the rat [43]. In patients with pancreatic tumours, infusion of this analogue suppressed release of growth hormone, glucagon, insulin and a variety of gastro-intestinal hormones, but was not long-acting [44]. Dramatic prolongation of action was observed (hormonal suppression for 5–10 h) when the peptide was administered subcutaneously to individuals with pancreatic endocrine tumours, suggesting that duration of action is mediated by a depot effect [45].

Attempts to synthesize analogues with the deletion of non-essential amino acids while simultaneously increasing the rigidity of the peptide resulted in the bicyclo structure (2), which demonstrated activity similar to that of somatosta-

```
       O
       ‖
      C-NH-Cys-Phe-D·Trp
     /         |       |
 (CH2)6        |       |
     \         |       |
      NH-C-Cys-Thr — Lys
         ‖
         O
```

(2)

tin in terms of growth hormone, insulin and glucagon suppression in rodents, and had an extended duration of action following subcutaneous administration [46]. Computer modelling efforts led to the synthesis of non-cysteine-containing analogues such as the hexapeptide, (3), which also had extended duration of action [47]. Further refinement of these structures yielded L-363, 586 (4), an analogue with growth hormone, glucagon- and insulin-suppressing

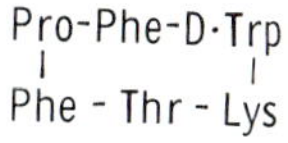

```
Pro-Phe-D·Trp              NMe·Ala-Thr-D·Trp
 |         |                    |         |
Phe - Thr - Lys                Phe - Val - Lys

     (3)                            (4)
```

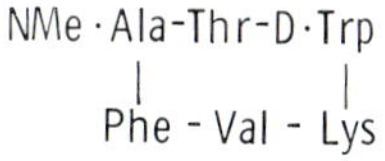

activity being 50–100-times that of somatostatin [48]. Oral administration of this analogue improved control of postprandial hyperglycaemia in insulin-treated alloxan diabetic dogs; however, bioavailability is only 1–3% [49]. Excellent glycaemic control was achieved in individuals with IDDM during infusions of this analogue. A 20-fold increase in potency and half-life was observed compared with somatostatin in these studies [50].

Another abbreviated analogue, the octapeptide, DPhe-Cys-Phe-DTrp-Lys-Thr-Cys-Thr(ol) (SMS201-995) also had both increased potency and prolonged action in rats and rhesus monkeys [51]. In normal humans, subcutaneous administration of this peptide, which has a long half-life (greater than 100 min versus 2 min for somatostatin), inhibited arginine-induced insulin and growth hormone, but not glucagon, release; however, it did inhibit the postprandial rise in glucagon levels [52]. Excellent control of postprandial glucose levels was achieved with single subcutaneous doses as low as 50 μg to subjects with IDDM [53]. Concomitant suppression of plasma glucagon, triacylglycerol and free fatty acid levels was observed. This analogue is an exceptionally potent and long-acting growth-hormone-suppressing agent, and a number of clinical studies have confirmed its utility in the treatment of acromegaly [54–56]. Unfortunately, prolonged suppression of growth hormone release may have adverse effects on growth rates in children and adolescents with IDDM, thus limiting its utility. Other long-acting analogues may have a similar liability.

CONCLUSIONS

Several somatostatin analogues improve glycaemic control in diabetic dogs and humans. They are of sufficient duration to suggest clinical efficacy. However, the high cost of production of these peptides and the development of alternative methods of achieving excellent glycaemic control, such as insulin pump and intensive subcutaneous insulin therapies, may limit the utility of these peptides.

CIGLITAZONE

One of the most promising novel hypoglycaemic agents for treatment of NIDDM is ciglitazone (5-(4-(1-methylcyclohexylmethoxy)benzyl)thiazolidine-2,4-dione) (5) [57–60]. It has effective hypoglycaemic activity in insulin-

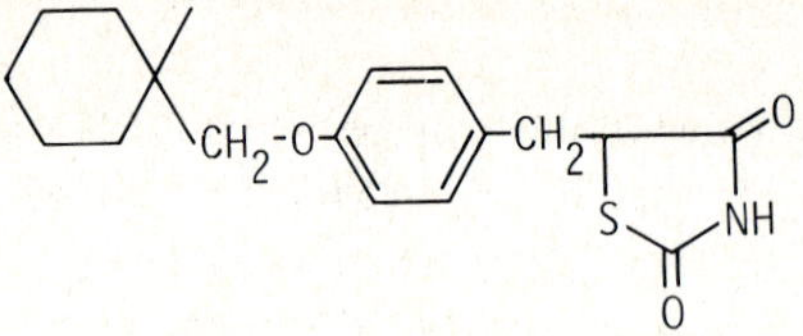

CH_2-O- -CH_2 O S NH O

(5)

resistant animal models, which seems to result from a potentiation of peripheral insulin action. However, the drug has no effect on the glycaemic level of normal or insulinopoenic animals. Insulin synthesis and storage are also improved with ciglitazone treatment, although it is unclear whether these are direct or indirect effects of the drug.

IN VIVO ACTIVITY

Ciglitazone was initially characterized in the obese diabetic yellow KK mouse [61]. While neither food intake nor body weight was altered by 4 days of dosing (30–186 mg/kg per day), blood glucose, plasma NEFA, triacylglycerols and cholesterol all were significantly reduced in a dose-dependent manner. *In vivo* assessment of insulin resistance in yellow KK mice using simultaneous subcutaneous injections of insulin and glucose [62] showed that ciglitazone treatment caused a decrease in insulin resistance [61]. In studies using other insulin-resistant animal models, such as the *ob*/*ob* and *db*/*db* mouse, ciglitazone was also an effective hypoglycaemic agent [63]. In mildly insulin-resistant 20-week-old rats [61] the drug, while maintaining normal glucose levels, caused a reduction of plasma insulin and triacylglycerol levels. Ciglitazone, at doses effective in insulin-resistant animal models, failed to have any measurable effects in normal Sprague-Dawley rats. When large doses of ciglitazone were administered (300 mg/kg per day for 9 days) to normal rats, some reduction of the glucose profiles normally observed during oral glucose loading was noted but other effects were not evident [63]. Likewise, in streptozotocin-diabetic rats (a model of insulin-dependent diabetes) ciglitazone (100 mg/kg), given for 12 days, had no observable effect on blood glucose or plasma triacylglycerols. These initial *in vivo* observations in normal, insulin-resistant and insulin-dependent diabetic animals indicated that ciglitazone was an effective hypoglycaemic agent only in insulin-resistant animals. Secondly, even large doses of the drug caused only a normalization of blood glucose; hypoglycaemia was not observed.

EFFECTS ON INSULIN BINDING

In an effort to determine the mechanism of action of ciglitazone, its effects on insulin binding have been studied in isolated adipocytes from the obese mouse (C57BL/6J–*ob/ob*) and its lean (+/?) counterpart [64]. Insulin binding in adipocytes from *ob/ob* mice normally differs from that observed in their lean counterparts in several ways. First, the number of low-affinity sites is reduced; second, the affinity of these sites is increased; third, the number of high-affinity sites is also reduced. Chronic treatment of *ob/ob* mice with ciglitazone caused a reversion of the numbers and affinities of these sites toward those normally observed in lean (+/?) mice. In lean mice, chronic drug treatment had no observed effect on the number of high- or low-affinity sites or their affinities. Insulin binding in normal lean, old obese, and streptozotocin-treated Sprague-Dawley rats was also compared [65]. In these rat studies, chronic treatment had no effect on insulin binding to isolated adipocytes from these animal models. Additionally, an inability of the drug to restore normal insulin binding after exposure of adipocytes to isoprenaline (which decreases binding) was also observed [66]. Collectively, these results indicate that ciglitazone does not exert its primary hypoglycaemic activity through alterations in insulin binding. It is possible that the changes in binding observed in adipocytes isolated from treated *ob/ob* mice resulted from relief of the normally observed hyperinsulinaemia rather than by any direct effects of the drug.

EFFECTS ON GLUCONEOGENESIS

Ciglitazone could lower plasma glucose by reducing hepatic glucose production (gluconeogenesis and glycogenolysis) and/or increasing glucose utilization by muscle and fat. Gluconeogenesis, as measured by [^{14}C]lactate conversion to glucose *in vivo* in the *ob/ob* mouse, was decreased by ciglitazone treatment [67]. In contrast, more recent studies using perfused livers of ciglitazone-treated *ob/ob* mice have shown gluconeogenesis from lactate to be unchanged; glucose formation from alanine was, however, decreased [68]. Even when gluconeogenesis was decreased by ciglitazone *in vivo*, the decreases which were noted were small in size and were probably not sufficient to account for the hypoglycaemic effects of ciglitazone. It is also important to note that the quantitation of all these measurements is confounded by the dilution of radioactivity by other substrates and metabolic intermediates as well as potential changes in their uptake by the liver.

EFFECTS ON GLUCOSE METABOLISM IN ADIPOSE TISSUE AND MUSCLE

The effects of chronic ciglitazone treatment on glucose metabolism in adipose tissue were studied in isolated fat pads from *ob/ob* and +/? mice incubated with [1-^{14}C]glucose [64]. Basal $^{14}CO_2$ production (a measure of pentose phosphate shunt activity) in fat pads from untreated lean animals was at least 3-fold higher than in pads from untreated obese mice. Treatment with ciglitazone caused only small, nonsignificant rises in basal $^{14}CO_2$ production in fat pads from either lean or obese mice. The stimulatory effect of insulin on $^{14}CO_2$ production in fat pads from treated obese mice was significantly enhanced when compared with untreated obese controls. On the other hand, insulin stimulation of $^{14}CO_2$ formation in fat pads from treated lean mice was not significantly different from that in untreated lean controls. Ciglitazone treatment in obese mice also significantly enhanced the stimulatory effects of insulin on [1-^{14}C]glucose incorporation into lipids, but had no effect on the insulin-stimulated increments in fat pads from lean mice. These observations indicate an overall increase in insulin responsiveness of adipose tissue in obese mice with ciglitazone treatment.

The effects of ciglitazone on muscle glucose metabolism have been studied in the perfused hindquarter of the obese mouse [68]. Following oral dosing of ciglitazone (150 mg/kg per day) for 9 days, 100 μU/ml of insulin caused a significant increase in the perfused hindquarter uptake of 2-deoxy-D-glucose in ciglitazone-treated mice, while in untreated mice, insulin at this dose had no effect. The enhancement of the insulin effect by drug treatment was still not of a sufficient magnitude to normalize muscle insulin sensitivity, since in perfused hindquarters from lean animals the maximal insulin stimulation of 2-deoxy-D-glucose uptake was significantly greater than that found in obese animals treated with ciglitazone. The basal rate of 2-deoxy-D-glucose uptake was unaffected by ciglitazone. Basal rates of $^{3}H_2O$ formation from [5-^{3}H]glucose in perfused hindlimbs of *ob/ob* mice were also unaffected by ciglitazone treatment. Whether ciglitazone potentiated insulin stimulation of this process was not reported. It is clear that glucose metabolism in both adipose tissue and muscle can be significantly affected by ciglitazone treatment and these changes are probably the primary cause of the potent hypoglycaemic effect of the drug.

EFFECTS ON THE PANCREAS

While the effects of ciglitazone do not appear to be a result of the drug's ability to enhance insulin secretion, as a fall in the plasma insulin levels accompanies the fall observed in the glucose level, ciglitazone treatment does appear to alter islet structure and perhaps function. Initial studies with ciglitazone showed that it promoted regranulation of islets in the *ob/ob* mouse [63]. A more thorough morphometric evaluation of islet surface area in *ob/ob* mice and their lean counterparts was later conducted [69]. In this study pancreatic islets from chronically-treated *db/db* mice were also studied. After 7 weeks of ciglitazone treatment, moderate to heavy granulation was observed in islets from both *db/db* and *ob/ob* mice, whereas the islets from untreated mice were extensively degranulated [69]. Untreated mice of both types showed extensive hypertrophy of the rough endoplasmic reticulum, Golgi and mitochondria indicative of stressed beta cells. Surface area measurements of islets showed ciglitazone caused a reduction in surface area in the *ob/ob* mice, but not to the level of the lean counterparts, suggesting that the islet hypertrophy indicated above had been partially ameliorated but not completely reversed. These chronic studies provide evidence for the beneficial effects of chronic treatment with ciglitazone on beta cell function. It is unclear whether these effects of increased regranulation and an improved pattern of insulin storage and synthesis are direct effects of the drug or simply a result of near normalization of the plasma glucose and consequently a weakened stimulus for continued insulin secretion.

EFFECTS ON RENAL LESIONS

The effects of chronic (12 and 20 weeks) administration of ciglitazone on blood glucose and renal lesions in diabetic (C57BL/KsJ – *db/db*) mice have also been studied [70]. Blood glucose levels were significantly improved in both studies. Deposition of fluorescein-conjugated IgM and IgG in the glomerular mesangium and renal tubules was evaluated by light microscopy and a significant reduction of IgG deposition was noted. Expansion of the mesangium which normally occurs in *db/db* mice was not retarded by ciglitazone treatment. Ciglitazone completely ameliorated glycogen vacuolization of tubular epithelial cells which normally occurs due to excessive reabsorption of glucose during hyperglycaemia.

EFFECTS ON THERMOGENIC IMPAIRMENT

Coincident with the development of insulin resistance in the *ob/ob* mouse is an impairment in thermogenic responsiveness of the brown adipose tissue of this animal [71, 72]. The activity of the thermogenic pathway is generally assessed by measurement of a GDP binding protein in isolated mitochondria [73]. Seven-day treatment of *ob/ob* mice with ciglitazone restored responsiveness of this protein to cold exposure, suggesting that the drug had ameliorated the thermogenic defect. At the same time, an almost complete restoration of the impaired insulin stimulatory effect on lipogenesis in brown adipose tissue was also observed [74]. These results serve to support the idea that ciglitazone causes a generalized therapeutic improvement in insulin resistance rather than improvement in one or more specific derangements caused by the resistance.

BIOAVAILABILITY AND DISPOSITION

The bioavailability and disposition of ciglitazone in both rats and dogs have been studied [75–77]. In dogs, a meal accompanying the administration of drug either in tablet or suspension (micro-pulverized ciglitazone in water) caused an increase in its bioavailability [75]. Experiments using agents such as propantheline bromide to increase GI residence time indicated that the effect of meals to increase bioavailability was in part due to an enhancement of GI residence time [76].

The disposition of ciglitazone is strikingly different in rats than in dogs [77]. In rats, a maximal plasma level was reached 2 h after oral dosing with an apparent half-life of 4.9 h, while in dogs a plateau was reached at 1 h, persisted for 10 h and had an apparent half-life of 23.5 h. In rats, plasma concentrations of metabolites were higher than those of the unchanged ciglitazone, while in dogs the reverse was observed. Several of the metabolites that have been identified also have hypoglycaemic activity. After 7 days of administration no accumulation of the drug was found in plasma or tissues and 97.5% of the dose was eliminated from the body within 96 h.

CONCLUSIONS

The ability of ciglitazone to normalize blood sugar levels and ameliorate most, if not all, degenerative changes which occur in non-insulin-dependent diabetes certainly make it a potentially exciting therapy. A recent report [78] indicates that analogues with a potency at least 10-fold that of ciglitazone have been synthesized. Studies on the toxicity of ciglitazone and/or its analogues have not yet been reported, nor have any studies on the clinical efficacy of these compounds.

FATTY ACID OXIDATION INHIBITORS

The use of fatty acid oxidation inhibitors as orally effective hypoglycaemic compounds has its roots in the Randle glucose–fatty acid cycle first proposed in the 1960's [79]. This hypothesis recognized the reciprocal relationship that exists between fat and carbohydrate metabolism. It followed that in individuals such as diabetics, who often have an increased utilization of fatty acids, a reduction in fatty acid oxidation should enhance carbohydrate utilization and consequently lower blood glucose levels. A variety of fatty acid oxidation inhibitors do have significant hypoglycaemic activity [80–83]; however, their activity may reside more in their potent inhibition of hepatic gluconeogenesis than in their enhancement of peripheral glucose utilization.

CARNITINE PALMITOYLTRANSFERASE (CPT)

Efforts to develop potent inhibitors of fatty acid oxidation have focused on agents that selectively inhibit the movement of long-chain fatty acyl CoA's into the mitochondrion by the inhibition of carnitine palmitoyltransferase (CPT). This enzyme has been widely accepted as the key regulatory enzyme in the oxidation of long-chain fatty acyl groups [84] and catalyzes the movement of long-chain fatty acyl CoA's across the inner mitochondrial membrane (see *Figure 6.1*). The long-chain fatty acyl CoA's initially undergo transesterification catalyzed by CPT I (located on the outer aspect of the inner membrane) to form an acyl-carnitine derivative. The acyl-carnitine derivative is rapidly translocated across the membrane where CPT II (located on the inner portion of the inner membrane) catalyzes a transesterification reaction in which carnitine is released simultaneously with resynthesis of fatty acyl CoA. The movement of medium-chain fatty acyl CoA's into mitochondria is carnitine-independent and is not regulated by CPT. CPT I and II can be differentiated not only on the basis of their locations but also based on their sensitivity to inhibition by malonyl-CoA (the product of the first committed step in fatty acid synthesis and an important regulator of fatty acid oxidation [84]). CPT I is inhibited by malonyl-CoA, whereas CPT II is not. Whether CPT I and CPT II are distinct

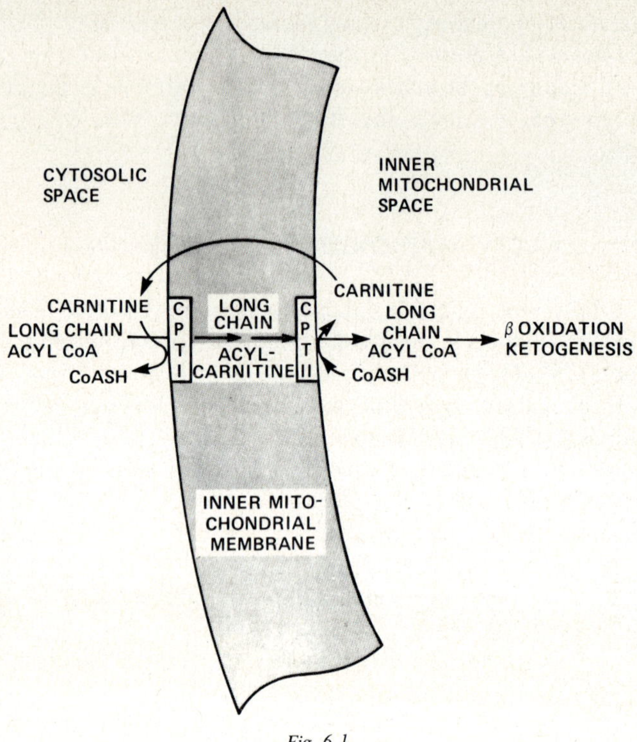

Fig. 6.1.

isozymes or the same enzyme with different affinities for malonyl-CoA (due to differences in their local membrane environments) has not yet been completely resolved [85].

CPT INHIBITORS

Most inhibitors of this enzyme reported in the literature are long-chain fatty acyl CoA analogues [80, 83, 86]. The two most thoroughly characterized CPT inhibitors are TDGA (2-tetradecylglycidate (6) and its methyl ester (MeTDGA)) and POCA (chlorophenylpentyloxiranecarboxylate (7)). These inhibitors affect only CPT I and not CPT II [87–89], with IC_{50} values reportedly in the nanomolar range. The following discussion will focus primarily on these two compounds, since they have been the most thoroughly characterized, and are probably closest to comprehensive clinical trials. Other inhibitors of CPT I should cause similar if not identical effects. Two other

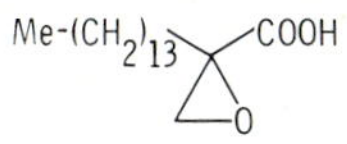

Me-$(CH_2)_{13}$ COOH O

(6)

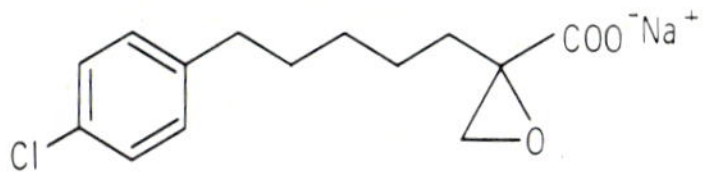

Cl COO$^-$Na$^+$ O

(7)

potentially useful fatty acid oxidation inhibitors have recently been reported. Emeriamine (8) is a carnitine analogue which has significant hypoglycaemic effects in fasted rats and several other animal models. Emeriamine inhibits CPT I in isolated mitochondria with an IC_{50} value in the micromolar range [90]. Where appropriate in the following discussion, comparisons will be made with emeriamine, although only limited data are available on the pharmacology of this compound. Since emeriamine is a carnitine analogue, it may suffer the disadvantage of inhibiting other carnitine-dependent enzymes, thus causing deleterious side-effects. Another novel compound, 2-(3-methylcinnamylhydrazono)propionate (MCHP) (9) apparently inhibits the translocation of long-

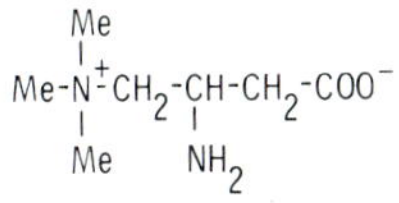

Me-N^+(Me)(Me)-CH_2-CH(NH_2)-CH_2-COO^-

(8)

C_6H_5-CH=C(Me)-CH_2-NH-N=C(Me)COOH

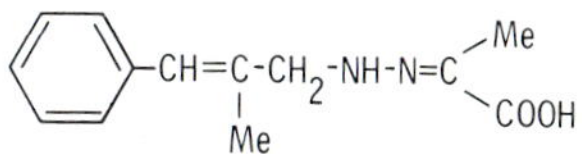

(9)

chain fatty acyl carnitines across the mitochondrial membrane but has little or no effect on either CPT I or CPT II [91]. Its derivation, mechanism of action and pharmacology will be discussed in a subsequent section.

The biologically active forms of TDGA and POCA are their CoA esters [87, 89, 92]. TDGA and MeTDGA were ineffective in the inhibition of CPT activity unless a suitable period of preincubation supplemented with free CoA and ATP was allowed for formation of the CoA derivative by endogenous acyl CoA synthase [87]. Similar studies have shown that the CoA ester of POCA is required for activity [89]. The effects of these inhibitors appeared to be irreversible in nature, since CPT activity was not regenerated by repeated washing of isolated mitochondria previously incubated with the CoA form of the inhibitors [92, 93]. Neither inhibitor prevented the oxidation of palmitoylcarnitine by isolated intact mitochondria, indicating that CPT I rather than CPT II is inhibited. The potencies of MeTDGA-CoA and POCA-CoA were similar, with IC_{50} values for rat liver CPT I of 26 and 20 nM, respectively [87, 89]. Preincubation with palmitoyl-CoA prior to addition of the inhibitor protected the enzyme [87, 89]. Emeriamine, the carnitine analogue, had an IC_{50} value of 62 μM [90] with 1.2 mM carnitine as a substrate.

EFFECTS OF FASTING ON CPT INHIBITION

CPT I is more strongly inhibited by POCA-CoA and TDGA-CoA in mitochondria isolated from the liver of fed animals than from that of 24-h-fasted animals [92]. This is analogous to the change in sensitivity of the enzyme to malonyl-CoA observed with fasting [94]. However, fasting did not affect inhibitor potency in muscle [92]. Although the hepatic enzyme is more sensitive to the effects of POCA-CoA and TDGA-CoA in the fed state, the inhibitors appear to have demonstrable effects on plasma glucose only in the fasted state [80, 95]. This is directly related to the contribution that gluconeogenesis makes to overall hepatic glucose production in the fasted state. Normally, only a small portion of total glucose output by the liver in the fed state is derived from gluconeogenesis, whereas in the fasted state this process accounts for the major portion of hepatic glucose output.

TISSUE SENSITIVITY

The sensitivity of CPT I from different tissues to POCA and MeTDGA varies, as does its sensitivity to malonyl-CoA [88, 92, 96, 97]. Studies in which MeTDGA was administered *in vivo* followed by isolation of mitochondria from various tissues showed that liver CPT I was most sensitive to the inhibitor, followed by heart and then diaphragm [97]. A similar pattern of tissue sensitivity was observed with POCA [88, 92]. Preliminary studies comparing the effects of emeriamine in isolated hepatocytes and cardiomyocytes showed the hepatocytes to be 10-fold more sensitive than the cardiomyocytes when palmitate oxidation or CPT I activity was used as a measure of effectiveness [90].

EFFECTS ON FATTY ACID OXIDATION AND SYNTHESIS

The inhibition of CPT I by either POCA or MeTDGA caused a decrease in [^{14}C]palmitate oxidation. MeTDGA effectively inhibited palmitate oxidation in kidney cortex slices [98], diaphragm [98], heart [99] and hepatocytes [100] from fasted rats. POCA produced an almost complete suppression of long-chain fatty acid oxidation in the perfused rat heart from either fed or fasted animals [101]. POCA also inhibited hepatocyte oxidation of oleate up to 85% with maximal effects of POCA observed at concentrations as low as 1 μM

[102]. Esterification of oleate was increased in the same proportion as its oxidation was inhibited, so oleate uptake was unchanged. While oxidation of palmitoyl-CoA was strongly decreased in both liver and muscle mitochondrial fractions from POCA-fed animals (0.5% and 0.2% in chow) oxidation of palmitoylcarnitine was unaffected [92]. Metabolism of [^{14}C]octanoate was also unaffected by POCA, indicating that ketogenic enzymes and medium-chain fatty acid oxidation also remained uninhibited [102].

One means of regulation of fatty acid biosynthesis is through end-product inhibition of acetyl-CoA carboxylase by long-chain acyl CoA's. Thus it might be expected that POCA-CoA, a long-chain fatty acyl CoA analogue, would also be inhibitory. In hepatocytes from fed rats, at concentrations above 5 μM, POCA caused an inhibition, while at a concentration of 2 μM a slight stimulation was observed [103]. Inhibition of fatty acid synthesis by 100 μM POCA-CoA (71%) was greater than that observed from 250 μM palmitoyl-CoA (30%). Since 100 μM concentrations of these CoA derivatives caused equivalent inhibition of acetyl-CoA carboxylase in liver extracts (30%), POCA-CoA must have a direct inhibitory effect on fatty acid synthetase. The slight stimulation of fatty acid synthesis observed at low levels (2 μM) of POCA may be related to an increased flux through pyruvate dehydrogenase and to a more oxidized state of the mitochondrial matrix, resulting in increased citrate formation.

EFFECTS ON GLUCONEOGENESIS

The inhibition of fatty acid oxidation potently inhibits gluconeogenesis (a major component of glucose production in the fasted state) [89]. The effects of MeTDGA and POCA on gluconeogenesis have been well studied *in vitro* [90, 100, 104–106], although their effects on gluconeogenesis *in vivo* have not been reported. In isolated rat hepatocytes, MeTDGA significantly inhibited gluconeogenesis from lactate, pyruvate, dihydroxyacetone phosphate and fructose [100]. At the same time, acetyl-CoA levels were depleted from 108 to 41 nmol/g and a corresponding rise in the free CoA levels occurred. These changes are consistent with the hypothesis that the inhibition of gluconeogenesis is the result of an increase in pyruvate utilization *via* increased flux through pyruvate dehydrogenase and a decrease in pyruvate carboxylase activity (a critical gluconeogenic enzyme) because of the decreased concentration of acetyl-CoA. Similar to the effects of MeTDGA, POCA inhibited oleate oxidation, gluconeogenesis, and ketogenesis in isolated hepatocytes from 24-h-fasted rats [105]. Full effects of POCA on fatty acid oxidation in hepatocytes have been observed at concentrations of 1 μM [102].

EFFECTS ON PYRUVATE DEHYDROGENASE

One major site of regulation of glucose oxidation is the intramitochondrial pyruvate dehydrogenase (PDH) complex. It exists in both an active (dephosphorylated) and inactive (phosphorylated) form. In situations of relative insulin deficiency such as starvation and diabetes, the relative proportion of PDH in the active form is decreased mainly as a result of increased PDH kinase activity. A number of factors regulate the kinase, such as the NADH/NAD and acetyl-CoA/CoA ratios [107, 108]. The effects of TDGA and POCA on PDH have also been studied. TDGA administered intravenously to normal fed rats had no effect on the active form of PDH from heart, diaphragm, liver, kidney and adipose tissue [109]. The drug was able to reverse the effects of a 48 h fast on heart PDH, which normally decreases the amount in the active form, and the drug also partially reversed the effect of starvation on kidney PDH [109]. PDH of the other tissues was unaffected. The compound has been studied in various animal models of diabetes with variable results. The compound failed to reverse the effects of alloxan diabetes on PDH from rat hearts perfused without insulin. The addition of insulin alone only partially normalized the proportion of PDH in the active form, while the combination of insulin and TDGA caused a complete normalization [109]. In mice treated with gold thioglucose to induce obesity and hyperinsulinemia, the proportion of PDH in the active form is decreased, but a single dose of TDGA (25 mg/kg) restored the PDH proportion back to that of the lean controls [110]. Perfusion of hearts from streptozotocin-diabetic rats with POCA caused a 3-fold increase in PDH, although PDH activity was not influenced directly by POCA [88].

EFFECTS ON GLUCOSE METABOLISM *IN VIVO*

In vivo there are two primary determinants of the plasma glucose level, the rate of hepatic glucose production and the rate of peripheral glucose utilization. Inhibition of fatty acid oxidation should have significant effects on both of these processes, at least in the fasted state. The effects of MeTDGA *in vivo* on glucose turnover have been studied in both normal and streptozotocin diabetic dogs using [6-^{3}H]glucose [111]. No effects of the compound were noted on glucose turnover in the normal animals, but the length of fast was not specified in these experiments. The dog normally absorbs the contents of a mixed meal at a much lower rate than either man or the rat [112] so that, unless the dogs were fasted at least 48 h, it is unlikely that the effects of a fatty acid oxidation inhibitor would be evident (thus explaining the lack of effects of MeTDGA on glucose turnover). In diabetic dogs, MeTDGA caused the plasma glucose level to fall from 155 to 113 mg/100 ml, but had no significant effect on glucose turnover. With MeTDGA treatment, restoration of the acute suppressive effect of insulin on glucose production that was absent in the diabetic condition was also observed. It is noteworthy that the normal counterregulatory increase in glucose production in response to insulin-induced hypoglycaemia was unimpaired in both the normal and diabetic MeTDGA-treated dogs.

As might be expected following administration of either inhibitor *in vivo*, non-esterified fatty acids (NEFA) rise, but with chronic administration of the drugs, their levels reportedly fall back to those normally found in the plasma [113]. The rise is probably due to the decreased utilization of the NEFA, but the reason for the transient nature of this rise has remained unexplained. Either lipolytic inhibition occurs to restrict inflow of NEFA into the plasma or the processes of esterification and re-esterification are increased to store excess NEFA as triacylglycerols in some tissue or tissues [95, 114]. Chronic use of the drug has resulted in a fatty liver, but it is unclear whether this explains the transient nature of the rise in NEFA.

CLINICAL STUDIES

While none of the fatty acid oxidation inhibitors has proven clinical efficacy, several reports indicate that they may have potential in the treatment of diabetes. MeTDGA was used with some success to treat a middle-aged, severely insulin-resistant diabetic who suffered from a syndrome of anti-insulin receptor antibodies, thus disallowing the use of insulin in any quantity for treatment of the disorder [115]. Insulin given up to 1280 units/day did not cause a lowering in the glucose level. The excessive hyperglycaemia from which this patient suffered (fasting plasma glucose of up to 724 mg/100 ml) was due almost exclusively to an overproduction of glucose by the liver. Decreased utilization of glucose did not seem to play a significant rôle. However, when MeTDGA was administered, the respiratory quotient increased from 0.71 to 0.84, indicating that the drug had caused enhanced utilization of glucose by peripheral tissues. Glucose production also fell, reflecting decreases in the rates of both hepatic glycogenolysis and gluconeogenesis. MeTDGA has also been used in other studies in insulin-dependent diabetics [116]. As a supplement to their normal insulin dose, six ketotic IDDM patients were given 50 mg/day MeTDGA for 11 consecutive days. Ketonuria was abolished and remained abolished. Glycosuria, and fasting and postprandial plasma glucose levels were also reduced, requiring the insulin dosage to be reduced in three patients. These studies suggest that the use of MeTDGA in IDDM may be a beneficial adjunct to current therapy. To date, there have been no studies reported in NIDDM patients with any of the fatty acid oxidation inhibitors.

EFFECTS ON EXERCISE CAPACITY AND HYPOGLYCAEMIC COUNTERREGULATION

There has been some speculation that, with the chronic use of a fatty acid oxidation inhibitor, an impairment of exercise capacity and/or the counter-regulatory mechanisms involved in glucose homeostasis might occur. In normal [117] or streptozotocin-diabetic [118] rats, MeTDGA caused no decrease in the ability of the animals to perform strenuous or prolonged, moderately strenuous exercise. However, based on a reduced ability by tissues to use NEFA, one might expect that the ability to sustain exhaustive exercise would be reduced, particularly once glycogen levels had been depleted.

Adrenaline-mediated increases in glucose production represent a major mechanism by which the body defends against hypoglycaemia. The effects of chronic MeTDGA dosing on adrenaline-induced hyperglycaemia were studied in the fed and fasted rat [119]. The hyperglycaemic response to adrenaline in fed rats was unaffected by MeTDGA treatment, although in fasted rats, drug treatment slightly impaired the induction of hyperglycaemia by the catecholamine.

EFFECTS ON CARDIAC STRUCTURE AND FUNCTION

Since the heart derives the majority of its energy from the oxidation of fatty acids, a major concern about the chronic use of fatty acid oxidation inhibitors relates to their potential for cardiotoxicity or at least impairment of normal cardiac function. The acute effects of POCA on cardiodynamics and myocardial metabolism were studied in normoxic and in underperfused canine hearts in anaesthetized animals [120, 121]. In the normoxic heart, although arterial NEFA levels rose continuously, myocardial uptake was almost completely inhibited; glucose uptake remained unchanged but lactate uptake increased markedly, suggesting a switch from oxidation of NEFA to carbohydrate. In the face of these changes, an initial increase in pulmonary vascular resistance was noted along with increasing vascular pressure up to 4 h; otherwise, all other haemodynamic parameters measured remained unchanged, including cardiac output, aortic pressure, and heart rate. When the effects of POCA were examined in the underperfused canine myocardium, the drug induced a transient decrease in left ventricular power and cardiac output. A transient increase in end-diastolic segment length and a sustained decrease in systolic shortening during the ejection phase in the underperfused area were also observed.

The normalization of some diabetic cardiac abnormalities by MeTDGA such as depression of calcium uptake, elevation of long-chain acyl carnitines in sarcoplasmic reticulum [122] and shifts in myosin isozymes have been observed [123]. However, other chronic studies in which MeTDGA was administered twice daily at either 10 mg/kg or 25 mg/kg to normal rats showed that after 5 weeks of treatment significant deleterious changes in a number of parameters had occurred [124]. The triacylglycerol and phospholipid content of the cardiac tissue was elevated and the hearts themselves were enlarged, being flabby and discoloured with dilated ventricles when compared to controls. Other chronic studies have shown an increase in heart size in *db/db* mice with TDGA treatment [125, 126]. However, these same studies showed a beneficial reversal of diabetic renal abnormalities such as immunoglobulin deposition in the glomerular mesangium and mesangial matrix expansion with chronic TDGA treatment [125, 126]. Clearly, longer-term, more detailed studies on the toxicity of these inhibitors are needed to evaluate both the beneficial and potentially deleterious aspects of these compounds. Compounds with enhanced hepatic versus cardiac reactivities may minimize cardiac toxicity.

MISCELLANEOUS FATTY ACID OXIDATION INHIBITORS

A number of other inhibitors of fatty acid oxidation have been reported in the literature [81, 86, 140–142] including 2-bromopalmitoyl-CoA, bromoacetyl-CoA and *S*-methanesulphonyl-CoA decanoylcarnitine, all inhibitors of carnitine palmitoyltransferase, 4-pentenoic acid, an inhibitor of 3-ketoacyl-CoA thiolase and acetoacetyl-CoA thiolase, and hypoglycin, a compound isolated from the Jamaican ackee fruit (*Blighia sapida*) which potently inhibits several acyl-CoA dehydrogenases. The toxicity associated with these compounds or their expense have precluded further development as hypoglycaemic agents.

CONCLUSIONS

The clinical efficacy of fatty acid oxidation inhibitors has not yet been proven. Some beneficial effects have been noted in insulin-dependent diabetics, but no controlled clinical trials in non-insulin-dependent diabetics have been reported. Although the inhibition of fatty acid oxidation causes hypoglycaemia in the fasted state, it is unclear whether these compounds will be effective in an individual with normal eating habits. The safety and efficacy of this class of compounds can be determined only by further studies.

HYDRAZONOPROPIONIC ACIDS

Over 20 years ago, monoamine oxidase inhibitors of the hydrazine type were proposed as supplementary hypoglycaemic agents for treatment of diabetes mellitus [130–133]. Due to their toxicity, this approach was abandoned [134], only to be rekindled by a group at the Institut für Klinische Chemie [135, 136]. They found two hydrazine analogues, (2-phenylethylhydrazono)- and 2-(cyclohexylethylhydrazono)propionic acids (PEHP (10) and CHEHP (11), respectively), with increased hypoglycaemic activity and reduced toxicity. Both compounds at doses of between 145 and 800 μmol/kg (30 and 170 mg/kg, respectively) significantly lowered blood glucose in 48-h-fasted guinea-pigs, rats and hamsters. Glucose-lowering effects were also observed in 12-h-fasted diabetic mice. CHEHP lowered ketone body levels in fasted guinea-pigs, while having no observable effects on plasma triacylglycerol or cholesterol levels. These compounds were ineffective hypoglycaemic agents in fed animals and appeared to cause no change in cellular ATP/ADP ratios, as has been reported for biguanides [137].

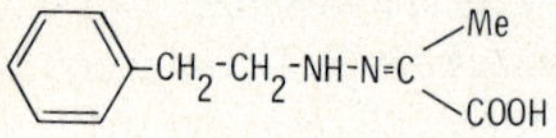

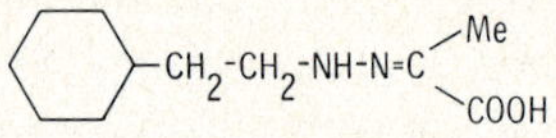

(10)

(11)

EFFECTS ON MONOAMINE OXIDASE

Since compounds of the hydrazine type also inhibit monoamine oxidase, the effects of PEHP and CHEHP on this enzyme were studied. While PEHP inhibited brain and liver monoamine oxidase by more than 80%, CHEHP had only slight inhibitory effects (14% at 145 μmol/kg) 3 h following intraperitoneal injection. *In vitro* characterization of these inhibitors supported these *in vivo* experiments. PEHP was found to behave as a non-competitive inhibitor with a K_i of 70 μmol/l, while the inhibitory effects of CHEHP were consistent with competitive inhibition and a K_i of 340 μmol/l [138]. Despite its inhibitory effects on monoamine oxidase, a 3 month treatment of guinea-pigs with PEHP (145 μmol/kg per day) caused no change in behaviour or observable condition of the animals; such chronic treatment did not alter the acute metabolic effects of the drug, suggesting that either PEHP or CHEHP might be suitable for use as a hypoglycaemic agent. However, a PEHP-CHEHP analogue, 2-(3-methylcinnamylhydrazono)propionate (MCHP, (9)) has recently been synthesized which has little inhibitory effect on monoamine oxidase, while retaining potent hypoglycaemic properties [91].

EFFECTS ON CARBOHYDRATE AND LIPID METABOLISM

Further characterization of MCHP demonstrated that doses as low as 20.5 μmol/kg (5 mg/kg) in the 48-h-fasted guinea-pig caused a significant decline in plasma glucose levels and a reduction in ketone body levels [139]. 7 day treatment with MCHP did not alter the acute effects of the drug, but caused a significant decline in plasma cholesterol levels (1.40 ± 0.55 to 0.59 ± 0.17 mM) [139]. This fall may be due to a reduction in the cholesterol's biosynthetic rate corresponding to a fall in acetyl-CoA levels which was observed. The compound was much less potent in the fasted rat than the fasted guinea-pig; a minimally effective hypoglycaemic dose was 212 mg/kg.

In vitro studies on the hepatic effects of MCHP showed that, in perfused guinea-pig livers, glucose formation from pyruvate or lactate was reduced by 60% [140]. A reduction of glucose formation from alanine by the drug was also observed. In isolated rat hepatocytes, similar effects of MCHP were noted, but again at much higher levels of the drug than in the guinea-pig, paralleling the already mentioned dose-dependent differences between the two species.

EFFECTS ON FATTY ACID TRANSPORT

Recent work [91] has shown MCHP to be an effective inhibitor of long-chain fatty acid transfer across the mitochondrial membrane. In perfused livers (guinea-pig), the inhibitory effect of MCHP on glucose production was abolished when octanoate was substituted for oleate. Ketogenesis from either palmitoylcarnitine or palmitoyl-CoA + carnitine was potently inhibited in isolated liver mitochondria, indicating that MCHP may either inhibit CPT I and/or II or may interfere with the translocation of palmitoylcarnitine across the inner mitochondrial membrane. The studies cited above indicated that this inhibitor is unlike either MeTDGA or the phenylalkyloxiranes, which inhibit only CPT I. A recent report [141] in which the investigators attempted to identify the exact inhibitory locus of MCHP action shows that the site is neither CPT I or CPT II but instead the actual translocation of palmitoylcarnitine across the membrane. Since octanonate relieves the hypoglycaemic effects of MCHP, it is likely that the translocation of only long-chain fatty acids is affected. It is currently unclear by what mechanism MCHP inhibits fatty acid transfer across the mitochondrial membrane.

EFFECTS ON GLUCOSE UPTAKE BY THE JEJUNUM

The influence of these compounds on jejunal glucose uptake was studied [142] to determine if effects similar to biguanide hypoglycaemics on glucose absorption contributed to the observed hypoglycaemia with MCHP. Absorption of glucose and alanine was measured in perfused jejuna isolated from 48-h-fasted rats. Glucose uptake was reduced by CHEHP and MCHP as well as by DBI (phenylethyl biguanide). MCHP was an effective inhibitor of glucose uptake when added either to the vascular or to the mucosal side. Thus, the effect of MCHP and CHEHP on jejunal glucose uptake is similar to the effects of biguanide hypoglycaemics and may play a rôle in the hypoglycaemic activity of MCHP and related hydrazones.

CONCLUSIONS

While hydrazonopropionic acids exert their primary hypoglycaemic action through an inhibition of fatty acid oxidation, other apparently unrelated effects are also evident (i.e., inhibition of monoamine oxidase and jejunal glucose uptake). The use of either TDGA or POCA as a hypoglycaemic agent would seem to offer a more directed and defined approach, with less risk of side-effects unrelated to the primary mechanism of action of the drug.

AS-6

Another novel hypoglycaemic agent, AS-6, (12), is derived from ascochlorin which was first discovered in the filter cake of the fermented broth of the fungus, *Ascochyta viviae* Libert [143]. AS-6, the 4-*O*-carboxymethylated derivative of ascochlorin, is a more potent hypoglycaemic agent and is more readily absorbed from the gut than ascochlorin [144, 145].

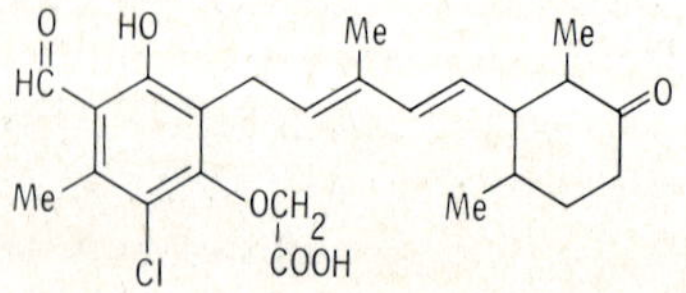

O HC HO Me Me O Me OCH2 COOH Me Cl

(12)

EFFECTS ON CARBOHYDRATE AND FAT METABOLISM

Initial characterization of AS-6 in streptozotocin-diabetic mice showed it to cause a reduction of about 26% in glucose, 31% in free fatty acids and 38% in insulin levels as compared with controls after 7 days of an oral-dosing regimen begun at the same time as the streptozotocin injection [145]. Similarly, in streptozotocin-injected rats, oral administration of the compound for 14 days reduced the level of hyperglycaemia by 24%. In glucose tolerance tests in normal mice, AS-6 treatment caused a more rapid attenuation of the hyperglycaemia than in untreated controls [145]. These early studies indicated that AS-6 might have promise as an antidiabetic drug.

In *db/db* mice, 7 day treatment with AS-6, while not affecting food consumption or weight gain, ameliorated polydipsia, polyuria and glucosuria [146]. This amelioration was associated with a marked decrease in serum glucose and triacylglycerol levels. In the *db/db* mouse, the effects of AS-6 on glucose levels appeared to be synergistic with insulin, since after 1 week of AS-6 treatment insulin alone caused a 20% fall in glucose, AS-6 alone caused a 34% fall, but together they caused a 67% fall.

In vitro studies [147] on isolated adipocytes from *db/db* mice treated for 1 week with AS-6 showed an increase in insulin binding (1.4–3.3-fold), basal glucose uptake (17.5%), insulin-stimulated glucose uptake (58%) and basal lipogenesis (3.4-fold greater than in untreated controls).

EFFECTS ON THE PANCREAS

Administration of AS-6 in *db/db* mice prevented degenerative endocrine pancreatic changes such as degranulation of beta cells, increased basophillic nature of islet exocrine border, infiltration of islets by small round cells, and fat necrosis occurring around the pancreas [146]. However, an increase in size and number of islets normally seen during *db/db* development was not prevented. Similar metabolic and pancreatic effects have been noted for ciglitazone, but the two compounds probably work through different mechanisms, since administration of AS-6 to insulinopoenic diabetic animals [145] had a hypoglycaemic effect while ciglitazone did not.

EFFECTS ON SUBCELLULAR ABNORMALITIES

The effects of AS-6 on certain subcellular abnormalities associated with *db/db* mouse adipocytes have also been examined [148, 149]. Separation of proteins by SDS gel electrophoresis from partially purified plasma membranes and mitochondria isolated from *db/db* and +/? adipocytes showed several significant differences in the pattern of bands observed. AS-6 treatment of *db/db* mice for 7 days caused a reversion of the band pattern of *db/db* membranes to that of +/? membranes [148]. The effects of insulin on ^{32}P incorporation into plasma membrane protein of adipocytes from *db/db* and +/? mice was also measured. When compared with membranes from untreated *db/db* animals on a milligram protein basis, insulin increased labelling 1.8-fold in lean (+/?) and 2.4-fold in AS-6 treated *db/db* membranes. Calcium binding to adipocyte plasma membranes, which is decreased in *db/db* mice, was returned to levels found in lean (+/?) mice with AS-6 treatment. Whether these subcellular effects of AS-6 are important in the restoration of insulin sensitivity in the *db/db* mouse is not currently known.

MISCELLANEOUS EFFECTS

Treatment with AS-6 caused a decrease in cholesterol levels [150, 151], but it is unclear whether this property relates to its ability to enhance insulin action on lipid metabolism or results from other actions. Although ascochlorin originally was associated with cytotoxic effects in HeLa cell cultures [143], chronic studies (10 weeks) in mice with AS-6 showed no effects on food consumption or toxic signs in liver [147].

AS-6 has been less well characterized than ciglitazone and many questions remain unanswered about its mechanism of action, toxicity and clinical efficacy, but it does at least provide a new and promising direction on which to base further novel synthetic efforts.

β-ADRENERGIC AGONISTS

The utility of β-adrenergic agonists as hypoglycaemic agents has been surprising, since acute administration of isoproterenol (isoprenaline) [152] or the more selective β_2 agonist, terbutaline [153], caused a deterioration in glycaemic control in humans. Hepatic glucose production was increased by isoproterenol even in the presence of elevated insulin levels [152]. In addition, the rate of glucose uptake may be suppressed by acute administration of β-agonists, since the adrenaline-mediated inhibition of glucose uptake was reversed by β-blocking agents [154, 155]. However, chronic administration of β-adrenergic agonists improved glucose homeostasis. Following 1 or 2 weeks of daily terbutaline therapy (5 mg, T.I.D.) to normal humans, insulin-stimulated glucose metabolism increased by 29% as measured in the euglycaemic, hyperinsulinemic clamp technique [156]. Glucose storage (non-oxidative glucose disposal) was increased, while oxidative glucose metabolism remained unchanged.

New β-adrenergic agonists have been designed for their utility in treatment of obesity and NIDDM, as opposed to the traditional antiasthmatic β_2 agonists. A subset of β-adrenergic receptors that does not fall clearly into either β_1 or β_2 has been described in rat brown adipose tissue [157]. Selective activation of this receptor, by chronic administration of BRL-26830 (13) to

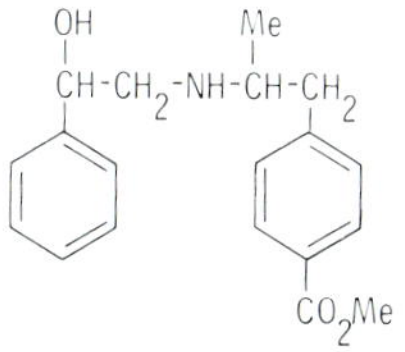

OH Me
CH-CH_2-NH-CH-CH_2
CO_2Me

(13)

genetically obese (57 BL/6, *ob/ob*) mice, stimulated the metabolic activity of brown adipose tissue, elevated caloric consumption (thermogenesis) and resulted in a highly significant reduction in weight gain. Experiments conducted in other models of obesity, including gold thioglucose-treated mice, cafeteria-fed obese mice and genetically obese Zucker (*fa/fa*) rats, demonstrated weight loss without significant reduction in food intake [158].

Although weight loss *per se* may improve glucose tolerance in obese individuals with NIDDM [159–161], BRL-26830 improved glycaemic control in situations where no weight loss occurred. For example, acute administration of this drug (with effective doses being between 0.5 and 10 mg/kg, p.o.) lowered basal glucose levels up to 30% and improved glucose tolerance tests (a maximal reduction of 50% of the area under the glucose curve) in 24-h-fasted normal mice [162]. BRL-26830 stimulated basal insulin release in 24-h-fasted rats and lowered glucose levels following a subcutaneous glucose load. In rats fasted only 5 h, this drug did not affect glucose levels even at doses as high as 186 mg/kg, p.o. Under these conditions, the drug stimulated glucose utilization but also stimulated hepatic glucose production to equivalent levels (measured by radiotracer techniques); thus no net change in glucose levels occurred [163]. Acute treatment of fasted C57BL/6 *ob/ob* mice with BRL-26830 also resulted in either no change or a deterioration of glucose tolerance [162].

Chronic administration of BRL-26830 improved glucose metabolism in a number of animal models of insulin resistance. For example, daily dosing of this β-agonist (1 mg/kg) for 6 weeks to C57BL/6 *ob/ob* mice produced substantial improvements in subcutaneous glucose tolerance tests [162]. A dose of the drug which allowed normal weight gain in *ob/ob* mice was chosen in this study to avoid the complications of an antiobesity effect of BRL-26830 on glucose tolerance (the minimal antiobesity dose in these mice is 5 mg/kg). Chronic treatment of genetically obese, diabetic (C57BL/KsJ, *db/db*) mice with BRL-26830 also produced beneficial effects [164]. Blood glucose and HbA_{1c} returned to near normal levels, while elevations in fed insulin levels were noted and the suppressed pancreatic insulin stores returned to normal levels. Paradoxically, *db/db* mice treated with BRL-26830 were heavier than were controls. Glucose turnover studies in Zucker fatty rats have provided some information concerning the mechanism of the normalization of glucose metabolism in previously insulin-resistant animals [165]. Chronic drug treatment increased total glucose disposal and metabolic clearance rates, but a counterproductive increase was observed in the endogenous glucose appearance rate.

At the cellular level, BRL-26830 increased insulin sensitivity in several tissues. The effects of 3 weeks of drug administration to Zucker rats were studied in isolated soleus muscle [166]. In lean animals (*Fa/?*), an enhanced sensitivity of the glycolytic rate to stimulation by insulin was noted. Similar treatment in obese animals (*fa/fa*) resulted in increased insulin sensitivity of both glycolysis and glycogen synthesis, returning to levels found in lean, untreated controls. Since weight changes in these studies were minimal, the effects of BRL-26830 were most likely due to direct actions on the target tissue. Changes in insulin sensitivity also occurred in adipose tissue following chronic BRL-26830 administration. Adipocyte insulin receptor number and glucose transport rate were increased in obese mice (*ob/ob*), with no change in receptor affinity [167]. In this study, basal glucose metabolism in the major thermogenic tissue, brown adipose tissue (BAT), was also stimulated [168]. Acute administration of BRL-26830 doubled BAT glucose utilization, while chronic dosing with BRL-26830 produced an 8-fold increase in BAT-glucose utilization (1.6% to 12.5% of whole body glucose utilization).

A second β-adrenergic agonist with potential utility in the improvement of insulin sensitivity is Ro-16-8714 (14). When obese mice (C57BL/6J, *ob/ob*) received this agent for 15 days, glycosuria rapidly diminished and blood glucose was normalized, while circulating insulin levels were not altered [169]. During this chronic dosing study, the *in vivo* rate of glucose oxidation was increased. These findings may be related to increased lipogenic activity of brown adipose tissue and subsequent oxidation of newly synthesized fatty acid. Chronic treatment of Zucker fatty rats also produced improvements in glucose metabolism, with normalization of glucose tolerance tests.

In conclusion, improvements in both obesity and hyperglycaemia are noted

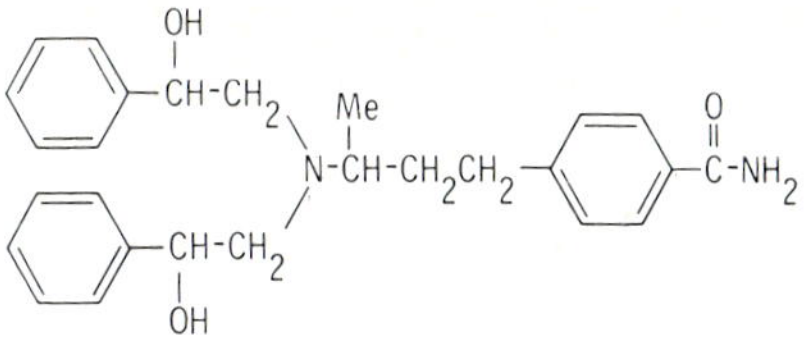

(14)

following chronic administration of β-adrenergic agonists. Brown adipose tissue appears to play an important rôle in some of these actions. Since the activity of brown adipose tissue is questionable in adult humans, the utility of these drugs remains to be established in clinical studies. Unless the β-agonists under study are very selective, undesirable properties associated with other β-agonists, such as tremor, may limit their usefulness. Such drugs would also have to have minimal lipolytic activity at the level of white adipocytes, or increased rates of lipolysis could occur, leading to elevated circulating levels of NEFA, and accelerated fatty acid oxidation.

ANORECTIC AGENTS

Weight loss in the treatment of obese NIDDM is often an effective means of achieving improved glycaemic control [159–161]. When diet therapy alone is inadequate to initiate a weight reduction programme, a variety of anorectic agents are available for short-term therapy [170]. Two of the agents, mazindol and fenfluramine, may also possess activities which improve glucose control independent of the weight loss they induce.

Mazindol, an imidazoisoindole (15) structurally unrelated to amphetamine

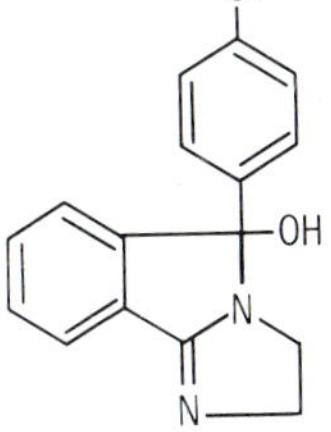

(15)

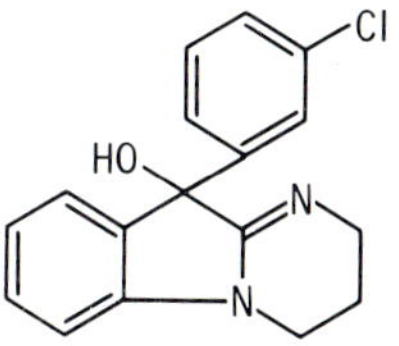

(16)

or fenfluramine, stimulated glucose uptake in isolated human skeletal muscle (gluteus maximus or gluteus medius) either in the presence or absence of insulin [171]. Unlike fenfluramine (see below), this activity is independent of a 5-hydroxytryptamine mechanism [172]. Acute administration of mazindol (4 mg, p.o.) to obese subjects resulted in an improvement in oral glucose tolerance and a reduction in glucose-stimulated insulin secretion. However, no alterations in glucose or insulin profiles were noted following an intravenous glucose challenge, suggesting that mazindol may have altered glucose absorption [173]. Ciclazindol (16), a drug structurally related to mazindol, also stimulated glucose uptake into human skeletal muscle in both the presence and absence of insulin [172]. Obese diabetics treated for 4 weeks with ciclazindol had improvements in oral glucose tolerance similar to those achieved with metformin, although this effect may be due to the modest weight loss induced by ciclazindol [174].

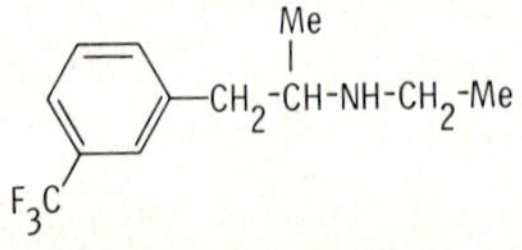

Me

F_3C–C_6H_4–CH_2-CH-NH-CH_2-Me

(17)

Fenfluramine (17) is an anorectic agent structurally similar to amphetamine, but with a mechanism of action relating to its 5-hydroxytryptamine-agonistic activity, which distinguishes it from the amphetamine class of drugs [175]. *In vitro* studies in both rat diaphragm [176] and human skeletal muscle [177] have demonstrated the ability of fenfluramine to stimulate glucose uptake. This effect was not limited to normal muscle, since studies in rat diaphragm from streptozotocin-diabetic animals showed the same effect [178] in the presence of insulin. This process was stereoselective, since the (−) isomer of fenfluramine had significantly greater effects than the (+) isomer [179]. Both the major metabolite of fenfluramine, norfenfluramine, and a closely related compound, flutiorex, also stimulate glucose influx in isolated muscle in the presence of insulin [179, 180]. Fenfluramine action on rat diaphragm is mediated *via* 5-hydroxytryptamine receptors, since fenfluramine-mediated glucose uptake was selectively reversed by the 5-hydroxytryptamine antagonist, methysergide [181]. In addition to its effects on muscle, fenfluramine stimulated glucose oxidation in adipocytes and adipose tissue pieces from humans in the presence or absence of insulin [182]. These effects were small, but significant (15–20% above basal) at therapeutic doses and were more pronounced at higher fenfluramine doses. Fenfluramine displaced ^{125}I-labelled insulin from specific binding sites, suggesting that it may interact with the insulin receptor to have an insulin-like effect. Fenfluramine also stimulated peripheral glucose uptake in canine hind limb [178] and in human forearm [183] perfusion studies.

The effect of fenfluramine on glucose tolerance in obese diabetics independent of weight loss has been investigated in several clinical studies. For example, fenfluramine (60 mg/day for 8 weeks) induced a 20% reduction in fasting glucose levels in obese female diabetics without the added variable of alterations in body weight [184]. Glucose profiles following a standard meal were improved compared with placebo control, although insulin release was similar in both groups. Glycosylated haemoglobin levels were decreased following chronic therapy with fenfluramine when compared with the control group. Improvements in glucose metabolism with fenfluramine independent of weight loss have also been noted in other clinical studies in diabetic [185, 186] and normal [187] subjects.

In conclusion, at least two anorectic agents have effects on glucose metabolism independent of their abilities to induce weight loss. Whether these, or related, agents may be useful hypoglycaemic agents merits further consideration.

STEROIDS

Dehydroepiandrosterone (DHEA, (18)) is a major secretory product of the adrenal cortex [188] which ameliorates several metabolic abnormalities found in obese, insulin-resistant rodents. The first metabolic alteration was noted in mice carrying the viable yellow (genetic obesity) mutation [189] where a 0.2% dietary addition or oral administration (150–500 mg/kg, three times per week, 28 weeks) of DHEA reduced weight gain with no alteration in food consumption. Decreased accumulation of carcass triacylglycerol and decreased

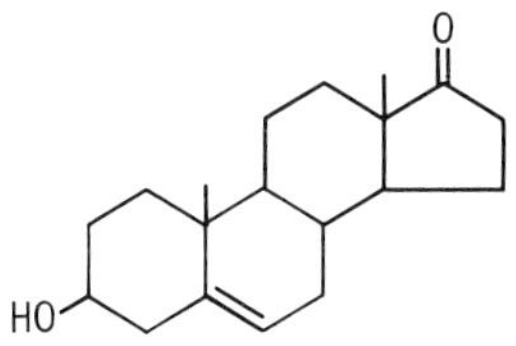

O

HO

(18)

rates of hepatic lipogenesis were also found. Subsequent studies have also demonstrated DHEA-mediated reductions in weight gain in several other animal models of obesity, such as Zucker fatty rats [190, 191] and *db/db* mice on a C57BL/6 background [192].

Substantial improvements in glucose metabolism in insulin-resistant rodents have been demonstrated with chronic dosing of DHEA and its metabolites. These alterations have been observed either in the presence or absence of steroid-induced alterations in body weight (an effect that is highly dependent on the animal strain employed). For example, DHEA prevented severe diabetes in C57BL/KsJ *db/db* mice treated continuously for 18 weeks with a diet containing 0.4% DHEA [192–194], although final weights of these animals were similar to those of untreated diabetic mice. Blood glucose levels were normalized, while plasma and pancreatic insulin levels rose dramatically; beta cell granulation increased 10-fold. In contrast, the *db/db* mutant on a BL/6 background was more sensitive to DHEA, exhibiting markedly reduced weight gain at both 0.1 and 0.4% DHEA in the diet. The transient hyperglycaemia normally observed in these mice was completely prevented. In addition to improvements in genetically obese mice, glucose metabolism in mildly insulin-resistant old mice was also improved by DHEA. 2-year-old mice demonstrated substantial improvements in glucose tolerance tests following 4 weeks of DHEA treatment [194].

Although DHEA elevated insulin levels in the circulation and pancreas of several strains of obese mice, it may also act to improve insulin sensitivity without alterations to insulin release. Prolonged treatment of Zucker fatty (*fa/fa*) rats produced a decrease in body weight coupled with normal glucose levels, a decrease in plasma insulin levels [191, 195] and a reduction in β-cell hyperplasia [195], suggesting improvements in insulin sensitivity in these insulin-resistant animals. These reductions in body weight were not coupled with substantial alterations in food consumption, suggesting that DHEA elevated metabolic rate, a result confirmed in normal rats [196]. Lipid accumulation is reduced in fat deposits [191, 195], as is fatty acid synthetic activity [191]. This action may be the result of an inhibition of glucose-6-phosphate dehydrogenase [191], the source of reducing equivalents for fatty acid biosynthesis. DHEA impaired basal and insulin-stimulated glucose oxidation and conversion to lipids when adipocytes were isolated from treated Zucker fatty rats [197], a result that is consistent with inhibition of glucose-6-phosphate dehydrogenase, but paradoxical in an animal with reduced body weight, fat pad weight, and improved insulin sensitivity (adipocyte glucose metabolism is usually elevated in such animals). DHEA may act on muscle or liver to improve glucose metabolism, possibilities that have not been examined to date.

Although DHEA produces pronounced changes in glucose metabolism and insulin sensitivity, three metabolic products of DHEA, 3α-hydroxyetiocholanolone (α-ET, (19)), 3β-hydroxyetiocholanolone (β-ET, (20)) and DHEA

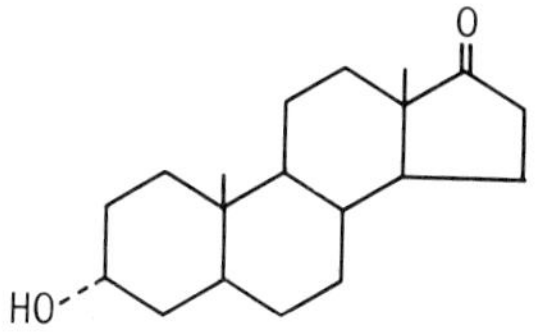

(19)

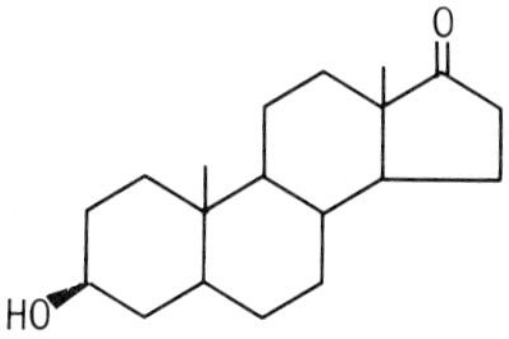

(20)

sulphate, are more potent hypoglycaemic agents [192]. All three of these steroids were approximately 4-times as effective as DHEA in normalizing blood glucose levels and increasing plasma insulin and pancreatic insulin levels in C57BL/KsJ *db/db* mice. In addition, œstradiol had the same effect in male mice but at much lower doses [192]. The benefits of both α-ET and β-ET were potentiated by low-dose œstradiol administration [192]. When studied in an obese moderately diabetic model, the viable yellow mouse (A^{vy}/A), blood glucose and insulin levels and body weight were reduced by these treatments to near normal levels [198].

This approach to therapy appears to suffer several disadvantages. DHEA may be considered a prohormone, since it may be converted not only to the etiocholanolones, but also to testosterone or œstradiol [188]. The possible androgenic or œstrogenic activity of DHEA and DHEA sulphate [199] seems to preclude any widespread use of this steroid. The etiocholanolones are more potent and are metabolic end products (not converted to steroids with androgenic or œstrogenic activity), but appear clinically to have substantial pyrogenic activity [200].

CONCLUSIONS

Diabetes mellitus is a heterogeneous disease that requires various therapeutic strategies, depending on the individual. Both exogenous insulin and sulphonylureas exert their actions by increasing local insulin concentrations at target tissues. However, in some cases, such as obese, insulin-resistant individuals, substantial circulating insulin may be present, but its actions may be impeded. In such situations, administration of hypoglycaemic agents which do not act by increasing insulin release may be the preferred therapeutic approach. Another drawback to the use of sulphonylureas is risk of severe hypoglycaemic episodes. While some of the approaches described in this review suffer from a similar liability, others such as ciglitazone appear to avoid this problem. Admittedly, the toxicities of the novel hypoglycaemic agents discussed herein are not yet well characterized, while the toxicities of the sulphonylureas are well defined as a result of their widespread and lengthy history of use. However, no severe liabilities for any of the novel agents discussed above have yet been reported. Thus, some of the approaches outlined in this review may prove to be valuable additions in controlling the aberrant glucose metabolism of diabetes.

ACKNOWLEDGEMENTS

We wish to thank Dr. Alan Corbin for his critical reading of the manuscript. We would also like to thank Ms. June Deitz for her skilled assistance in the typing and preparation of this text.

REFERENCES

1 W.H. Herman, P. Sinnock, E. Brenner, J.L. Brimberry, D. Langford, A. Nakashima, S.J. Sepe, S.M. Teutsch and R.S. Mazze, Diabetes Care, 7 (1984) 367.
2 H.E. Lebovitz and M.N. Feinglos, Diabetes Mellitus (Medical Examination Publishing Co., New Hyde Park, NY 1983) 3rd Edn. p. 591.
3 G. Schäfer, Diabete Metab., 9 (1983) 148.
4 W. Puls, H.P. Krause, L. Muller, H. Schutt, R. Sitt, G. Thomas, Int. J. Obes. 8 Suppl., 1 (1984) 181.
5 P. Brazeau, W. Vale, R. Burgus, N. Ling, M. Butcher, J. Rivier and R. Guillemin, Science, 179 (1983) 77.
6 D.J. Koerker, W. Ruch, E. Chideckel, J. Palmer, C.J. Goodner, J. Ensinck and C.C. Gale, Science, 184 (1974) 482.
7 M. Brown, J. Rivier and W. Vale, Endocrinology, 98 (1976) 336.
8 K.G.M.M. Alberti, S.E. Christensen, J. Eversen, K. Seyer-Hansen, N.J. Christensen, A.P. Hansen, K. Lundbeak and H. Orskov, Lancet, ii. (1973) 1299.
9 H. Leblanc, J.R., Anderson, M.B. Sigel and S.S.C. Chen, J. Clin. Endocrinol. Metab., 40 (1975) 568.
10 J.E. Gerich, M. Lorenzi, V. Schneider, J. Karam, J. Rivier, R. Guillemin and P.H. Forsham, N. Engl. J. Med., 291 (1974) 544.
11 C. Meissner, C. Thum, W. Beischer, G. Winkler, K.E. Schroder and E.F. Pfeiffer, Diabetes, 24 (1975) 988.
12 J.E. Gerich, M. Lorenzi, D.M. Bier, E. Tsalikian, V. Schneider, J.H. Karam and P.H. Forsham, J. Clin. Invest., 57 (1976) 875.

13 J.E. Gerich, T.A. Schultz, S.B. Lewis and J.H. Karam, Diabetologia, 13 (1977) 537.
14 P. Raskin and R.H. Unger, N. Engl. J. Med., 299 (1978) 433.
15 S.E. Christensen, A.P. Hansen, J. Weeke and K. Lundbeak, Diabetes, 27 (1978) 300.
16 J. Wahren and P. Felig, Lancet, ii (1976) 1213.
17 W. Creutzfeldt, P.G. Lankisch and U.R. Folsch, Dtsch. Med. Wochenschr., 100 (1975) 1135.
18 M. Sheppard, B. Shapiro, B. Pimstone, S. Kronheim, M. Berelowitz and M. Gregory, J. Clin. Endocrinol. Metab., 48 (1979) 50.
19 W.V. Tamborlane, R.S. Sherwin, R. Hendler and P. Felig, N. Engl. J. Med., 297 (1977) 181.
20 W.V. Tamborlane, R.S. Sherwin, R. Hendler and P. Felig, Metabolism, 27 (1978) 849.
21 S. Madsbad, K.G.M.M. Alberti, C. Binder, J.M. Burrin, O.K. Faber, T. Krarup, L. Regeur, Br. Med. J., 2 (1979) 1257.
22 D.F. Veber and R. Saperstein, Annu. Rep. Med. Chem., 14 (1979) 209.
23 J.E. Gerich, Am. J. Med., 70 (1981) 619.
24 P. Marbach, M. Neufeld and J. Pless, Adv. Exp. Med. Biol., 188 (1985) 339.
25 J. Rivier, M. Brown and W. Vale, Biochem. Biophys. Res. Commun., 65 (1975) 746.
26 M. Brown, J. Rivier and W. Vale, Endocrinology, 98 (1976) 336.
27 M. Brown, J. Rivier and W. Vale, Science, 196 (1977) 1467.
28 C. Meyers, A. Arimura, A. Gordin, R. Fernandez-Durango, D.H. Coy, A.V. Schally, J. Drouin, L. Ferland, M. Beaulieu and F. Labrie, Biochem. Biophys. Res. Commun., 74 (1977) 630.
29 E.L. Lien, Metabolism, 27 (1978) 1095.
30 P. Lins, S. Efendic, C.A. Meyers, D.H. Coy, A. Schally and R. Luft, Metabolism, 29 (1980) 728.
31 V. Schusdziarra, J. Rivier, R. Dobbs, M. Brown, W. Vale and R. Unger, Horm. Metab. Res., 10 (1978) 563.
32 G. Lenti, M. Trovati, R. Lorenzati, F. Vitelli, V. Tagliaferro, A. Marocco and G. Pangano, Acta Diabetol. Lat., 17 (1980) 9.
33 C. Neuhaus, W. Kerner, V. Maier and E.F. Pfeiffer, Horm. Met. Res. 8, Suppl., (1979) 115.
34 H. Fukushima, H. Maeda, K. Suzaki, M. Shimada, M. Sakakida, H. Kishikawa, N. Nakamura and H. Uzawa, Endocrinol. Jpn., 2 (1985) 241–48.
35 Y. Harano, H. Hidaka, T. Nakano, J. Emura, T. Kimura, S. Sakakara and Y. Skigeta, Biomed. Res., 1 (1980) 560.
36 D. Sarantakis, J. Teichman, R. Fenichel, E. Lien, FEBS Lett., 92 (1978) 153.
37 E. Lien and D. Sarantakis, Diabetologia, 17 (1979) 59.
38 E.L. Lien, J. Greenwood and D. Sarantakis, Diabetes, 28 (1979) 491.
39 G. Dimitriadis, P. Tessari and J. Gerich, Metabolism, 32 (1983) 987.
40 G. Dimitriadis and J. Gerich, Horm. Metab. Res., 17 (1985) 510.
41 J.L. DeYoung and E.L. Lien, J. Parenter. Sci. Technol., 6 (1984) 234.
42 C. Martin, B. Wallum, B. Drom, L. Hall and J. Gerich, Life Sci., 35 (1984) 2627.
43 W. Vale, J. Rivier, N. Ling and M. Brown, Metabolism, 27 Suppl. 1 (1978) 1391.
44 T.E. Adrian, A.J. Barnes, R.G. Long, D.J. O'Shaughnessy, M.R. Brown, J. Rivier, W. Vale, A.M. Blackburn and S.R. Bloom, J. Clin. Endocrinol. Metab., 53 (1981) 675.
45 A.J. Barnes, R.G. Long, T.G. Adrian, W. Vale, M.R. Brown, J.E. Rivier, J. Hanley, M Ghatei, D.L. Sarson and S.R. Bloom, Clin. Sci., 61 (1981) 653.
46 D.F. Veber, F.W. Holly, R.F. Nutt, S.J. Bergstrand, S.F. Brady, R. Hirschmann, M.S. Glitzer and R. Saperstein, Nature (London), 280 (1979) 512.

47 D.F. Veber, R.M. Freidinger, D.S. Perlow, W.J. Paleveda Jr., F.W. Holly, R.G. Strachan, R.F. Nutt, B.H. Arison, C. Homnick, W.C. Randall, M.S. Glitzer, R. Saperstein and R. Hirschmann, Nature (London), 292 (1981) 55.
48 D.F. Veber, R. Saperstein, R.F. Nutt, R.M. Freidinger, S.F. Brady, P. Curley, D.S. Perlow, W.J. Paleveda, C.D. Colton, A.G. Zacchei, D.J. Tocco, D.R. Hoff, R.L. Vandlen, J.E. Gerich, L. Hall, L. Mandarino, E.H. Cordes, P.S. Anderson and R. Hirschmann, Life Sci., 34 (1984) 1371.
49 L. Hall, D. Stenner, W. Blanchard, D. Veber, R. Saperstein, and J. Gerich, Diabetes, 30 (Suppl. 1) (1981) 44A.
50 I. Gottesman, J. Tobert, R. Vandlen and J. Gerich, Life Sci., 38 (1986) 2211.
51 W. Barnes, U. Briner, W. Doepfner, R. Haller, R. Huguenin, P. Marbach, T.J. Petcher and J. Pless, Life Sci., 31 (1982) 1133.
52 E. del-Pozo, M. Neufeld, K. Schluter, F. Tortosa, P. Clarenbach, E. Bieder, L. Wendel, E. Nuesch, P. Marbach, H. Cramer and L. Kerp, Acta Endocrinol. (Copenhagen), 111 (1986) 433.
53 G.A. Spinas, A. Bock and U. Keller, Diabetes Care, 8 (1985) 429.
54 G. Plewe, J. Beyer, U. Krause, M. Neufeld, E. del-Pozo, Lancet, ii (1984) 782.
55 S.W. Lamberts, R. Dosterom, M. Neufeld, E. del-Pozo, J. Clin. Endocrinol. Metab., 60 (1985) 1161.
56 S.W. Lamberts, M. Zweens, L. Verschoor, E. del-Pozo, J. Clin. Endocrinol. Metab., 63 (1986) 16.
57 T. Sohda, K. Mizuno, H. Tawada, Y. Sugiyama, T. Fujita and Y. Kawamatsu, Chem. Pharm. Bull., 30 (1982) 3563.
58 T. Sohda, K. Mizuno, E. Imamiya, Y. Sugiyama, T. Fujita and Y. Kawamatsu, Chem. Pharm. Bull., 30 (1982) 3580.
59 T. Sohda, K. Mizuno, E. Imamiya, H. Tawada, K. Meguro, Y. Kawamatsu and Y. Yamamoto, Chem. Pharm. Bull., 30 (1982) 3601.
60 T. Sohda, K. Meguro and Y. Kawamatsu, Chem. Pharm. Bull., 32 (1984) 2267.
61 T. Fujita, Y. Sugiyama, S. Taketomi, T. Sohda, Y. Kawamatsu, H. Iwatsuka and Z. Suzouk, Diabetes, 32 (1983) 804.
62 S. Taketomi, H. Ikeda, E. Ishikawa and H. Iwatsuka, Horm. Metab. Res., 14 (1982) 14.
63 A.Y. Chang, B.M. Wyse, B.J. Gilchrist, T. Peterson and A.R. Diani, Diabetes, 32 (1983) 830.
64 A.Y. Chang, B.M. Wyse and B.J. Gilchrist, Diabetes, 32 (1983) 839.
65 M. Kobayashi, M. Iwasaki, S. Ohgaku, H. Maegawa, N. Watanabe and Y. Shigeta, FEBS Lett., 163 (1983) 50.
66 D.M. Kirsch, W. Bachmann and H.U. Haering, FEBS Lett., 176 (1984) 49.
67 A.Y. Chang, B.J. Gilchrist and B.M. Wyse, Diabetologia, 25 (1983) 514.
68 N.S. Shargill, A. Tatoyan, M. Fukushima, D. Antwi, G.A. Bray and T.M. Chan, Metabolism, 35 (1986) 64.
69 A.R. Diani, T. Peterson, G.A. Sawada, B.M. Wyse, B.J. Gilchrist, A.E. Hearron and A.Y. Chang, Diabetologia, 27 (1984) 225.
70 A.R. Diani, T. Peterson, G. Sawada, K. Jodelis, B.M. Wyse, B.J. Gilchrist, A.E. Hearron and A.Y. Chang, Nephron, 42 (1986) 72.
71 S.W. Mercer and P. Trayhurn, Biosci. Rep., 4 (1984) 933.
72 S.W. Mercer and P. Trayhurn, Biochem. J., 212 (1983) 393.
73 D.G. Nicholls, Biochim. Biophys. Acta, 549 (1979) 1.
74 S.W. Mercer and P. Trayhurn, FEBS Lett., 195 (1986) 12.
75 S.R. Cox, E.L. Harrington, R.A. Hill, V.J. Capponi and A.C. Shah, Biopharm. Drug Dispos., 6 (1985) 67.

76 S.R. Cox, E.L. Harrington and V.J. Capponi, Biopharm. Drug Dispos., 6 (1985) 81.
77 H. Torii, K. Yoshida, T. Tsukamoto and S. Tanayama, Xenobiotica, 14 (1984) 259.
78 K. Meguro and T. Fujita, Eur. Pat. Appl., 85 307 084.5 (1985).
79 P.J. Randle, P.B. Garland, C.N. Hales and E.A. Newsholme, Lancet, i (1963) 785.
80 G.F. Tutweiler, T. Kirsch, R.J. Mohrbacher and W. Ho, Metabolism, 27 (1978) 1539.
81 H.S.A. Sherratt, Br. Med. Bull., 25 (1969) 250.
82 D. Billington, H. Osmundsen and H.S.A. Sherratt, Biochem. Pharmacol., 27 (1978) 2891.
83 K. Eistetter and H.P.O. Wolf, J. Med. Chem., 25 (1982) 109.
84 J.D. McGarry and D. Foster, Annu. Rev. Biochem., 49 (1980) 395.
85 P.E. Declerq, M.D. Venincasa, S.E. Mills, D.W. Foster and J.D. McGarry, J. Biol. Chem., 260 (1985) 12516.
86 M.R. Edwards, M.I. Bird and E.D. Saggerson, Biochem., J., 230 (1985) 169.
87 T.C. Kiorpes, D. Hoerr, W. Ho, L.E. Weaner, M.G. Inman and G.F. Tutweiler, J. Biol. Chem., 259 (1984) 9750.
88 P. Roesen and H. Reinauer, Metabolism 33 (1984) 177.
89 H.P.O. Wolf and D.W. Engel, Eur. J. Biochem., 146 (1985) 359.
90 K. Kanamaru, S. Shinagawa, M. Asai, H. Okazaki, Y. Sugiyama, T. Fujita, H. Iwatsuka and Y. Yoneda, Life Sci., 37 (1985) 217.
91 I.V. Deaciuc, H.F. Kuehnle, K.M. Strauss and F.H. Schmidt, Biochem. Pharmacol., 32 (1983) 3405.
92 D.M. Turnbull, K. Bartlett, S.I.M. Younan and H.S.A. Sherratt, Biochem. Pharmacol. 33 (1984) 475.
93 G.F. Tutweiler and M.T. Ryzlak, Life Sci., 26 (1980) 393.
94 E.D. Saggerson and C.A. Carpenter, FEBS Lett., 129 (1981) 225.
95 P.P. Koundakjian, D.M. Turnbull, A.J. Bone, M.P. Rogers, S.I.M. Younan and H.S.A. Sherratt, Biochem. Pharmacol., 33 (1984) 465.
96 S.E. Mills, D.W. Foster and J.D. McGarry, Biochem. J., 214 (1983) 83.
97 G.F. Tutweiler, H.J. Brentzel and T.C. Kiorpes, Proc. Soc. Exp. Biol. Med., 178 (1985) 288.
98 G.F. Tutweiler, R. Mohrbacher and W. Ho, Diabetes, 28 (1979) 42.
99 F.J. Pearce, J. Forster, G. DeLeeuw, J.R. Williamson and G.F. Tutweiler, J. Mol. Cell. Cardiol., 11 (1979) 893.
100 G.F. Tutweiler and P. Dellevigne, J. Biol. Chem., 254 (1979) 2935.
101 H.S.A. Sherratt, S.J. Gatley, T.R. DeGrado, C.K. Ng and J.E. Holden, Biochem. Biophys. Res. Commun., 117 (1983) 653.
102 F. Demaugre, H.A. Buc, C. Cepanac, A. Moncion and J.P. Leroux, Biochem. J., 222 (1984) 343.
103 L. Agius, D. Pillay, K.G.M.M. Alberti and H.S.A. Sherratt, Biochem. Pharmacol., 34 (1985) 2651.
104 G.F. Tutweiler and H.J. Brentzel, Eur. J. Biochem., 124 (1982) 465.
105 C. Schudt and A. Simon, Biochem. Pharmacol., 33 (1984) 3357.
106 F. Demaugre, H.A. Buc, C. Cepanec, A. Moncion and J.P. Leroux, Biochem. J., 222 (1984) 343.
107 A.L. Kerbey, P.J. Randle, R.H. Cooper, S. Whitehouse, H.T. Pask and R.M. Denton, Biochem. J., 154 (1976) 327.
108 A.L. Kerbey, P.M. Radcliffe and P.J. Randle, Biochem. J., 164 (1977) 509.
109 I.D. Catterson, S.J. Fuller and P.J. Randle, Biochem. J, 208 (1982) 53.
110 I.D. Catterson, P.F. Williams, A.L. Kerbey, L.D. Astbury, W.E. Plehwe and J.R. Turtle, Biochem. J., 224 (1984) 787.

111 N. Altszuler, B. Winkler, R.W. Tuman and C.R. Bowden, Diabetes, 33 Suppl. 1 (1984) 169A.
112 M.A. Davis, P.E. Williams and A.D. Cherrington, Am. J. Physiol., 247 (1984) E362.
113 H.P.O. Wolf, K. Eistetter and G. Ludwig, Diabetologia, 22 (1982) 456.
114 S.C. Frost and M.A. Wells, Arch. Biochem. Biophys., 211 (1981) 547.
115 L. Mandarino, E. Tsalikian, S. Bartold, H. Marsh, A. Carney, E. Buerklin, G. Tutweiler, M. Haymond, B. Handwerger and R. Rizza, J. Clin. Endocrinol. Metab., 59 (1984) 658.
116 J. Verhaegen, J. Leempoels, J. Brugmans and G.F. Tutweiler, Diabetes, 33 Suppl. 1 (1984) 180A.
117 J.C. Young, J.E. Bryan, S.H. Constable, G.F. Tutweiler and J.O. Holloszy, Can. J. Physiol. Pharmacol., 62 (1984) 815.
118 J.C. Young, J.L. Treadway, E.I. Fader and R.F. Caslin, Diabetes, 35 (1986) 744.
119 G.F. Tutweiler, J. Joseph and N. Wallace, Biochem. Pharmacol., 34 (1985) 2217.
120 R. Seitelberger, O. Kraupp, A. Beck, S. Bacher and G. Raberger, J. Cardiovas. Pharmacol., 6 (1984) 902.
121 R. Seitelberger, O. Kraupp, M. Winkler, G. Brugger and G. Raberger, J. Cardiovas. Pharmacol., 7 (1985) 273.
122 A.G. Tahiliani and J.H. McNeil, Can J. Physiol. Pharmacol., 63 (1985) 925.
123 W.H. Dillman, Am. J. Physiol., 248 (1985) E602.
124 E. Bachmann, E. Weber and G. Zbinden, Biochem. Pharmacol., 33 (1984) 1947.
125 S.M. Lee, G. Tutweiler, R. Bressler and C. Kircher, Diabetes, 31 (1982) 12.
126 S.M. Lee, J.J. Bull and R. Bressler, Biochem. Med., 33 (1985) 104.
127 J.C. Fong and H. Schula, J. Biol. Chem., 253 (1978) 6917.
128 G. Delisle and I.B. Fritz, Proc. Natl. Acad. Sci. USA, 58 (1967) 790.
129 G.A. Stewart and T. Hanley, in Oral Hypoglycemic Agents, ed. G.D. Campbell (Academic Press, New York, 1969), pp. 347–407.
130 H.M. Van Praag and B. Leijnse, Clin. Chim. Acta, 8 (1963) 466.
131 A.J. Cooper and K.M.G. Keddie, Lancet, i (1964) 1133.
132 L. Wickstroem and K. Petterson, Lancet, ii (1964) 995.
133 P.I. Adnitt, Diabetes, 17 (1968) 628.
134 L.I. Goldberg, J. Am. Med. Assoc., 190 (1964) 132.
135 R. Haeckel and M. Oellerich, Eur. J. Clin. Invest., 7 (1977) 393.
136 R. Haeckel and M. Oellerich, Horm. Metab. Res., 11 (1979) 606.
137 R. Haeckel and H. Haeckel, Diabetologia, 8 (1972) 117.
138 M. Oellerich and R. Haeckel, Horm. Metab. Res., 12 (1980) 182.
139 M. Oellerich, R. Haeckel, K.H. Wirries, G. Schumann and M. Beneking, Horm. Metab. Res., 16 (1984) 619.
140 R. Haeckel, M. Oellerich, G. Schumann and M. Beneking, Horm. Metab. Res., 17 (1985) 115.
141 F.H. Schmidt, I.V. Deaciuc and H.F. Kuehnle, Life Sci., 36 (1985) 63.
142 R. Haeckel, H. Terbutter, G. Schuman and M. Oellerich, Horm. Metab. Res., 16 (1984) 423.
143 G. Tamura, S. Suzuki, A. Takatsuki, K. Ando and K. Aruma, J. Antibiot., 21 (1968) 539.
144 Y. Nawata, K. Ando, G. Tamura, K. Arima and Y. Iitaka, J. Antibiot., 22 (1969) 511.
145 T. Hosokawa, M. Sawada, K. Ando and G. Tamura, Agric. Biol. Chem., 46 (1982) 775.
146 T. Hosokawa, K. Ando and G. Tamura, Diabetes, 34 (1985) 267.
147 T. Hosokawa, K. Ando and G. Tamura, Biochem. Biophys. Res. Commun., 126 (1985) 471.

148 T. Hosokawa, K. Ando and G. Tamura, Biochem. Biophys. Res. Commun., 125 (1984) 64.
149 T. Hosokawa, K. Ando and G. Tamura, Biochem. Biophys. Res. Commun., 127 (1985) 247.
150 T. Hosokawa, M. Sawada, K. Ando and G. Tamura, Agric. Biol. Chem., 44 (1980) 2461.
151 T. Hosokawa, K. Ando and G. Tamura, Agric. Biol. Chem., 46 (1982) 775.
152 L. Sacca, G. Morrone, M. Cicala, G. Corso and B. Ungaro, J. Clin. Endocrinol. Metab., 50 (1980) 680.
153 S. Carlstrom and H. Westling, Acta Med. Scand., 512 (1970) 33.
154 D.C. Deibert and R.A. DeFronzo, J. Clin. Invest., 65 (1980) 717.
155 R.A. Rizza, P.E. Cryer, M.W. Haymond and J.E. Gerich, J. Clin. Invest., 65 (1980) 682.
156 K. Scheidegger, D.C. Robbins and E. Danforth, Jr., Diabetes, 33 (1984) 1144.
157 J.R.S. Arch, A.T. Ainsworth, M.A. Cawthorne, V. Piercy, M.V. Sennitt, V.E. Thody, C. Wilson and S. Wilson, Nature (London), 309 (1984) 163.
158 J.R.S. Arch and A.T. Ainsworth, Am. J. Clin. Nutr., 38 (1983) 549.
159 L.H. Nerbergh, Ann. Intern. Med., 17 (1952) 935.
160 J.M. Olefsky, J. Clin. Invest., 57 (1976) 842.
161 J.M. Olefsky and O.G. Kolterman, Am. J. Med., 70 (1981) 151.
162 M.V. Sennitt, J.R.S. Arch. A.L. Levy, D.L. Simson, S.A. Smith and M.A. Cawthorne, Biochem. Pharmacol., 34 (1985) 1279.
163 M.A. Cawthorne, M.J. Carroll, A.L. Levy, C.A. Lister, M.V. Sennitt, S.A. Smith and P. Young, Int. J. Obes, 8 Suppl. 1 (1984) 93.
164 M.J. Carroll, C.A. Lister, M.V. Sennitt, N. Steward-Long and M.A. Cawthorne, Diabetes, 34 (1985) 1198.
165 S.A. Smith, A.L. Levy, M.V. Sennitt, D.L. Simson and M.A. Cawthorne, Biochem. Pharmacol., 34 (1985) 2425.
166 R.A.J. Challis, L. Budohoski, E.A. Newsholme, M.V. Sennitt and M.A. Cawthorne, Biochem. Biophys. Res. Commun., 128 (1985) 928.
167 P. Young, L. King and M.A. Cawthorne, Biochem. Biophys. Res. Commun., 133 (1985) 457.
168 P. Young, M.A. Cawthorne and S.A. Smith, Biochem. Biophys. Res. Commun. 130 (1985) 241.
169 M.K. Meier, L. Alig, M.E. Börgi-Saville and M. Müller, Int. J. Obes., 8 (1984), Suppl. 1, 215.
170 J.G. Douglas and J.F. Munro, Drugs, 21 (1981) 362.
171 M.J. Kirby and P. Turner, J. Pharm. Pharmacol., 28 (1976) 163.
172 M.J. Kirby and P. Turner, Br. J. Clin. Pharmacol., 4 (1977) 459.
173 L.C. Harrison, A.P. King-Roach and K.C. Sandy, Metabolism, 24 (1975) 1353.
174 M.E.J. Lean and L.J. Borthwick, Eur. J. Clin. Pharmacol., 25 (1983) 41.
175 S. Garathni, A. Bizzi, G. deGaetano, A. Jori and R. Saminin, in Recent Advances in Obesity Research, ed. A. Howard (J. Libbey, London 1975), p. 354.
176 J.S. Bajaj and J. Vallance-Owen, Horm. Metab. Res., 6 (1974) 85.
177 M.F. Kirby and P. Turner, Postgrad. Med. J., 51 Suppl. 1 (1975) 73.
178 K.D. Jorgensen, Acta Pharmacol. Toxicol., 40 (1977) 401.
179 M.J. Kirby and P. Turner, Postgrad. Med. J., 51 (1975) 73.
180 M.J. Kirby, H. Cara-Georgious-Markomihalakis and P. Turner, Br. J. Clin. Pharmacol., 2 (1975) 541.
181 M.J. Kirby and P. Turner, Arch. Int. Pharmacodyn. Ther., 225 (1977) 25.
182 L.C. Harrison, F.I.R. Martin, A. King-Roach and R.A. Melick, Postgrad, Med. J. 51 (1975) 110.

183 W.H.J. Butterfield, and M.J. Whichelow, Lancet, i (1968) 109.
184 M. Verdy, L. Charbonneau, I. Verdy, R. Belanger, E. Bolte and J.L. Chiasson, Int. J. Obes., 7 (1983) 289.
185 J.K. Wales, Acta Endocrinol., 90 (1979) 616.
186 P. Hamet, I. Verdy, J.L. Chiasson and M. Verdy, Clin. Invest. Med., 9 (1986) 12.
187 J.R.W. Dykes, Postgrad. Med. J., 49 (1973) 314.
188 R.H. Williams, Textbook of Endocrinology (W.B. Saunders, Philadelphia, PA) 5th Edn. (1974).
189 T.T. Yen, J.A. Allan, D.V. Pearson, J.M. Acton and M.M. Greenberg, Lipids, 12 (1977) 409.
190 M.P. Cleary, A. Shepherd and B. Jenks, J. Nutr., 114 (1984) 1242.
191 A. Shepherd and M.P. Cleary, Am. J. Physiol., 246 (1984) E123.
192 D.L. Coleman, E.H. Leiter and N. Applezweig, Endocrinology, 115 (1984) 239.
193 D.L. Coleman, E.H. Leiter and R.W. Schwizer, Diabetes, 31 (1982) 830.
194 D.L. Coleman, R.W. Schwizer and E.H. Leiter, Diabetes, 33 (1984) 26.
195 T.S. Gansler, S. Muller and M.P. Cleary, Proc. Soc. Exp. Biol. Med., 180 (1985) 155.
196 A.R. Tagliaferro and J.R. Davis, Fed. Proc., 42 (1983) 326.
197 S. Muller and M.P. Cleary, Metabolism, 34 (1985) 278.
198 D.L. Coleman, Endocrinology, 117 (1985) 2279.
199 M.J.K. Harper, Endocrinology, 84 (1969) 229.
200 S.M. Wolff, Ann. Intern. Med., 67 (1967) 1268.

Progress in Medicinal Chemistry – Vol. 24, edited by G.P. Ellis and G.B. West

7 Calcium Channel Blocking Drugs

HELEN E. KENDALL, Ph.D., M.P.S. and DAVID K. LUSCOMBE, Ph.D., F.P.S.

Medicines Research Unit, Division of Clinical Pharmacy, Welsh School of Pharmacy, UWIST, P.O. Box 13, Cardiff, CF1 3XF, United Kingdom

INTRODUCTION

Over the last two decades, many new drugs have been developed for the treatment of cardiovascular disorders. One of the most important discoveries has been the existence of drugs which inhibit the entry of calcium ions into cells by blocking the voltage-dependent calcium channels. Fleckenstein first introduced the term 'calcium antagonist' to describe compounds which interfere directly with the transmembrane calcium supply [1]. He considered that calcium antagonists should be divided into specific and non-specific inhibitors, depending on whether the calcium antagonism is responsible for the major action of the compound or whether the antagonism is merely a side-effect which becomes apparent only when high doses are administered. A further subdivision of calcium antagonists has recently been suggested [2], which divides calcium antagonists into calcium availability inhibitors and calcium effect inhibitors. However, since drugs may inhibit the availability of calcium by blocking voltage-dependent calcium channels, by stimulating the efflux of calcium from the cell, or by stimulating the uptake of calcium into storage sites

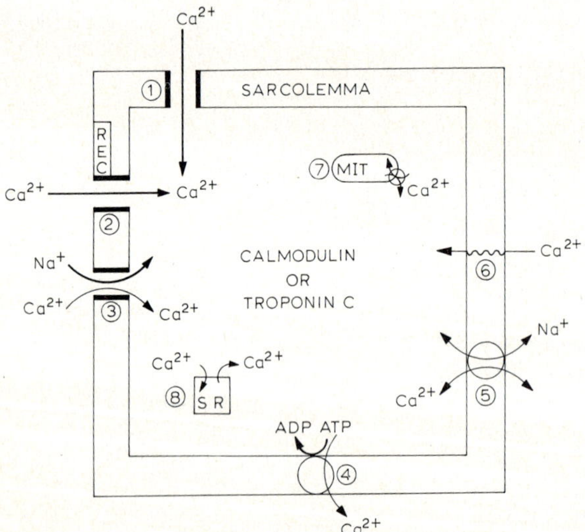

Figure 7.1. Diagrammatic representation of cellular Ca^{2+} movements. Calcium ions (Ca^{2+}) may cross the sarcolemma by the following routes: (1) voltage-operated calcium channels; (2) receptor-operated calcium channels; (3) fast sodium channels; (4) Ca^{2+}-ATPase pump; (5) Na^+–Ca^{2+} exchange; and (6) Ca^{2+} 'leak' pathways. Inside the cell, Ca^{2+} may be taken up into and released from mitochondria (MIT) (7) and sarcoplasmic reticulum (SR) (8). Ca^{2+} also binds to the intracellular proteins, calmodulin and troponin C. REC represents a specific membrane receptor site.

(see *Figure 7.1*), the term calcium antagonist is clearly a rather imprecise one in describing the mechanism of action of such drugs. The use of other terms such as calcium entry blocker, calcium inhibitor or slow channel blocking drug is similarly inexact. Currently, it is considered that drugs which act through inhibition of the inward slow calcium current by preventing the passage of Ca^{2+} ions through the calcium channel in the cell membrane are best described as calcium channel blocking drugs.

These calcium channel blocking drugs form a diverse group of substances. Although their major mode of action is inhibition of the slow calcium current in cardiac and smooth muscle cells, many drugs in this category may also exert other effects, as for example, on the fast sodium current or as calmodulin antagonists. An expert committee appointed by the World Health Organisation has recently divided the calcium channel blocking drugs into six separate classes [2a]. The phenylalkylamines such as verapamil (1) constitute class I, the 1,4-dihydropyridines, for example, nifedipine (2) form class II, whereas diltiazem (3) and KB944 (fostedil, 4) are in class III. Drugs such as flunarizine

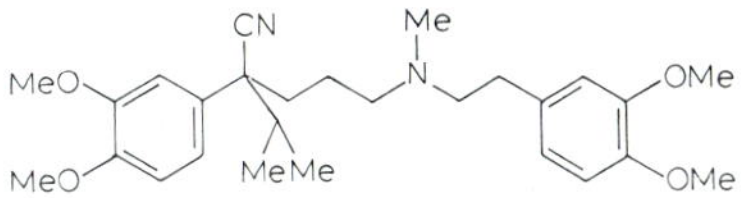

(1)

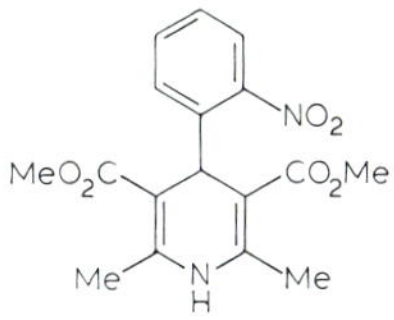

(2)

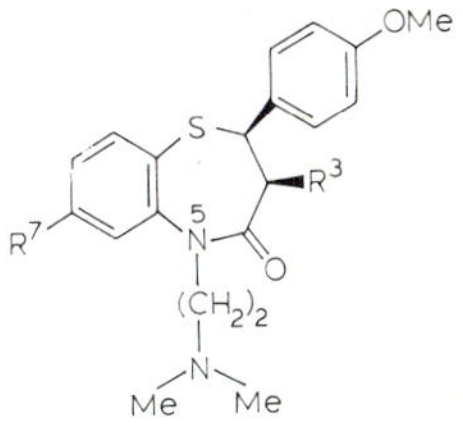

(3) R^3 = OCOMe
R^7 = H

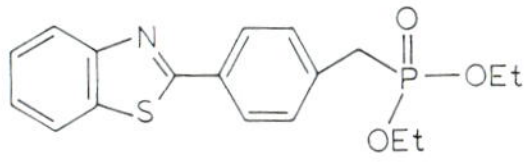

(4)

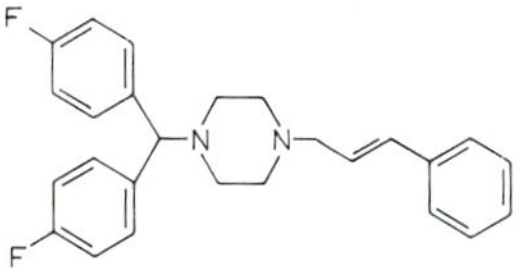

(5)

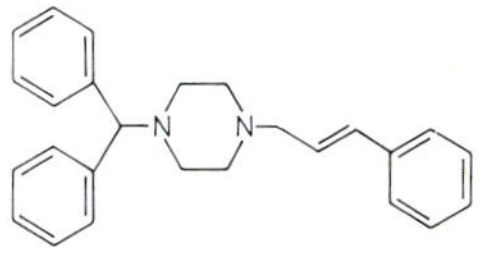

(6)

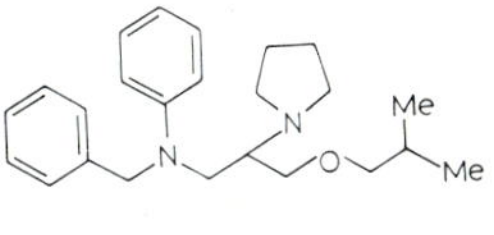

(7)

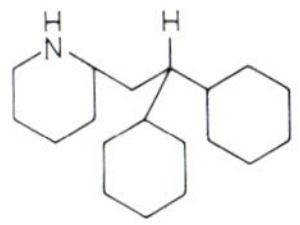

(8)

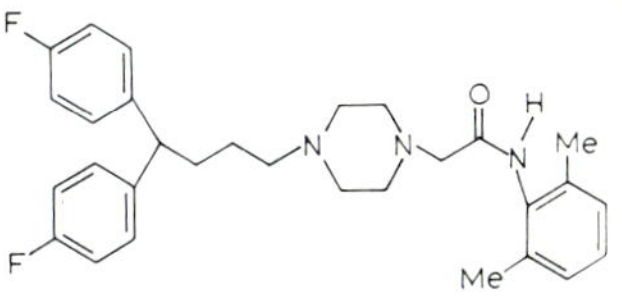

(9)

(5), and cinnarizine (6) are included in class IV and prenylamine-like compounds form class V. The last class (class VI) contains all other compounds which have some calcium channel blocking activity, for example, bepridil (7), perhexiline (8) and lidoflazine (9). This review discusses mainly the drugs in classes I, II and III. Although the class IV, V and VI drugs do block calcium channels, their effects are much less selective; for example, many of these drugs bind to calmodulin and inhibit its activity at similar concentrations [3, 4]. Thus it is uncertain whether these drugs cause vasodilation as a result of calcium channel blockade or as a result of calmodulin antagonism. Another difference between the drugs in classes IV, V and VI compared with those in classes I, II and III is that drugs such as flunarizine, cinnarizine and pimozide inhibit Ca^{2+}-induced contractions in skinned guinea-pig taenia preparations, whereas nifedipine, verapamil and diltiazem do not inhibit such contractions [5]. This suggests that an intracellular site of action may contribute to the calcium antagonist effect of class IV, V and VI drugs. Bepridil and lidoflazine have been shown to block the fast sodium channels in Purkinje fibres, atrial fibres and papillary muscles [6–9] and this may also complicate the interpretation of their pharmacological effects.

EFFECTS OF CALCIUM CHANNEL BLOCKADE

EXCITATION-CONTRACTION COUPLING

The passage of calcium into cardiac and smooth muscle cells acts as the major trigger for their contraction, although there are differences in the way in which actin and myosin interact in the two types of muscle cell. The transmembrane influx of Ca^{2+} is itself too small to initiate a contraction, but it has been suggested that the inward Ca^{2+} current releases more Ca^{2+} from stores in the sarcoplasmic reticulum [10–13]. Ca^{2+} ions then bind to regulatory proteins, which bring about an interaction between actin and myosin. In cardiac cells, the regulatory protein is troponin C. Binding of Ca^{2+} to troponin C displaces tropomyosin (which inhibits actin-myosin cross-bridging) and allows the thin actin filaments to slide over the thick myosin filaments, resulting in contraction [10, 13]. The regulatory protein in smooth muscle is calmodulin. In the presence of sufficient intracellular Ca^{2+}, calmodulin binds to and activates myosin light-chain kinase [11–13]. This produces phosphorylation of myosin and activates cross-bridging, which again results in contraction [14, 15]. Thus the influx of Ca^{2+} into muscle cells is essential for the contraction of cardiac and smooth muscle.

Calcium channel blockers act to uncouple excitation-contraction coupling in myocardium and smooth muscle by inhibiting the influx of Ca^{2+}. Vasodilation [1, 16] and depression of atrial and ventricular activity [17, 18] occur as a direct result. The calcium channel blockers prevent contraction in response to depolarization of the cell membrane by electrical stimulation [17, 19–21] or by agonists such as high potassium ion concentrations [22–33], noradrenaline [22–24, 28, 29, 32, 33], 5-hydroxytryptamine [21, 24] or prostaglandin $F_{2\alpha}$ [21, 24].

RECEPTOR-OPERATED AND VOLTAGE-OPERATED CHANNELS

Differences exist in the ability of calcium channel blockers to inhibit contractions induced by high potassium ion (K^+) depolarization compared with contractions resulting from the interaction of agonists with their receptor sites. In the majority of tissues, K^+-invoked responses are blocked to a greater extent than responses to noradrenaline (NA) (norepinephrine) [22–24, 33]. However, in the rat portal vein, both diltiazem and nifedipine inhibit contractions induced by NA significantly more than those induced by K^+ [29]. Furthermore, in rabbit mesenteric resistance vessels, Ca^{2+} influx stimulated by NA is more sensitive to inhibition by diltiazem than is Ca^{2+} influx stimulated by K^+ [22].

This difference between agonist-induced responses and K^+-induced responses has been attributed to the fact that some agonist responses appear to be mediated partially by release of Ca^{2+} from intracellular stores as well as by influx of Ca^{2+} *via* calcium channels [22, 24, 34], whereas K^+-induced responses are mediated by activation of calcium channels due to depolarization of the cell membrane [24]. Some agonists activate Ca^{2+} influx and therefore contraction in the absence of changes in membrane potential [22, 24, 25, 35] and this suggests that Ca^{2+} may enter the cell *via* calcium channels that are not dependent on depolarization. These channels have been designated receptor-operated channels (ROC's) to distinguish them from the voltage-operated channels (VOC's).

Evidence for two separate populations of calcium channels in smooth muscle has come from the observation that the influx of $^{45}Ca^{2+}$ stimulated by K^+ and by NA are additive when the two modes of activation are applied simultaneously to rabbit aorta [22, 24] or to rabbit mesenteric resistance vessels [22]. The mechanism of Ca^{2+} influx *via* the ROC's is not known, but it has been suggested that the release of intracellular Ca^{2+} is involved in the opening of the ROC's by NA [22]. The ROC's appear to be less sensitive to calcium channel blockade than the VOC's, but, as has already been noted, this is not the case in all tissues. The variable sensitivity to calcium channel blockers throughout the vascular system has been attributed to differences in the Ca^{2+} sources for contraction and to different types of calcium channel mediating Ca^{2+} influx [24].

MOVEMENT OF RADIOLABELLED Ca^{2+} ACROSS THE CELL MEMBRANE

Uptake of $^{45}Ca^{2+}$ has been measured to provide direct evidence that the calcium channel blocking drugs inhibit the influx of Ca^{2+} into muscle cells. Tracer studies are difficult to interpret because a small influx of Ca^{2+} into the cell may trigger intracellular Ca^{2+} release or the influx of Ca^{2+} may then be rapidly extruded back into the extracellular fluid [36]. Chelation of extracellular Ca^{2+} with EGTA (ethylene glycol bis(2-aminoethyl)-*N*,*N*,*N*′,*N*′-tetraacetic acid or replacing the Ca^{2+} with La^{3+} is necessary before the small amount of intracellular Ca^{2+} can be measured [24, 36, 37]. In rabbit aorta, the inhibition of $^{45}Ca^{2+}$ uptake by calcium channel blockers has been shown to closely parallel the inhibition of K^+-induced contractions [22, 23, 33, 38]. On the other hand, contractile responses to high concentrations of NA have been found to be considerably more resistant to calcium channel blockers than NA-stimulated $^{45}Ca^{2+}$ influx [22, 23, 33, 38]. The lack of correlation between NA-induced uptake of $^{45}Ca^{2+}$ and degree of contraction suggests that intracellular release of Ca^{2+} is important for contraction in response to NA [38].

It has been reported that the resting influx of Ca^{2+} into cells is not affected by calcium channel blocking drugs [24, 29]. However, a recent study found that verapamil increases $^{45}Ca^{2+}$ uptake and the total intracellular Ca^{2+} content of the resting tissue [39]. Verapamil may stimulate Ca^{2+}-ATPase in the sarcoplasmic reticulum, thereby increasing Ca^{2+} uptake. Although Church and Zsoter [29] have found that the resting influx of $^{45}Ca^{2+}$ is unaffected by diltiazem, nifedipine or verapamil in rat aorta and rabbit mesenteric vein, they also failed to demonstrate any significant effect of these drugs on K^+-stimulated uptake of $^{45}Ca^{2+}$ [29]. In addition, they showed that the efflux of $^{45}Ca^{2+}$ from rabbit mesenteric vein is increased by nifedipine but is not affected by verapamil or diltiazem. Most studies have found that the calcium channel blocking drugs inhibit the influx of $^{45}Ca^{2+}$ on depolarization of the cell membrane but they do not stimulate the efflux of $^{45}Ca^{2+}$ [24].

ACTION POTENTIALS

The effects of calcium channel blocking drugs on the ionic currents of cardiac cell membranes can be studied by recording action potentials. In most cells, a large, fast inward Na^+ current is responsible for the early spike component and a small, slow inward Ca^{2+} current maintains the action potential plateau [40]. Diltiazem, D600 (gallopamil), nisoldipine and nilvadipine have all been found to shorten the plateau phase of the action potential [41–43a]. These drugs plus verapamil also depress the height of the plateau in intact hearts [40], papillary muscles [41, 42, 44], Purkinje fibres [43] and ventricular trabeculae [41].

The maximum rate of rise of the action potential (V_{max}) is reduced by calcium channel blocking drugs only in certain circumstances. Diltiazem, KB944, (−)-verapamil and (−)-D600 have no effect on V_{max} in papillary muscles or ventricular trabeculae [42, 44, 45], which suggests that these drugs do not affect the fast Na^+ current. However, (+)-D600 and (+)-verapamil have been found to depress V_{max}, and this may be attributed to blockade of the fast Na^+ current [44]. In young chick hearts, V_{max} is decreased by nifedipine, verapamil, diltiazem and bepridil [40, 46], but the decrease in V_{max} is not due to blockade of the fast Na^+ current because young hearts do not possess fast sodium channels [40]. A slow Na^+ current was shown to be the main inward current [40], which demonstrates that calcium channel blocking drugs inhibit the passage of Na^+ ions as well as Ca^{2+} ions. The slow action potentials observed in young chick hearts can be completely abolished by the calcium channel blockers [40, 46].

Diltiazem, D600 and KB944 have been found to slow the rate of repolarization of cell membranes and hence prolong action potential duration [41, 42, 45]. This has been attributed to inhibition of the outward currents [41, 45]. Verapamil [44] and nisoldipine [43] also prolong action potential duration.

VOLTAGE CLAMP STUDIES

Investigation of the effects of the calcium channel blocking drugs on the calcium current in cardiac tissues has been greatly facilitated by the development of the voltage clamp technique. Initially the voltage clamp could be applied only to groups of cells such as Purkinje fibre bundles [43, 47], papillary muscles [41, 42, 48, 49] or ventricular trabeculae [41, 48, 50, 51], but more recently recordings of the calcium current in single ventricular cells have been obtained by using a patch clamp [52–57]. The voltage clamp controls the membrane potential by holding it at values chosen by the experimenter and keeping it uniform over an area of membrane through which the recorded current flows. There are two methods of achieving this: (a) by inserting two fine glass microelectrodes inside the cell and (b) by establishing a double sucrose-gap, which greatly increases the resistance between two or more regions of a cardiac cell so that the regions on each side of the gap are effectively only connected by the intracellular fluid [58].

In order to study solely the calcium current, other membrane currents must be inactivated. Na^+ current can be eliminated by addition of tetrodotoxin [43, 47] or by holding the membrane potential at −40 to −50 mV before the clamp step to the voltage yielding the maximum calcium current [46, 47, 49]. One of the outward K^+ currents is activated by an increase in intracellular Ca^{2+} concentration and since the net sum of the inward and outward currents is measured the calcium current might be considerably underestimated [36]. This can be overcome by blocking the K^+ current with tetrabutylammonium chloride [43, 47] or by using Ba^{2+} as the charge carrier instead of Ca^{2+} [36, 52].

Nifedipine [59], nisoldipine [43], verapamil [48, 50], D600 [50] and diltiazem [42] have all been shown to inhibit calcium current by the voltage clamp technique. D600 [41] and diltiazem [42] also prolong the time to peak inward current. Verapamil, diltiazem and nitrendipine block both inward and outward current through calcium channels [52], which implies that the drugs act by channel blockade rather than by depleting a pool of extracellular charge carriers. Recently two distinct components of calcium channel current have been identified in cardiac cells and these may correspond to two types of calcium channel [53, 54]. One component corresponds to the calcium current previously reported in cardiac cells. This current persists even at relatively positive holding potentials and inactivates slowly [54]. The second component is present only when cells are held at negative potentials and it inactivates more rapidly [53, 54]. The conductance when Ba^{2+} is used as the charge carrier is also different between the two components. The new calcium channel is insensitive to block by nimodipine [53] or nitrendipine [54], unlike the previously reported calcium channel. The rôle of this new channel in cardiac conduction has yet to be established.

A use-dependent effect has been found with the calcium channel blocking drugs. This can be seen in a number of ways. Firstly, there is more block of the calcium current for voltage clamp steps arising from less negative voltages (holding potentials) [42, 47–49]. Secondly, blockade increases only after repeated depolarizations from the more positive holding potential when verapamil, D600 or diltiazem are used to block the calcium channel [47, 48, 52]. On the other hand, the 1,4-dihydropyridines, nitrendipine and nisoldipine, block a single depolarizing pulse to the same extent as a train of brief pulses [47, 52]. However, voltage-dependent block is seen with nisoldipine and nitrendipine [47]. Frequency-dependent effects can also be demonstrated in isotonically contracting cat papillary muscles [20]. (±)-Verapamil and (±)-D600 cause inversion of the positive staircase which normally results from increasing the frequency of stimulation in stages, so that every increase in the frequency of stimulation is accompanied by a more pronounced negative inotropic effect. Verapamil, nifedipine, diltiazem and bepridil are able to block slow action potentials induced in cultured chick heart cell reaggregates in a frequency-dependent manner, but the effect of nifedipine (and of bepridil to a lesser extent) is also due to an apparent use-independent component [46].

A modulated receptor hypothesis has been developed to account for the use-dependent effects of the sodium channel blocking antiarrhythmic drugs and the calcium channel blockers [60]. According to this hypothesis, the calcium channel blocking drug binds to a receptor site on or very close to the trans-membrane calcium channel and the activity of the receptor for the drug is modulated by the channel state and/or potential. Three different channel states have been proposed: rested, activated and inactivated. When the drug associates with the channel, the channel no longer conducts Ca^{2+} ions and its voltage dependence is shifted to a more negative potential.

Conflicting evidence has been obtained as to whether the calcium channel blocking drug binds to the channel in its activated or inactivated state. Diltiazem, D600 and verapamil have been shown to block both inactivated [49, 52] and activated (open) channels [47, 52]. Sanguinetti and Kass [47] have proposed a model which predicts that ionized drugs can gain access to the channel-associated receptor only *via* a hydrophilic pathway that is available only when channels are in the active state configuration. Since verapamil is almost completely ionized at pH 7.4, this scheme predicts that it can only block channels by first binding to channels in the activated state. For verapamil, block develops only during repetitive pulsing to more positive voltages from a holding potential of -45 mV, where the probability of channel opening is much greater. Further evidence in support of this model has been obtained by studying contractures of frog toe muscles produced by 123 mM K^+ [27]. Verapamil, in concentrations between 0.1 μM and 30 μM, causes little or no reduction in the first test with high K^+ after drug application, but when tests are repeated at 10–15 min intervals, block is produced.

The 1,4-dihydropyridines appear to interact with both the activated and inactivated channel states [47, 52, 57]. Nisoldipine and nitrendipine are neutral drugs at pH 7.4 and therefore, according to the model of Sanguinetti and Kass [47], they can bind to activated channels via a hydrophilic pathway (as for verapamil) but they can also reach the same receptor in the inactivated state *via* the lipid membrane surrounding the channel. Sanguinetti and Kass suggested that, in the absence of repetitive pulsing from holding potentials less than -45 mV, inactivated channel block is more important to the action of nisoldipine. However, in the presence of repetitive depolarization, nisoldipine can produce activated channel block and this will contribute to the overall reduction of calcium channel current under these conditions.

Different modes of channel gating have been found by studying the activity of single calcium channels recorded from cell-attached patches on guinea-pig ventricular cells [57]. The terms mode 0, mode 1 and mode 2 were proposed to distinguish the three distinct forms of gating behaviour. Mode 1 is characterized by brief channel openings occurring in rapid bursts. In mode 2 behaviour, long openings of the channel occur and in mode 0 the channel is unavailable for opening. The effects of 1,4-dihydropyridines on channel gating have been investigated by Hess, Lansman and Tsien [57]. Nimodipine promotes mode 0 behaviour; that is, the reduction of the calcium current is due to an increase of the proportion of sweeps with no openings. Nitrendipine, however, alters channel activity in several ways. The proportion of sweeps with detectable openings is greatly decreased but there is also a larger percentage of sweeps that contain long openings and the distribution of open times is prolonged. This suggests that nitrendipine has agonist as well as antagonist activity (*vide infra*).

NON-COMPETITIVE INTERACTION BETWEEN CALCIUM AND THE CALCIUM CHANNEL BLOCKING DRUGS

Originally it was suggested that inorganic cations and the calcium channel blocking drugs compete with Ca^{2+} for a common site within the calcium channel, the so-called metal cation coordination site [61]. However, it has now been shown that the interaction between calcium and the calcium channel blocking drugs cannot be explained by simple competition for a binding site. After exposure of isotonically contracting cat papillary muscles to verapamil or D600, increasing the Ca^{2+} concentration fails to restore the amplitude-frequency relationships seen under control conditions; instead, the amplitude-frequency relationships are simply shifted to higher amplitudes [19]. The effect of varying the calcium concentration on the inhibition of the calcium current produced by D600, verapamil or Mn^{2+} has been studied using voltage clamp conditions [50]. The inhibition of the calcium conductance by the drugs is non-competitive, whereas Mn^{2+} competes with Ca^{2+} for the same site. In guinea-pig hearts, verapamil blocks the slow response induced by electrical stimulation following inhibition of the fast sodium channels. This slow response is not blocked completely by verapamil and the inhibition can be only partially reversed even when the calcium concentration is elevated to 8 mM [40].

Diltiazem, D600 and nitrendipine are less effective in blocking Ba^{2+} currents than in blocking Ca^{2+} currents [52]. Agents which act at the metal cation co-ordination site should be more effective in reducing Ba^{2+} currents than

Ca^{2+} currents, because Ba^{2+} is known to bind less strongly to the cation site. Hess and Tsien have suggested that there may be two cation-binding sites per channel controlling the passage of inorganic ions [55]. If calcium channel blocking drugs bind to sites within the channel other than the cation-binding sites, they may possibly influence the cation site allosterically.

STRUCTURE-ACTIVITY RELATIONSHIPS

Although the calcium channel blocking drugs are a fairly diverse group of compounds, they can be divided (see p. 251) into phenylalkylamines (class I), 1,4-dihydropyridines (class II), benzothiazepines (class III) and benzothiazolylbenzylphosphonates (class III). Qualitative structure-activity relationships have been determined for each group of compounds and some quantitative structure-activity relationships are also available for the phenylalkylamines and 1,4-dihydropyridines. These investigations have led to a better understanding of the interaction between a drug and the calcium channel and to the development of drugs with improved clinical profiles, i.e., greater selectivity and fewer adverse effects.

PHENYLALKYLAMINES

The general structure of the phenylalkylamines is shown in *Table 7.1*. Both aromatic ring systems are essential for the typical inotropic effect of verapamil (1.1), since compound D619 (in which the *N*-homoveratryl substituent is replaced by an *N*-methyl group) shows a completely different negative inotropic action [62]. The optimal pattern for the substituents R^1, R^2 and R^3 has been investigated. The 3,4-Cl_2, 3-CF_3 and 3,4,5-$(OCH_3)_3$ substituted derivatives are more active than verapamil [62, 63]. By Hansch analysis, the electronic parameters, sigma (Hammett constant), and *F* (field constant) give the best correlation with biological effectiveness [62, 63a]. Potency is therefore increased when the pi-electron density in the benzene ring next to the asymmetric carbon is diminished. Compound T13 (1.2) is an exception to this rule, since its actions are weaker than would be predicted from its rather high positive sigma value. This might be due to the absence of a substituent in the *para*-position (R^2) [62].

Table 7.1. PHENYLALKYLAMINE STRUCTURES

	Name	R^1	R^2	R^3	R^4	R^5	R^6	R^7	R^8
1.1	verapamil	H	OMe	OMe	Pr^i	CN	Me	OMe	OMe
1.2	T13	H	H	CF_3	Pr^i	CN	Me	OMe	OMe
1.3	D525	H	OMe	OMe	Et	CN	Me	OMe	OMe
1.4	D586	H	OMe	OMe	$(CH_2)_7Me$	CN	Me	OMe	OMe
1.5	D490	H	OMe	OMe	CH_2Ph	CN	Me	OMe	OMe
1.6	carboxyverapamil	H	OMe	OMe	Pr^i	COOH	Me	OMe	OMe
1.7	tiapamil	H	OMe	OMe	(*)	(*)	Me	OMe	OMe
1.8	D600 (gallopamil)	OMe	OMe	OMe	Pr^i	CN	Me	OMe	OMe
1.9	D594	H	OMe	OMe	Pr^i	CN	Pr^i	OMe	OMe
1.10	demethoxyverapamil (D888)	H	OMe	OMe	Pr^i	CN	Me	H	OMe
1.11	anipamil	H	H	OMe	Pr^i	CN	Me	H	OMe
1.12	ronipamil	H	H	H	Pr^i	CN	Me	H	H

(*), R^4, R^5 = $-SO_2(CH_2)_3SO_2-$.

Replacement of the isopropyl group at R^4 reduces the potency of the compounds. D525 (1.3), D586 (1.4) and D490 (1.5) are all less active than verapamil, although the nature of the frequency-dependent negative inotropic effect remains unchanged [62]. Carboxyverapamil (1.6), in which COOH has been substituted for the nitrile group at R^5, is about 10-fold less potent than verapamil in blocking isoprenaline-induced slow action potentials in guinea-pig papillary muscle [64]. Replacement of both the nitrile group and the isopropyl group is possible. Tiapamil (1.7) is a non-chiral analogue of verapamil which has a similar profile of action to verapamil and D600 (1.8) but is not as potent [63]. The substituted phenyl ring, the asymmetric carbon and groups R^4 and R^5 can be replaced completely, as is shown by AQA39 (10) and UL-FS49 (11).

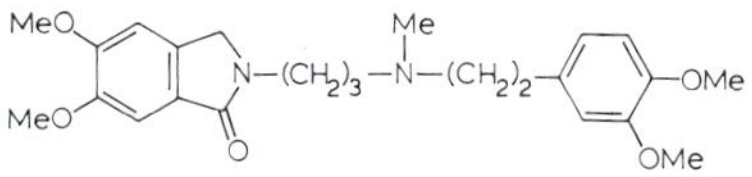

(10)

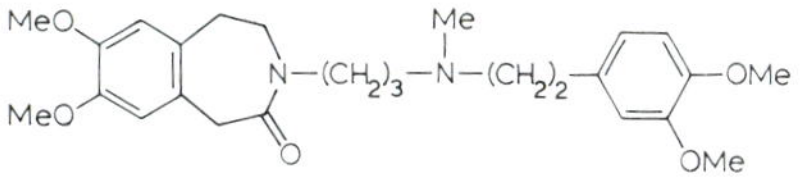

(11)

These two drugs differ from verapamil, nifedipine and diltiazem because they have a specific bradycardia effect without prominent vasodilating activity [65, 66]. Calcium channel blockade has been shown with AQA39 [67] but AQA39 and UL-FS49 also have cardiac anticholinergic effects [65]; clearly, further work is required to determine their mechanisms of action more fully.

Dealkylation of the nitrogen atom produces a compound with inotropic effects completely different to those of verapamil [62]. If the R^6 methyl group is replaced with an isopropyl group as in D594 (1.9) then potency is reduced but the compound acts qualitatively as does verapamil [62]. Quaternization of the nitrogen results in complete loss of efficacy [62]. Groups R^7 or R^8 may be replaced by hydrogen atoms as in demethoxyverapamil (also known as desmethoxyverapamil or D888) (1.10), anipamil (1.11) and ronipamil (1.12). Demethoxyverapamil is 12- to 18-times more potent than verapamil in inhibiting Ca^{2+}-induced contractions of potassium-depolarized aortic smooth muscle or force of contraction in cat papillary muscle [68]. Both benzene rings in ronipamil are unsubstituted.

As already stated, verapamil and D600 contain an asymmetric carbon atom and therefore exist in two optically active forms. In general, the *S*(−)-enantiomers of verapamil and D600 are more potent than the *R*(+)-enantiomers [19, 28, 69].

1,4-DIHYDROPYRIDINES

The structural features leading to optimal activity of the 1,4-dihydropyridines have been studied in detail. This group comprises the largest number of calcium channel blocking drugs and the individual compounds vary both in potency and selectivity. 1,4-Dihydropyridine compounds have been developed which show agonist effects at the calcium channel.

The 1,4-dihydropyridine ring is essential for optimal activity; the oxidized (pyridine) derivatives are only weakly active [40]. An unsubstituted NH group at R^1 on the dihydropyridine ring (see *Table 7.2*) is also important [70–73]. Flordipine (2.2) is an interesting new drug which is substituted at R^1. Flordipine and nifedipine (2.1) show similar potency *in vitro* in relaxing potassium-depolarized canine veins, but flordipine is more than 100-times less potent in causing relaxation of arterial strips [74]. In addition to calcium entry blockade, flordipine has been found to inhibit cyclic nucleotide phosphodiesterase [74]. Substitution at R^2 and R^6 by lower alkyl groups also confers optimal activity. 2,6-Dimethyl substitution is best [71], although replacement of an alkyl group by amino, cyano (e.g., nilvadipine (2.17)) or formyl is tolerated [73–75]. Amlodipine (2.20) has an aminoethoxymethyl group at R^2.

The simple 2,6-dimethyl-3,5-alkoxycarbonyl-1,4-dihydropyridines have some hypotensive activity in the anaesthetized animal, but good activity is generally only observed with compounds having a cyclic substituent at R^4 [71]. A substituted phenyl ring at R^4 is most common and the effects of different substituents have been widely studied. The most potent derivatives are the *ortho*-phenyl-substituted compounds, whereas *meta*-substituted derivatives are less potent and *para* substitution at the benzene ring causes a further decrease in potency [71, 75–77]. By Hansch analysis, Mahmoudian and Richards [77a] have determined that the requirements for optimal activity include (1) a bulky substituent at the *ortho*-position, (2) a wide nut not lengthy substituent at the *meta*-position and (3) a small (preferably H) substituent at the *para*-position. Electronegativity of the substituent does not appear to be important, since compounds possessing both electron-withdrawing and electron-donating substituents in the *ortho*-position of the 4-phenyl ring are active [71]. However, Meyer has suggested that electron-withdrawing substituents such as NO_2, CN, CF_3, SO_2-alkyl and Cl increase the activity compared with those of donor substituents [75].

Steric factors appear to be of major importance in determining activity. *Ortho* substitution in the phenyl ring of the 1,4-dihydropyridines restricts the phenyl group orientation such that its ring plane is perpendicular to the dihydropyridine ring to minimize steric interactions between this substituent and the

Table 7.2. 1,4-DIHYDROPYRIDINE STRUCTURES

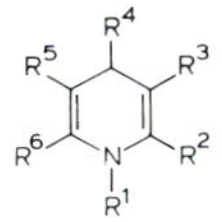

	Name	R^1	R^2	R^3	R^4	R^5	R^6
2.1	nifedipine	H	Me	COOMe	$2\text{-}NO_2C_6H_4$	COOMe	Me
2.2	flordipine	(a)	Me	COOEt	$2\text{-}CF_3C_6H_4$	COOEt	Me
2.3	nicardipine	H	Me	$COO(CH_2)_2NMeCH_2Ph$	$3\text{-}NO_2C_6H_4$	COOMe	Me
2.4	nitrendipine	H	Me	COOEt	$3\text{-}NO_2C_6H_4$	COOMe	Me
2.5	nimodipine	H	Me	$COO(CH_2)_2OMe$	$3\text{-}NO_2C_6H_4$	$COOPr^i$	Me
2.6	nisoldipine	H	Me	COOMe	$2\text{-}NO_2C_6H_4$	$COOCH_2Pr^i$	Me
2.7	felodipine	H	Me	COOEt	$2,3\text{-}Cl_2C_6H_3$	COOMe	Me
2.8	PY 108-068 (darodipine)	H	Me	COOEt	(b)	COOEt	Me
2.9	PN 200-110 (isradipine)	H	Me	COOMe	(b)	$COOPr^i$	Me
2.10		H	Me	COOMe	$3\text{-}NO_2C_6H_4$	$COO\text{-c-}C_3H_5$	Me
2.11		H	Me	COOMe	$2,3\text{-}Cl_2C_6H_3$	$COOCH_2COMe$	Me
2.12	WY-44705	H	Me	COOMe	(c)	(d)	(d)
2.13	Bay K 8644	H	Me	COOMe	$2\text{-}CF_3C_6H_4$	NO_2	Me
2.14	CGP 29392	H	(e)	(e)	$2\text{-}CHF_2OC_6H_4$	COOEt	Me
2.15	YC-170	H	Me	CONHPh	$2\text{-}ClC_6H_4$	(f)	Me
2.16	202-791	H	Me	$COOPr^i$	(b)	NO_2	Me
2.17	nilvadipine (FR34235)	H	CN	COOMe	$3\text{-}NO_2C_6H_4$	$COOPr^i$	Me
2.18	ryodipine (ryosidine)	H	Me	COOMe	$2\text{-}CHF_2OC_6H_4$	COOMe	Me
2.19	mesudipine	H	Me	COOEt	(g)	COOEt	Me
2.20	amlodipine	H	(h)	COOEt	$2\text{-}ClC_6H_4$	COOMe	Me
2.21	niludipine	H	Me	$COO(CH_2)_2OPr^n$	$3\text{-}NO_2C_6H_4$	$COO(CH_2)_2OPr^n$	Me

(a) = $-(CH_2)_2-N$ (b) = 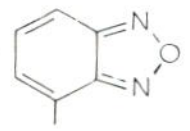 (c) =  (d) = 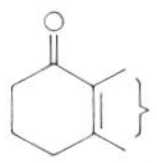 (e) = 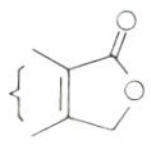 (f) = $COO(CH_2)_2$ (g) = SCH_3

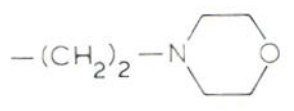

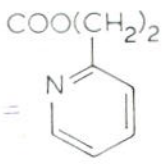

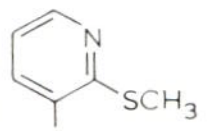

(h) = $-CH_2O(CH_2)_2NH_2$

3,5-diester substituents [70]. The dihydropyridine ring has been shown to exist in the boat-type conformation and those compounds containing an *ortho* substituent other than hydrogen on the phenyl ring exhibit the least amount of ring puckering [77]. A good correlation has been found between the degree of ring puckering and the relative activity of the parent unsubstituted compound and *ortho* and *meta* derivatives [77]. Disubstitution of the phenyl ring is acceptable when in the *ortho*- and/or *meta*-positions, as, for example, in felodipine (2.7) [75]. PY 108-068 (2.8) and PN 200-110 (2.9) have a condensed aromatic system at R^4, whereas mesudipine (2.19) contains a substituted pyridine ring.

The ester substituents at R^3 and R^5 also greatly influence the potency and selectivity of the 1,4-dihydropyridines. Compounds with non-identical ester functions appear to have greater vasodilatory activity than the symmetrically substituted derivatives [63, 70, 75]. The presence of non-identical substituents at R^3 and R^5 also makes the 1,4-dihydropyridine molecule chiral. The two stereoisomers may show very different activities [63, 78, 79]. Most of the 1,4-dihydropyridine analogues have one ester group in the *cis* orientation (ester carbonyl group is *cis* to a 1,4-dihydropyridine double bond) and one ester group in the *trans* orientation (carbonyl group is *trans* to a 1,4-dihydropyridine double bond) [70, 80]. This arrangement automatically causes a dissymmetric orientation of the ester groups relative to the 1,4-dihydropyridine ring. There appears to be considerable tolerance for the size of the ester groups. In anaesthetized animals, similar potency is found for the methyl and ethyl esters, but the ethyl esters usually have greater oral potency [71]. Some of the newest 1,4-dihydropyridines to be reported contain cyclopropyl (2.10) or 2-oxopropyl esters (2.11). The ester groups can be replaced by other electron-withdrawing substituents such as acyl or nitrile groups, but this usually leads to decreased activity [70–72, 75]. A perfluorophenyldihydropyridine (2.12) (WY-44705) has been developed with only one ester group. WY-44705 is hypotensive orally in spontaneously hypertensive rats and possesses 4-times less atrial depressant activity than nifedipine [66].

Quantitative structure-activity relationships have been determined for some of the 1,4-dihydropyridines. Hansch analysis reveals significant correlations between negative inotropy and (a) minimum width (i.e., *ortho*-substituted phenyl derivatives), (b) van der Waals's volume (ester-substituted derivatives) and (c) lipophilicity (ester derivatives) [76].

Interestingly, 1,4-dihydropyridines, such as Bay K 8644 (2.13), CGP 28392 (2.14) and YC-170 (2.15), have been discovered which show calcium agonist effects. For instance, these compounds demonstrate positive cardiac inotropic and vasoconstrictor activities [81]. Bay K 8644 increases twitch tension in guinea-pig atria in a frequency-dependent manner and prolongs action potential duration in calf ventricular muscle [82]. Voltage clamp investigations have shown that Bay K 8644 enhances calcium current, especially at negative membrane potentials [35]. However, at high concentrations (1 to 10 μM), Bay K 8644 results in coronary vasodilation and decreased cardiac contractility [83]. Hence agonist and antagonist effects are found in the same compound. It has also been shown that low concentrations of nifedipine, nitrendipine (2.4) and nicardipine (2.3) exhibit small but definite positive inotropic effects [83]. At higher concentrations, these three drugs exert pronounced negative inotropic effects, as would be expected with calcium channel blockers. It would therefore appear that several of the 1,4-dihydropyridines may be partial agonists.

Evidence has also been found to suggest that the two enantiomers of a molecule may show separate agonist and antagonist actions. (+)-Bay K 8644 in μM concentrations has vasodilating and negative inotropic properties, whereas (−)-Bay K 8644 in nanomolar concentrations increases coronary perfusion pressure and contractility [79]. Similarly, (−)-202-791 (2.16) inhibits the maximum rate of rise of the slow response action potential (V_{max}) and exerts a negative inotropic effect in guinea-pig papillary muscle, but (+)-202-791 enhances V_{max} and increases contractile force [78, 84]. The partial agonist effects can not always be attributed to different actions of the stereoisomers, however, since partial agonism has been shown with nifedipine, which is an achiral molecule. Since agonistic and antagonistic effects can be shown for the same molecular structure, small differences in binding site interactions may contribute to the initiation of agonist or antagonist responses [85]. The structural differences which may be important include the planarity of the 1,4-dihydropyridine ring, the hydrogen bonding capability of the N-H function and the orientation of the ester groups [85].

BENZOTHIAZEPINES

The diltiazem molecule ((3), R^3 = OCOMe, R^7 = H) has two chiral centres and it is also capable of *cis-trans* isomerism at these two carbon atoms. In general, the *trans* compounds do not cause vasodilation. Diltiazem is the dextrorotatory *cis* enantiomer. The laevorotatory *cis* enantiomer has a 10-fold longer duration of activity than diltiazem in increasing blood flow in the coronary sinus [75].

Variation of the substituents on the aromatic rings is possible. In the noncondensed aromatic ring, further alkoxy substituents or replacement of a methoxy group by hydroxy is accompanied by decreased potency. Only replacement of methoxy by *p*-methyl is tolerated without any marked drop in activity. Substitution with Cl at R^7 of the condensed aromatic ring is possible.

However, other substitution patterns result in decreased activity [75]. The dialkylaminoalkyl group on the N-5 atom is essential for activity. Dealkylation or quaternization at the terminal nitrogen leads to almost inactive compounds. Substitution by hydroxy, alkoxy or aralkoxy for the acetoxy function at R^3 produces compounds with activity similar to that of diltiazem, as does the replacement of the acetoxy group by longer-chain aliphatic or aromatic acyloxy groups [75].

BENZOTHIAZOLYLBENZYLPHOSPHONATES

KB944 (4) is a recently developed calcium channel blocking drug. It has a similar pharmacological profile to diltiazem [86]. Few structure-activity data are available, but it has been reported that the introduction of a pyridine nitrogen into the phenyl ring and variations in the alkoxy substitution at the phosphorus atom are possible without loss of activity [75].

BINDING STUDIES

Calcium channel blocking drugs exert dose-dependent electrophysiological and pharmacological actions, and therefore it is possible that specific recognition sites exist either on cell membranes or within cells, to mediate the actions of these drugs. Moreover, the discrete structure-activity relationships which have been determined for each group of drugs suggest that there may be different sites for the different classes of drugs. The first evidence of a specific membrane binding site was reported in 1981 by Bellemann, Ferry, Lubbecke and Glossmann [87], who found that tritiated nitrendipine bound in a reversible and saturable manner to partially purified guinea-pig heart membranes. Following this initial publication, many other studies have investigated the binding of a variety of calcium channel blocking drugs in a wide range of tissues such as heart, brain, skeletal muscle, vascular smooth muscle and ileum. The results of these studies have been very important in elucidating the nature of the calcium channel and of receptor sites for calcium channel blocking drugs. Classification of the calcium channel blockers is also possible by studying the allosteric interactions which occur when binding of one drug is inhibited or potentiated by another drug.

BINDING OF 1,4-DIHYDROPYRIDINES

Several 1,4-dihydropyridine calcium channel blockers have been tritiated and used to investigate binding of this class of drugs. The majority of studies utilized [^{3}H]nitrendipine [26, 87–115], but [^{3}H]nimodipine [115–123], [^{3}H]nifedipine [124] and [^{3}H]PN 200-110 [97, 106, 125, 126] have also been employed in some studies. Drug binding has been shown to be specific, saturable, rapid and reversible. Scatchard plots of the specific binding at equilibrium are linear, consistent with mass action behaviour. The apparent dissociation constant (K_D) and the number of binding sites (B_{max}) may be determined from the Scatchard plot.

[^{3}H]Nitrendipine binds with high affinity to sites in several different tissues. It is interesting to compare the binding data from the available studies both within a particular tissue and between different types of tissue. In the heart, K_D for binding of [^{3}H]nitrendipine varies from 0.048 nM [97] up to about 3.23 nM [104]. Similarly, there is a wide range for the number of binding sites (1800 fmol/mg protein [88] down to 6.7 fmol/mg protein [102]). This variability in the data may be due to differences in the preparation of cardiac membranes for the binding experiments. Some studies have used membranes consisting of both sarcolemma and sarcoplasmic reticulum, whereas other studies have used purified sarcolemmal membranes alone. Data for binding of [^{3}H]nitrendipine to purified canine cardiac sarcolemma are presented in *Table 7.3*. It can be seen that, although the K_D values are less variable, there is still a wide range of values for B_{max}. Hence other factors must influence the binding of [^{3}H]nitrendipine besides the purity of the membrane preparation.

Glossmann and Ferry [115] have found that [^{3}H]nitrendipine binding to guinea-pig heart membranes increases with time; 50 to 60 min are required to reach steady-state binding at 37 °C. Binding in rabbit ventricular membranes plateaus after 20–60 min at 25 °C depending on the concentration of

Table 7.3. K_D AND B_{max} VALUES FOR [^{3}H]NITRENDIPINE HIGH-AFFINITY BINDING IN CANINE CARDIAC SARCOLEMMA

K_D *(nM)*	B_{max} *(fmol/mg protein)*	*Reference*
0.32	1330	89
0.3	1800	88
0.25	665	99
0.19	571	99
0.14	960	103
0.11	230	101

[^{3}H]nitrendipine [127]. Specific binding of [^{3}H]nitrendipine increases linearly with increasing membrane protein and the pH optimum for specific binding is 7.0 to 7.6 at 25°C [127]. Since these various factors (temperature, incubation time, pH and membrane protein concentration) differ between binding assays, it is not surprising that the B_{max} and K_D values are also variable. Glossmann and Ferry [115] have shown that binding can also be affected by the ionic strength of the buffer used and by the presence of various anions. When the nitrate anion is included in the buffer, specific binding increases [115]. Several cations have the ability to inhibit or enhance binding (*vide infra*) and the buffer used for the binding assay will probably contain at least trace amounts of these cations.

Species differences do not appear to be very important for the binding of [^{3}H]nitrendipine to cardiac membranes. K_D values between 0.1 and 0.3 nM have been found for dog [88, 89, 99, 101, 103], guinea-pig [91, 87], rabbit [92], rat [94, 100, 102, 104, 107] and chick [96]. Considerable variation is found in B_{max} values for a number of species and there is no evidence to suggest that the number of binding sites is especially high in any one species. However, the number of binding sites does appear to be higher in certain tissues compared to the other organs for a particular species. The tissue distribution of [^{3}H]nimodipine binding in guinea-pigs has been investigated [122]. Skeletal muscle, heart and brain membranes show the highest amounts of [^{3}H]nimodipine binding, respectively. Confirmation that skeletal muscle contains large amounts of [^{3}H]nitrendipine binding sites has been obtained by several groups [91, 107, 112]. Dissociation constants in brain, heart and ileum are similar (generally about 0.1 to 0.2 nM), whilst K_D values in skeletal muscle are about 2 nM [91].

The subcellular distribution of [^{3}H]nitrendipine and [^{3}H]nimodipine sites has been studied in several tissues. In skeletal muscle preparations, the binding sites appear to be localized in the T-tubules [107, 112, 122]. This portion of the sarcoplasmic reticulum is thought to contain the principal sites for storage and release of calcium [107, 112]. However, in smooth muscle such as rat gastric fundus [111] or guinea-pig ileum [109, 26] the [^{3}H]nitrendipine binding sites are associated with the plasma membrane. Conflicting evidence has been found as to the location of the binding sites in heart. Purification of cardiac sarcolemma increases the number of binding sites (B_{max}) about 3–8-fold [95, 108, 120] and this suggests that the sarcolemma is one of the major binding fractions. Williams and Jones [128] found that although cardiac sarcolemma contains a significant number of [^{3}H]nitrendipine binding sites, the highest density of sites (1.5 pmol/mg protein) is in a subfraction enriched in ryanodine-sensitive sarcoplasmic reticulum vesicles [128]. [^{3}H]Nimodipine binds to both sarcoplasmic reticulum and sarcolemmal fractions from dog heart [129]. On the other hand, no specific binding of [^{3}H]nitrendipine to sarcoplasmic reticulum has been found by Sarmiento, Janis, Colvin, Triggle and Katz [103].

As would be expected, the binding of tritiated 1,4-dihydropyridines can be inhibited by unlabelled 1,4-dihydropyridines in a competitive manner [26, 91, 102–104, 106, 108, 112, 115, 116, 120, 127, 130–132]. The competition curves exhibit parallel slopes with apparent Hill slope, n_H, values near unity [127]. This implies that the 1,4-dihydropyridines compete for the same receptor site. This inhibition of binding also shows a large degree of stereoselectivity [26, 104, 115, 116, 127]. A good correlation has been found between binding activities and inhibition of contraction in guinea-pig ileum for a series of 1,4-dihydropyridines [26, 104]. The relative order of potency is the same for inhibition of binding as it is for the pharmacological response in mammalian heart [127]. Calcium channel agonist compounds, Bay K 8644, CGP28392 and YC-170, also inhibit the binding of [^{3}H]nitrendipine or [^{3}H]PN 200-110 in a similar manner as do the calcium channel blocking 1,4-dihydropyridines [91, 106, 127, 131, 133].

Binding of 1,4-dihydropyridines to high-affinity sites in the heart has been well characterized, but the existence of another low-affinity binding site has also been reported, as shown in *Table 7.4* [87, 88, 98, 100, 120, 125]. This site has not been studied as extensively and in fact only one high-affinity binding site has been found in several studies [90–93, 108, 116]. For instance, only one high-affinity site has been found in cat myocardium, although an additional component of binding corresponding to the low-affinity site is present in the

Table 7.4. DATA FOR BINDING OF TRITIATED 1,4-DIHYDROPYRIDINES TO TWO SITES IN CARDIAC TISSUE

Tritiated 1,4-dihydro-pyridine	*Species*	*High-affinity binding*		*Low-affinity binding*		*Ref.*
		K_D *(nM)*	B_{max} *(fmol/mg protein)*	K_D *(nM)*	B_{max} *(fmol/mg protein)*	
Nitrendipine	dog	0.17	99	12.4	520	98
Nitrendipine	dog	0.3	1,400– 1,800	140– 217	14,000– 32,000	88
Nitrendipine	guinea-pig	0.1	300	67	35,000	87
Nitrendipine	chick	0.1	270	5.2	1,560	125
Nitrendipine	rat	0.343		47.6	12,400	100
PN 200-110	chick	0.09	190	2.7	700	125
Nimodipine	cow	0.35	300	33	8,200	120

dog [98]. Higher concentrations of [^{3}H]nitrendipine or [^{3}H]nimodipine are required to reveal the low-affinity site compared to the high-affinity site [88, 98, 120]. Calcium concentration can also influence the number of binding sites. In the presence of low calcium concentrations (0.2 to 10 μM), two affinity states for binding of [^{3}H]PN 200-110 have been observed, but in the presence of higher concentrations of calcium (0.2 to 1 mM) all the receptors are converted to a single high-affinity state [125]. Therefore, failure to find two binding sites may be due to using inappropriate conditions.

BINDING OF PHENYLALKYLAMINES

Binding sites have been found in heart, brain and skeletal muscle for the radioactive ligands, [^{3}H]verapamil [95, 118, 122, 134–136] and [^{3}H]demethoxyverapamil [68, 120, 125, 126, 137–140]. High- and low-affinity sites have been identified for [^{3}H]demethoxyverapamil [120, 125, 140] and there is also evidence for two sites for [^{3}H]verapamil binding [119, 136, 141], but because of the high values for non-specific binding at high concentrations of verapamil, the low-affinity site is difficult to detect by Scatchard analysis [141]. The K_D values for high-affinity [^{3}H]demethoxyverapamil binding are between 0.17 [125] and 7.2 nM [140], whereas for verapamil these values are between 16 [136] and 93.8 nM [134]. The largest B_{max} values for the high-affinity site are found in skeletal muscle T-tubule membranes [118, 135, 138, 139] as for the 1,4-dihydropyridines. In cardiac homogenates, binding sites for demethoxyverapamil and verapamil are present in both the sarcolemma and sarcoplasmic reticulum [95, 120, 137]. The sarcoplasmic reticulum appears to have an excess of sites for the phenylalkylamines compared with those for the 1,4-dihydropyridines [95]. It has been proposed that the sarcoplasmic reticulum contains a low-affinity site for demethoxyverapamil which is not related to the high-affinity site located on the calcium channel [137].

The specific binding of [^{3}H]verapamil to T-tubule membranes increases with increasing pH [141]. The optimum pH is between 7.8 and 8.2 [118]. [^{3}H]Demethoxyverapamil binding to cardiac sarcolemma is also increased at higher pH values and is optimal between pH 8.0 and 8.5 [120]. Decreasing the temperature of the binding assay from 37°C to 2°C increases the number of verapamil binding sites [118, 126]. Other phenylalkylamines such as D600 and tiapamil inhibit the binding of verapamil or demethoxyverapamil in a competitive manner [68, 120, 140, 141]. This inhibition is stereoselective and for [^{3}H]-demethoxyverapamil the potency series has been determined as follows: (–)-demethoxyverapamil > tiapamil > (–)-D600 > (–)-verapamil > (+)-demethoxyverapamil > (+)-D600 > (+)-verapamil [120].

BINDING OF DILTIAZEM

The binding of (+)-*cis*-[^{3}H]diltiazem to skeletal muscle T-tubule membranes [118, 122, 138, 142] or to purified cardiac sarcolemma [95] has been reported, but extensive binding studies have not yet been performed. Like verapamil, the binding of (+)-*cis*-[^{3}H]diltiazem is temperature-dependent. Increasing the temperature from 2 to 37°C leads to a 5-fold decrease in specific binding [142]. Reported K_D values are about 35 to 50 nM in T-tubule membranes and cardiac sarcolemma [95, 138, 142]. The number of binding sites (B_{max}) is greater in T-tubule membranes (11 000 to 50 000 fmol/mg protein) than in cardiac sarcolemma (220 to 600 fmol/mg protein) [95, 138, 142]. Binding is stereoselective, since the IC_{50} values (concentration at which 50% inhibition of (+)-*cis*-[^{3}H]diltiazem binding occurs) are very different for unlabelled (+)-*cis*-diltiazem (54 nM) and (−)-*cis*-diltiazem (6,680 nM) [142].

ALLOSTERIC REGULATION OF BINDING

The interaction between the phenylalkylamines, the 1,4-dihydropyridines and diltiazem has been investigated using binding techniques. A drug from any one of these three different classes of calcium channel blockers does not inhibit the binding of a drug from another class in a competitive manner, unlike the inhibition of nitrendipine binding by other 1,4-dihydropyridines or the inhibition of demethoxyverapamil binding by phenylalkylamine compounds. Thus it has been proposed that there are three main drug-receptor sites on the calcium channel and binding of a drug to its receptor site influences the binding of drugs to the other sites by allosteric interactions [115, 118, 143]. Evidence to support this theory has been widely obtained.

Verapamil and D600 inhibit the binding of [^{3}H]nitrendipine in a non-competitive manner in heart, brain, ileum, aorta, coronary artery and skeletal muscle [26, 91, 93, 96, 101, 102, 105, 108, 113, 130, 132, 144, 145]. The maximum inhibition of binding is not 100%, and less than 50% inhibition has been found in some studies [93, 108, 130, 132, 145]. The binding of other 1,4-dihydropyridines is inhibited in a similar manner by verapamil or D600 [106, 117, 120, 124]. When Hill plots are used to determine the effect of verapamil or D600 on the binding of [^{3}H]nifedipine in rabbit papillary muscles, the slope factors (n_H values) are significantly less than one [124], confirming the non-competitive nature of the inhibition. Tiapamil, however, gives an n_H value close to unity under the same experimental conditions [124] and tiapamil has also been shown to produce 100% inhibition of [^{3}H]nitrendipine binding

[26, 127, 131, 145]. Despite this apparent competitive inhibition of 1,4-dihydropyridine binding, tiapamil is probably an allosteric modulator like verapamil, since the tiapamil inhibition curve of [^{3}H]nitrendipine binding in guinea-pig ileum is biphasic [26], and at higher concentrations of [^{3}H]nitrendipine, binding to rabbit cardiac membranes is only partially inhibited (80%) by tiapamil [127].

The effect of diltiazem on 1,4-dihydropyridine binding is temperature-dependent. At low temperatures, 0 to 4°C, diltiazem inhibits [^{3}H]nitrendipine binding to heart and brain membranes in a non-competitive fashion [102, 131, 132]. However, at 37°C, potentiation of both [^{3}H]nitrendipine and [^{3}H]nimodipine binding by diltiazem has been observed in coronary arteries [101], heart [117, 120, 123, 127, 131, 146] and skeletal muscle T-tubules [147]. At inbetween temperatures such as 20 to 25°C, diltiazem shows variable effects on binding of 1,4-dihydropyridines. In some studies, diltiazem has had no effect on the binding of [^{3}H]nitrendipine [131, 132] or [^{3}H]nifedipine [124] to cardiac membranes. At 20°C, high concentrations of diltiazem (10 μM) inhibit [^{3}H]nimodipine binding to bovine cardiac sarcolemma but at lower concentrations no effect is seen [120]. The effect of diltiazem on [^{3}H]nitrendipine binding to guinea-pig ileum is also concentration-dependent at 25°C. At low concentrations (5 to 10 nM), a modest inhibitory effect is apparent but at 50 μM a potentiation is observed which becomes inhibitory at the highest concentration (100 μM) studied [26].

1,4-Dihydropyridines inhibit the binding of [^{3}H]verapamil or [^{3}H]demethoxyverapamil to membranes from heart, brain and skeletal muscle (68, 95, 118, 120, 122, 135, 136, 139]. The inhibition is again non-competitive and 100% inhibition is not achieved [68, 95, 120], and therefore it can be described as a negative heterotropic allosteric interaction. Binding of tritiated phenylalkylamines is also inhibited by diltiazem in skeletal muscle [118, 122, 134, 135, 139], brain [68, 134] and heart [95, 120]. The effect of other calcium channel blocking drugs on the binding of (+)-*cis*-[^{3}H]diltiazem to guinea-pig skeletal muscle microsomes has been studied [142]. The phenylalkylamines, verapamil, D600 and tiapamil, inhibit the binding of tritiated diltiazem. The effect of the 1,4-dihydropyridines is, however, dependent on the temperature. At 30°C, 1,4-dihydropyridines stimulate (+)-*cis*-[^{3}H]diltiazem binding but at 2°C the same drugs inhibit (+)-*cis*-[^{3}H]diltiazem binding in a concentration-dependent manner [142].

How are these allosteric interactions brought about between the different classes of calcium channel blocking drugs? Verapamil increases the dissociation constant (K_D) for binding of [^{3}H]nitrendipine [91, 102, 127, 132]; that is, the affinity of the binding site for [^{3}H]nitrendipine is reduced. This reduced affinity is probably due to acceleration of the rate of dissociation of [^{3}H]nitrendipine from its binding site [132]. The effect of verapamil on the association kinetics of [^{3}H]nitrendipine is to increase the rate of association [132]; however, it has been found that verapamil decreases the association rate constant for binding of [^{3}H]nimodipine [117]. Hence the contribution of a change in association kinetics to the reduced binding affinity for 1,4-dihydropyridines in the presence of verapamil is not clear.

In skeletal muscle T-tubules, ileum and heart, the number of binding sites for [^{3}H]nitrendipine or [^{3}H]nimodipine is increased by diltiazem [26, 117, 123, 127, 146, 147]. This has been attributed to the conversion of low-affinity sites into high-affinity sites in the presence of diltiazem [123, 146, 147]. In rat brain, the number of binding sites (B_{max}) for [^{3}H]nitrendipine is not increased by diltiazem, but the K_D is decreased, i.e., the affinity of the binding site for nitrendipine is greater [113]. A decrease in K_D is also found when diltiazem potentiates the binding of [^{3}H]nitrendipine to rabbit skeletal T-tubule membranes, although under these conditions B_{max} is increased [147]. Diltiazem decreases the dissociation rate constant for [^{3}H]nitrendipine or [^{3}H]-nimodipine binding. In other words, diltiazem slows the dissociation of the 1,4-dihydropyridines from their receptors [113, 120, 127].

Nitrendipine, PN 200-110 and diltiazem have all been found to decrease the number of binding sites for [^{3}H]demethoxyverapamil, thereby inhibiting the binding of the phenylalkylamine [68, 95, 135, 139]. Usually the K_D values have either been unchanged [95, 135] or only increased slightly [68, 139] in the same assays. However, in one case, diltiazem has been observed to increase the equilibrium dissociation constant (K_D) without changing the B_{max} [135]. The effect of nitrendipine on the binding of (+)-*cis*-[^{3}H]diltiazem to guinea-pig skeletal muscle microsomes has been studied [142]. At 30°C, nitrendipine increases the density of sites labelled by (+)-*cis*-[^{3}H]diltiazem, but at 2°C the density is decreased by nitrendipine.

INFLUENCE OF CATIONS ON THE BINDING OF CALCIUM CHANNEL BLOCKING DRUGS

Current through the calcium channel can be carried by divalent cations besides Ca^{2+} such as Ba^{2+} and Sr^{2+} [47, 52, 54, 55, 148], whilst other inorganic cations, for example Ni^{2+}, La^{3+}, Mn^{2+}, Co^{2+} and Cd^{2+}, block the calcium channels [50, 56, 61]. It has been suggested that these cations compete with Ca^{2+} for a binding site associated with the calcium channel [1, 50, 52, 61, 148]. Hess and Tsien proposed the presence of two cation-binding sites per channel [55] and Glossmann, Ferry, Goll and Rombusch have indicated that there may be from two to four interacting divalent cation sites, which have positive and negative co-operativity [118]. The presence of inorganic cations has been found to affect the binding of calcium channel blocking drugs and therefore it is reasonable to assume that allosteric interactions may occur between the cation-binding sites and the receptor sites for the different calcium channel blocking drugs.

In the presence of EDTA (ethylenediaminetetraacetic acid) or EGTA (ethylene glycol bis(2-aminoethyl)-*N*,*N*,*N′*,*N′*-tetraacetic acid), binding of [^{3}H]nitrendipine and [^{3}H]nimodipine to membranes prepared from heart, brain or ileum is inhibited [90–92, 105, 107, 115, 120, 143]. The number of binding sites (B_{max}) is reduced by EDTA [90, 92]. However, in skeletal muscle membranes, binding of [^{3}H]nitrendipine is not influenced by 1 mM EGTA [107], and EDTA (at 0.1 to 1 mM) stimulates [^{3}H]nitrendipine binding [91]. The effect of EDTA on binding can be reversed by replacing the divalent cations. Thus, Ca^{2+} ions reverse the EDTA-induced reduction in [^{3}H]nitrendipine or [^{3}H]nimodipine binding in brain, heart and ileum [91, 92, 105, 115]. The EDTA stimulation of [^{3}H]nitrendipine binding in skeletal muscle is also reversed by Ca^{2+} [91]. The ability of a series of cations to restore 1,4-dihydropyridine binding inhibited by EDTA has been investigated in guinea-pig ileum [92] and guinea-pig brain [115]. Although a slightly different order of potency was obtained in the two studies, it has been shown that Ca^{2+}, Sr^{2+}, Mg^{2+} and Mn^{2+} can fully reverse the EDTA-induced inhibition of [^{3}H]nitrendipine or [^{3}H]nimodipine binding, whilst Ba^{2+}, Ni^{2+} and Zn^{2+} only partially restore [^{3}H]nitrendipine binding. At higher concentrations, Zn^{2+}, Co^{2+} and Ni^{2+} inhibit [^{3}H]nimodipine binding [115].

The binding of 1,4-dihydropyridines is also influenced by cations without the prior application of EDTA or EGTA. Ca^{2+} ions have been found to both stimulate and inhibit binding in a concentration-dependent manner. Below 2 mM, Ca^{2+} stimulates binding of [^{3}H]nitrendipine or [^{3}H]nimodipine in brain [105], heart [120] and ileum [26], whereas at higher concentrations of Ca^{2+}, inhibition of [^{3}H]nitrendipine binding is observed [102, 112, 132]. The cations La^{3+}, Cd^{2+}, Cu^{2+}, Zn^{2+}, Pb^{2+}, Mg^{2+}, Mn^{2+} and Ba^{2+} have all been shown to inhibit [^{3}H]nitrendipine binding in skeletal muscle, brain, heart and ileum [26, 90, 91, 102, 112, 127, 132]. The most potent inhibitors include Cd^{2+}, La^{3+}, Cu^{2+}, Zn^{2+} and Pb^{2+} [26, 127, 132], whereas Ca^{2+}, Ni^{2+} and Mn^{2+} are less potent [26, 112, 127, 132].

[^{3}H]Demethoxyverapamil binding is also inhibited by cations. In bovine cardiac sarcolemma the order of potency for inhibition is $Sr^{2+} > Ba^{2+} > Ca^{2+}$ and EGTA does not affect [^{3}H]demethoxyverapamil binding [120]. On the other hand, the order of potency for inhibition of binding to skeletal muscle T-tubule membranes is $Ca^{2+} > Sr^{2+} > Ba^{2+} > Mg^{2+}$ [138]. As the Ca^{2+} concentration is increased, the maximal binding capacity for [^{3}H]demethoxyverapamil decreases and the apparent equilibrium dissociation constant increases [138]. The effect of Ca^{2+} on [^{3}H]verapamil binding in skeletal muscle is concentration-dependent: at 0.1 mM Ca^{2+}, binding is stimulated, but above 1 mM Ca^{2+}, inhibition is seen [134]. The rank order of potency for inhibition of [^{3}H]verapamil binding to T-tubule membranes has been determined as $Ca^{2+} = Mn^{2+} > Mg^{2+} > Sr^{2+} > Ba^{2+} > Co^{2+} > Ni^{2+}$ [141], which is similar, but not identical, to the order of potency for inhibition of [^{3}H]demethoxyverapamil binding. Again Ca^{2+} induces a decrease of the maximal binding capacity for the radioligand, whilst K_D is increased [141]. Ca^{2+} and La^{3+} have also been shown to inhibit binding of (+)-*cis*-[^{3}H]diltiazem [138, 142].

RELATIONSHIP BETWEEN MEMBRANE BINDING SITES FOR CALCIUM CHANNEL BLOCKING DRUGS AND EFFECTS ON CALCIUM CHANNELS IN INTACT CELLS

The pharmacological relevance of the binding sites for tritiated calcium channel blocking drugs has yet to be fully established. It has been assumed that the binding sites are part of the calcium channel, but is this assumption justified? In smooth muscle such as ileum and coronary arteries, the dissociation constants for [^{3}H]nitrendipine binding to membrane preparations are in the same concentration range as the IC_{50} values (concentration producing 50% inhibition) for inhibition of K^+ depolarization-induced responses; that is, they are both about 0.1 to 2 nM [26, 70, 98, 101]. No such one-to-one correlation for binding and pharmacological effect is seen in the heart. Although the K_D for binding is similar in heart to the K_D for binding in ileum or coronary arteries, the concentration of nitrendipine which produces a 50% reduction in the cardiac inotropic response to electrical stimulation is at least 100-times the K_D for binding [89, 98, 101].

However, the potency of a series of 1,4-dihydropyridines to inhibit [^{3}H]nitrendipine or [^{3}H]nimodipine binding correlates well with the rank order of these compounds for inhibition of cardiac inotropic responses and for inhibition of smooth muscle contraction [26, 70, 85, 90, 104, 127]. This implies that the structure-activity dependence is similar for both binding and pharmacological activities, even if there is a difference in the concentrations at which effects are found [70]. Furthermore, diltiazem not only stimulates the binding of [^{3}H]nimodipine to canine cardiac sarcolemma but also potentiates the negative inotropic response of perfused rat hearts to nimodipine [146]. Thus, the ability of diltiazem to double the number of binding sites for [^{3}H]nimodipine [146] is accompanied by an increase in response.

There are binding sites for the calcium channel blocking drugs in brain and skeletal muscle, but the effects of calcium channel blockade cannot always be readily demonstrated in these tissues [34, 38]. The distribution of [^{3}H]demethoxyverapamil binding sites in the brain has been investigated using autoradiography [68]. The highest density of sites is found in the hippocampus and dentate gyrus, followed by the cerebral cortex and thalamic nuclei. Very few sites are present in the white matter or cerebellum. The pharmacological effects of the calcium channel blockers in the brain and neuronal tissue have been reviewed [34, 149, 150]. Transmitter release from the adrenal glands, cultured phaeochromocytoma and cultured neuronal cells is sensitive to low concentrations of the calcium channel blockers [149]. Also, Bay K 8644 augments the uptake of $^{45}Ca^{2+}$ into striatal synaptosomes [149] and increases K^+-evoked 5HT release in cerebral cortex slices [149, 150]. These two effects can be blocked by antagonistic 1,4-dihydropyridines [150]. Thus, the calcium channel blocking drugs do exert pharmacological actions on the brain, but the relevance of those actions to the binding sites for the drugs has yet to be established.

Contraction of skeletal muscle is not acutely dependent on the entry of extracellular Ca^{2+} [38]. However, calcium channels are present in skeletal muscle and appear to be located in the T-tubules [34, 38]. Calcium channel blocking drugs do affect contraction of skeletal muscle, but their effects on muscle tension do not correlate with the relative calcium channel blocking ability of the drugs or with the ability of the drugs to cross the cell membrane [151]. For example, verapamil and D600 depress twitch and tetanus tension, whereas nifedipine causes marked potentiation of twitch tension but does not alter tetanus tension [151]. Thus, despite the high density of binding sites for calcium channel blockers in skeletal muscle, the pharmacological relevance of these sites is still unknown.

A number of suggestions have been put forward to account for the disparity in the concentrations at which calcium channel blocking drugs bind to sites in the heart and at which they produce negative inotropic responses and to account for the lack of effect of the calcium channel blockers in brain and skeletal muscle. Firstly, the properties of the binding sites may be altered during the preparation of membranes [85], since calcium channels are not functional in isolated membrane vesicles [56]. However, phosphorylation of ventricular sarcolemmal membranes does not alter the binding properties of nitrendipine [99]. Binding studies have been performed using intact cultured chick ventricular cells [96] and also intact rat cardiac myocytes [129, 144]. The dissociation constants for [^{3}H]nitrendipine binding were 0.26 nM, 0.587 nM and 1.07 nM, respectively, in the three studies. Hence, even in intact cells, the K_D values are much lower than the concentration range for negative inotropic effects [96]. Depolarization of rat cardiac myocytes with 50 nM K^+ increases the maximum number of binding sites but does not alter the K_D for binding [144], and therefore the nitrendipine receptor density is voltage-dependent.

Binding to low-affinity sites may be important for the calcium channel blocking effects of 1,4-dihydropyridines. In chick hearts, the dissociation constants (K_D values) for binding of [^{3}H]nitrendipine to low-affinity sites are similar to the concentrations at which nitrendipine inhibits contractile force [95, 120, 125]. The concentration of nitrendipine which produces a half-maximal decrease in contractile force of dog heart trabecular muscles (328 nM) is in the same range as the K_D for binding of nitrendipine to low-affinity sites in canine cardiac sarcolemma (140 nM) [152]. [^{3}H]Bay K 8644 has been shown to bind to cultured myocardial cells with a K_D of 35 nM and a B_{max} of 1.07 pmol/mg protein [136]. The K_D for binding is very close to the concentrations of Bay K 8644 that produce half-maximal increases in developed force in isolated dog atrial (29.5 nM) and ventricular trabecular muscles (30.5 nM) [89]. Bay K 8644 displaces [^{3}H]nitrendipine binding in canine cardiac sarcolemma in a manner consistent with binding to two classes of sites [152]. The dissociation constant for inhibition of binding at the high-affinity site correlates with the concentration of Bay K 8644 producing a positive inotropic effect, while the K_D for inhibition of binding at the low-affinity site correlates with the concentration at which the negative inotropic effect of Bay K 8644 is seen in the dog heart. It has been suggested that binding of the 1,4-dihydropyridine drugs to the high-affinity site mediates their agonist effects and binding to the low-affinity site is responsible for their antagonist effects [88, 152].

If some tissues have an excess of binding sites, not all the binding sites will necessarily be coupled to functional calcium channels [85]. Thus, blockade of 50% of the sites might not be sufficient to reduce by half the calcium-dependent effects. A half-maximal decrease in response would be achieved only when considerably more of the binding sites are occupied [98, 143]. According to this hypothesis, the most sensitive tissues to channel blockade must be those with the lowest density of binding sites, e.g., coronary arteries; those of intermediate sensitivity have an intermediate density of binding sites (e.g., heart), and skeletal muscle, which has the highest density of sites, is virtually insensitive to calcium channel blockade [143]. There is evidence of spare binding sites in rat cardiac myocytes. Green, Farmer, Wiseman, Jose and Watanabe [144] estimated that there are 160,000 sites per polarized cell and 382,000 sites per depolarized cell, which is greater than the density of 2,000 to 10,000 functional calcium channels per cultured rat cardiac cell determined by Reuter [56]. An even larger number of binding sites (10^6) was calculated by De Pover, Lee, Matlib, Whitmer, Davis, Powell and Schwartz [129].

EFFECTS ON OTHER SUBCELLULAR SITES

The calcium channel blockers have been found to affect several different subcellular processes (see *Figure 7.2*) and these effects may contribute to the pharmacological profile of the drugs. However, these actions are mainly seen at higher concentrations of the drugs than the concentrations required for calcium channel blockade. For example, high concentrations of D600, nitrendipine and verapamil have been shown to block the fast sodium channels present in cardiac tissue [44, 153–156]. On the other hand, lower concentrations of verapamil do not decrease the maximum rate of depolarization of phase 0 of the action potential (V_{max}) [156, 157]. Neither diltiazem nor nisoldipine reduces V_{max} [7, 42]. Since influx of Na^+ into the cell *via* the fast sodium

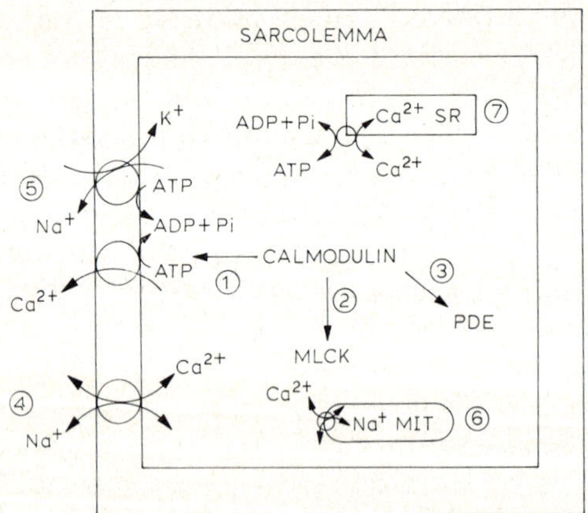

Figure 7.2. Some intracellular processes that may be affected by calcium channel blocking drugs. Calcium channel blocking drugs inhibit calmodulin-dependent sarcolemmal Ca^{2+}-ATPase ①, myosin light-chain kinase (MLCK) ② and phosphodiesterase (PDE) ③. Passive Na^+–Ca^{2+} exchange ④ may also be inhibited, whilst (Na^+ + K^+)-ATPase ⑤ is stimulated. Ca^{2+} release from mitochondria (MIT) in exchange for Na^+ ⑥ may be inhibited, but the effect of calcium channel blocking drugs on Ca^{2+} uptake into sarcoplasmic reticulum (SR) via *Ca^{2+}-ATPase ⑦ is variable.*

channels is responsible for phase 0 of the action potential [158], drugs which do not decrease V_{max} probably do not inhibit the sodium channels.

In cardiac cells, there are two time-dependent outward currents which contribute to repolarization of the cell. Both of these outward currents may be decreased by calcium channel blocking drugs. Since an increase in the intracellular Ca^{2+} concentration activates one of the outward currents (I_{to}), all the calcium channel blocking drugs should reduce I_{to} [159]. This has been shown with D600, nisoldipine and diltiazem [41–43, 159]. The other time-dependent outward current (I_x, delayed rectifier) is not calcium-activated [160]. Hume has found that D600, diltiazem and nisoldipine inhibit I_x in frog atrial cells [160]. However, a similar concentration of nisoldipine does not affect I_x in calf Purkinje fibres [43].

Diltiazem, nifedipine and verapamil decrease the Na^+-induced Ca^{2+} release from heart mitochondria without affecting the uptake of Ca^{2+} into this organelle [161, 162]. Diltiazem is the most potent inhibitor of Ca^{2+} release [161]. Nimodipine and nitrendipine increase (Na^+ + K^+)-ATPase activity in microsomal membranes from aorta and vas deferens but do not influence the activity of this enzyme in ventricular or brain membranes [163]. No stimulation of (Na^+ + K^+)-ATPase activity is found with nifedipine, diltiazem or verapamil [163]. Blocking the Na^+ pump reduces intracellular levels of Na^+, which would increase Na^+–Ca^{2+} exchange and therefore decrease intracellular Ca^{2+} [163]. Thus, the nimodipine-induced increase in (Na^+ + K^+)-ATPase activity might contribute to the vasodilator effect of the drug. Increased (Na^+ + K^+)-ATPase activity has been found in cardiac sarcolemma-enriched vesicles with 1 μM nicardipine, but not with a higher concentration (10 μM) [164]. Verapamil in μM concentrations inhibits the uptake of Ca^{2+} into cardiac membrane vesicles *via* the Na^+–Ca^{2+} exchange mechanism [165, 165a]. In addition, verapamil inhibits the binding of Ca^{2+} to cardiac sarcolemma [166, 167]. Both the stereoisomers of verapamil are equipotent inhibitors of binding [167] and verapamil has been shown to decrease the capacity of the low-affinity calcium binding sites [166]. Nifedipine, D600 and diltiazem do not affect cardiac sarcolemmal binding of calcium [166, 168].

Many of the calcium channel blocking drugs bind to calmodulin, which is a calcium-binding protein. Calmodulin mediates the activation by Ca^{2+} of a large number of enzymes and it appears to act as a multifunctional intracellular Ca^{2+} receptor [3]. Felodipine, nitrendipine, verapamil, D600 and diltiazem all bind to calmodulin [169–174] and their binding is increased in the presence of Ca^{2+} [169, 170, 172]. The binding of felodipine to calmodulin is also increased by diltiazem [169, 172]. It has therefore been suggested that the two drugs bind to different sites on calmodulin and that there are allosteric interactions between the binding sites [169, 172]. Verapamil can also increase felodipine binding, but is much less potent than diltiazem in this respect [172]. Calmodulin activates the Ca^{2+}-ATPases which are present in plasma membranes of heart and smooth muscle cells [3, 14]. Felodipine and verapamil inhibit the activity of cardiac sarcolemma Ca^{2+}-pumping ATPase [165a, 167, 175]. Nifedipine and nimodipine, however, stimulate ATPase activity [165a, 168, 175], although inhibition has been observed in one study at higher concentrations of these two drugs [165a].

In most tissues, a calmodulin-dependent phosphodiesterase has been found which hydrolyses both cyclic AMP and cyclic GMP [3]. The activity of the phosphodiesterase in the presence of calmodulin is inhibited by several of the calcium channel blockers. Nimodipine, nicardipine, felodipine, nisoldipine and nitrendipine are much more potent inhibitors than verapamil, D600 or diltiazem [171, 175–177]. Half-maximal inhibition of phosphodiesterase activity is seen with 2–20 μM concentrations of the 1,4-dihydropyridines, whereas more than 100 μM verapamil or diltiazem is required to produce 50% inhibition [171, 176]. The effects of nifedipine in different studies have been inconsistent. In some studies, low concentrations of nifedipine have been found to inhibit calmodulin-activated phosphodiesterase [176, 177], but in other investigations nifedipine has had only a slight inhibitory effect on phosphodiesterase at up to 10 μM [175, 178]. This discrepancy may be due to different sensitivities of phosphodiesterases from varying sources to nifedipine. There are also calmodulin-insensitive forms of phosphodiesterase whose activity is inhibited by verapamil, nimodipine, nicardipine, nifedipine and diltiazem [171, 177].

Antagonism of calmodulin may explain why calcium channel blocking drugs such as felodipine, nitrendipine and verapamil inhibit the phosphorylation of myosin light chains [174, 179, 180]. High concentrations of nitrendipine and verapamil are required to affect myosin light-chain kinase activity [180]. The effects of the calcium channel blocking drugs on the uptake of Ca^{2+} into sarcoplasmic reticulum (SR) and on Ca^{2+} release are varied. However, these effects do not appear to involve calmodulin, since the phosphorylation of phospholamban in cardiac SR is unaffected by felodipine, nitrendipine or verapamil [180]. Binding of Ca^{2+} to cardiac SR is decreased in the presence of high concentrations of nisoldipine, verapamil, D600 and diltiazem [181–183]. In skeletal muscle SR, high concentrations of verapamil, diltiazem, nisoldipine and felodipine also inhibit Ca^{2+} binding, but at lower concentrations nisoldipine and felodipine stimulate binding [181].

The effect of these drugs on Ca^{2+} uptake into SR again appears to be concentration-dependent. Felodipine, nitrendipine, nimodipine and nisoldipine stimulate Ca^{2+} uptake into both skeletal muscle and cardiac SR

[180, 181, 183a]. Decreased Ca^{2+} uptake is seen with verapamil, diltiazem and D600, particularly at high concentrations [180–182]. Ca^{2+} is taken up into SR by active transport. The energy for this process is provided by the action of Ca^{2+}-ATPase on the terminal phosphate of ATP [184]. In several studies, the calcium channel blocking drugs have been found not to affect the activity of Ca^{2+}-ATPase in cardiac SR [17, 180, 182]. However, Wang, Tsai and Schwartz [181] have shown that the calcium channel blockers can alter Ca^{2+}-ATPase activity and that some of these drugs have different effects on the Ca^{2+}-ATPase from cardiac SR compared to skeletal muscle SR. For instance, diltiazem stimulates Ca^{2+}-ATPase activity in skeletal muscle SR but decreases the enzyme's activity in cardiac SR. Verapamil also increases Ca^{2+}-ATPase activity in skeletal muscle SR but has no effect on cardiac SR Ca^{2+}-ATPase. The effects of nisoldipine and felodipine on Ca^{2+}-ATPase are similar for both cardiac and skeletal muscle SR. Low concentrations of the 1,4-dihydropyridines stimulate Ca^{2+}-ATPase activity, whereas at higher concentrations the activity is decreased.

Surprisingly, the calcium channel blockers have been found to inhibit the binding of several drugs to their receptors. For example, binding of agonists and antagonists to alpha-1 and alpha-2 adrenoreceptors, muscarinic receptors and D-2 dopamine receptors is inhibited by verapamil, D600, diltiazem and nicardipine [28, 185–188]. Verapamil and D600 are the most potent antagonists of binding [185–187]. This inhibition of binding suggests that the calcium channel blocking drugs may affect the functioning of the sympathetic and parasympathetic nervous systems in the intact animal.

Thus, the individual pharmacological effects exerted by each calcium channel blocker may depend upon the extent to which the drug affects other intracellular systems as well as its potency at the calcium channel. Many of the effects described here would tend to increase the vasodilator action of the drugs, such as inhibition of calmodulin-dependent enzymes. However, these other effects are subsidiary to blockade of the calcium channel, as they occur mainly at concentrations higher than those required to block the channels; therefore, at low concentrations the actions of the calcium channel blocking drugs are relatively specific.

THERAPEUTIC APPLICATIONS

As a result of their ability to relax smooth muscle, calcium channel blocking drugs have numerous therapeutic applications, mainly in the treatment of cardiovascular disorders but possibly also in therapy for bronchial asthma, gastrointestinal muscle spasms and uterine disorders. The *in vivo* effects of the calcium channel blockers vary depending on the properties of the individual drug. Tissue selectivity plays a very important rôle in determining their clinical usefulness. For example, calcium channel blocking drugs such as the 1,4-dihydropyridines have more pronounced effects on the vascular system than the heart. Some of the factors which determine the tissue selectivity of a particular drug have been described by Rampe, Su, Yousif and Triggle [72] and by Nayler and Dillon [189]. These include:

(1) pharmacokinetic factors affecting tissue distribution;
(2) the relative importance of Ca^{2+} entry through the slow channel to the functioning of a particular tissue;
(3) activation of cardiovascular reflex mechanisms;
(4) the number of drug-binding sites and differences in binding sites between different tissues;
(5) the extent to which a drug shows use-dependency;
(6) other properties of the drug besides calcium channel blockade;
(7) partial agonist activity of some of the 1,4-dihydropyridines;
(8) the pathology of the tissue.

A calcium channel blocking drug may have a different pharmacological profile *in vivo* than seen when using isolated tissue preparations. This is particularly the case in the cardiovascular system due to the operation of reflex mechanisms. For example, diltiazem, verapamil and nifedipine suppress sinoatrial (SA) nodal function in the excised rabbit heart; however, after i.v. dosing of these drugs to patients, sinus rate increases [190]. This increase in heart rate found after the administration of verapamil or nifedipine to volunteers can be prevented by autonomic blockade with atropine and propranolol [191], which confirms that reflex activation of sympathetic and vagal nervous activity modifies the cardiac response to calcium channel blocking drugs. In the excised rabbit atrioventricular (AV) nodal preparation, diltiazem, verapamil and nifedipine equally prolong the effective (ERP) and functional (FRP) refractory periods of the AV node. Clinically, diltiazem or verapamil still prolong ERP and FRP, but nifedipine shortens these two values [190]. Similarly, the atrial-His interval is lengthened by i.v. diltiazem or verapamil, while nifedipine shortens the AH interval [190]. The extent to which the calcium channel blocking drugs activate cardiovascular reflexes depends mainly on their potency as peripheral vasodilator drugs. Nifedipine has more potent vasodilator effects than suppressor effects on AV nodal conduction [16] and therefore the reflex increase in heart rate in response to vasodilation predominates over the negative inotropic effect.

LICENSED THERAPEUTIC INDICATIONS

Calcium channel blocking drugs which are currently licensed for use in the United Kingdom are presented in *Table 7.5*, nicardipine being the most recent addition to the list. To date, all the indications are of a cardiovascular nature and, indeed, the calcium channel blocking drugs have now become established as an important group of cardiovascular medicines.

In coronary artery disease, the calcium channel blocking drugs can be used to treat both Prinzmetal's (variant) angina and chronic stable angina pectoris. Calcium channel blocking drugs increase coronary blood flow by dilating the coronary arteries, and this is probably their major mode of action in Prinzmetal's angina, where coronary blood flow is reduced by vasoconstriction

Table 7.5. CALCIUM CHANNEL BLOCKING DRUGS LICENSED FOR USE IN THE UNITED KINGDOM

Non-proprietary name	*Proprietary name*	*Manufacturer*	*Indications*
Verapamil hydrochloride*	Securon Cordilox Berkatens	Knoll Abbott Laboratories Berk Pharmaceuticals	(1) treatment and prophylaxis of angina pectoris (2) treatment and prophylaxis of supraventricular tachycardia and paroxysmal supraventricular tachycardia of the reciprocating type, associated with the Wolff-Parkinson-White syndrome (3) treatment of mild to moderate hypertension and of renal hypertension
Nifedipine*	Adalat	Bayer U.K.	(1) treatment and prophylaxis of angina pectoris (2) treatment of hypertension (3) treatment of Raynaud's phenomenon
Ditiazem hydrochloride*	Tildiem	Lorex Pharmaceuticals	(1) prophylaxis and treatment of angina pectoris
Nicardipine hydrochloride	Cardene	Syntex Pharmaceuticals	(1) treatment of chronic stable angina pectoris (2) treatment of mild to moderate hypertension

* These drugs are also available in the U.S.A.

[192, 193]. However, in chronic stable angina, coronary blood flow is reduced due to stenosis of one or more coronary arteries (usually as a result of atherosclerosis) and in this situation the major mode of action of the calcium channel blockers is in reducing myocardial oxygen consumption. As calcium channel blocking drugs reduce peripheral resistance, they decrease the work of the heart [16]. Verapamil decreases myocardial contractility directly, but after administration of nifedipine this direct effect is counteracted by reflex stimulation of the heart and the heart rate may even be increased [194]. Verapamil, nifedipine and diltiazem appear to be equieffective in the treatment of angina [194]. Several clinical trials have been carried out to examine the effectiveness of some of the newer calcium channel blocking drugs. Nisoldipine [195], felodipine [196], PN 200-110 [197], gallopamil (D600) [198] and tiapamil [199] may all be useful as antianginal agents, but further assessment is required to establish their safety and efficacy.

All currently available calcium channel blocking drugs are licensed for the treatment of mild to moderate hypertension, except for diltiazem. Their antihypertensive effect is due to decreasing elevated peripheral resistance, which results directly from blockade of calcium channels in vascular smooth muscle [200]. The higher the vascular resistance, the more effective the calcium channel blocking drug is as a vasodilator. Therefore, these drugs reduce blood pressure to a greater extent in severe hypertension than they do in normotensive volunteers [200]. Verapamil and nifedipine are effective as sole therapy for mild to moderate hypertension and they can also be used in combination with several other antihypertensive drugs such as beta-adrenoceptor antagonists, methyldopa and thiazide diuretics [201]. However, verapamil should not be given intravenously to patients who are receiving treatment with a beta-adrenoceptor antagonist, since heart block may develop [201]. Although diltiazem has not been licensed for the treatment of hypertension in the U.K., it has been shown that diltiazem is as effective as nifedipine [200], hydrochlorothiazide [202] or metoprolol [203] when used as single drug therapy for mild to moderate hypertension. Felodipine [204, 205], nitrendipine [206] and tiapamil [207] have also been investigated as possible antihypertensive drugs.

The only calcium channel blocking drug to have been licensed for the treatment of cardiac arrhythmias is verapamil. Its main uses are in the treatment of supraventricular tachycardia (SVT) and paroxysmal SVT. Because verapamil lengthens the ERP and FRP of the AV node and prolongs AV nodal conduction time [16], it can be used to control the ventricular rate in atrial fibrillation or atrial flutter and it usually terminates re-entry arrhythmias involving the AV node [208, 209]. However, intravenous verapamil should not be given to patients who have the Wolff-Parkinson-White syndrome and atrial fibrillation, since there is an increased risk of precipitating ventricular tachycardia and fibrillation in these patients [201]. In general, ventricular tachycardia is not terminated by verapamil [210] although it is effective in suppressing ventricular extrasystoles [208]. Diltiazem shows similar electrophysiological effects on AV nodal conduction to verapamil, but it is not currently used to treat SVT [16]. Clinical trials have shown that diltiazem is effective in the management of paroxysmal SVT [211, 212] and the use of diltiazem for the treatment of cardiac arrhythmias deserves further investigation. Nifedipine does not have direct antiarrhythmic effects, probably because it does not influence AV nodal conduction time in the same way as verapamil [208].

Double-blind placebo-controlled trials have shown that both nifedipine and diltiazem are effective in controlling the symptoms of Raynaud's phenomenon [213–215]. Nifedipine increases fingertip blood flow by decreasing fingertip vascular resistance [216]. At present, the sole calcium channel blocking drug to have been licensed in the U.K. for the treatment of Raynaud's phenomenon is nifedipine. Since all calcium channel blockers produce vasodilation, they may all eventually prove to be effective in counteracting vasoconstriction in patients with Raynaud's phenomenon, but this has yet to be established.

OTHER CARDIOVASCULAR INDICATIONS UNDER INVESTIGATION

Because calcium channel blocking drugs show such varied activity within the cardiovascular system, it is not surprising that interest in their effectiveness as treatment for many different cardiovascular disorders has been aroused. The main disorders to have been investigated will be considered briefly. In the treatment of hypertensive emergencies, nifedipine has been shown to be very effective and it produces minimal adverse effects [217–219]. Verapamil has also been used successfully to control hypertensive crises but it needs to be given intravenously, unlike nifedipine, which can be administered sublingually [217]. Whether the administration of calcium channel blocking drugs increases survival rate following an acute myocardial infarction is controversial. Reduction of the extent of myocardial injury was first shown in animals with experimentally-induced myocardial ischaemia or infarction [220]. By measuring creatine kinase values, myocardial infarct size has been found to be decreased in patients given verapamil intravenously [221]. Verapamil also improves the mechanical performance of ischaemic myocardial segments (assessed by M-mode echocardiography) in patients with acute myocardial infarction [222]. However, the administration of calcium channel blocking drugs has not been shown to reduce the incidence of re-infarction [223, 224] or to decrease mortality [223, 225] when the drugs are given for a longer period (for 10 days or up to 6 months following myocardial infarction).

Several studies have shown that verapamil and nifedipine modify favourably the abnormal left ventricular diastolic function found in patients with hypertrophic cardiomyopathy [226–228]. Symptomatic status and exercise tolerance improve in most patients receiving long-term verapamil therapy [227, 228]. As the calcium channel blocking drugs are vasodilators, it has been suggested that they may be useful as drugs to decrease after-load in patients with congestive heart failure [229]. However, these drugs are also potentially negative inotropic agents and therefore they may cause a deterioration in cardiac function. Clinical studies have shown that nifedipine usually produces beneficial haemodynamic changes in patients with congestive heart failure, although haemodynamic deterioration following nifedipine administration has occasionally been reported [229, 230]. Verapamil has a greater negative inotropic action *in vivo* than nifedipine and doubts have been expressed as to whether verapamil should be used in patients with ventricular dysfunction [231]. Ferlinz and Citron [232] have shown, however, that intravenous administration of verapamil to patients with congestive heart failure can improve ventricular function. Long-term studies are required to determine which subgroups of patients will benefit from verapamil therapy and to identify subgroups at risk of haemodynamic deterioration following verapamil. At present, the calcium channel blocking drugs are unlikely to replace existing drugs used in the treatment of congestive heart failure.

In patients with pulmonary hypertension, calcium channel blocking drugs have been found to produce pulmonary artery dilation and therefore improvement in cardiac output and relief of hypoxia [233, 234]. Most of the studies have used short-term administration to carefully selected patients. Thus, the beneficial effects need to be confirmed in longer studies using more patients with advanced pulmonary hypertension. Detrimental deterioration in right ventricular performance has been found in some patients with severe right ventricular dysfunction following the administration of verapamil or nifedipine [233].

The calcium channel blocking drugs may prove to be very useful in the treatment of patients following acute stroke. In particular, nimodipine has been found to have selective actions on cerebral vascular smooth muscle without affecting systemic arterial pressure [16]. After intravenous administration, nimodipine increases hemispheric cerebral blood flow in patients with acute ischaemic stroke [235]. A placebo-controlled double-blind trial has shown that nimodipine significantly decreases the occurrence of severe neurologic deficits from spasm alone in patients who have had subarachnoid haemorrhage [236].

Cerebrovascular spasm also appears to be important in the initiation of migraine attacks; hence, clinical trials have been performed to assess the effectiveness of the calcium channel blocking drugs as prophylaxis for migraine. Nimodipine, nifedipine and verapamil are all effective in reducing the frequency and severity of vascular headaches [237]. Nimodipine may again be particularly useful because of its selective action on the cerebral blood vessels. Clinical trials have shown that it reduces migraine frequency and duration in a similar number of patients as the currently available drugs used for migraine prophylaxis [238].

POTENTIAL NON-CARDIOVASCULAR INDICATIONS

The ability of the calcium channel blocking drugs to relax non-vascular smooth muscle suggests that these drugs should be useful in a number of disorders involving spasm of smooth muscle, such as bronchial asthma or achalasia. However, the results of clinical trials have not always been encouraging and more research is required to establish which types of disorder respond to treatment with calcium channel blockers. For example, in bronchial asthma, calcium channel blocking drugs have been shown to provide a small protective effect against bronchoconstriction provoked by exercise, cold air and allergen challenge [239]. However, voltage-dependent calcium entry into airway cells does not appear to be very important in the pathogenesis of asthmatic bronchoconstriction [239] and therefore calcium channel blocking drugs are unlikely to play an important rôle in the treatment of asthma.

Nifedipine [240–242] and verapamil [243] decrease lower oesophageal sphincter pressure in patients with achalasia, thus the symptoms of achalasia are improved by treatment with a calcium channel blocking drug. Very few double-blind clinical trials have been performed to investigate this effect, but one recent study has shown that nifedipine does decrease symptoms in patients with achalasia when the drug is given sublingually before meals [242]. Although further clinical studies are required, it is likely that calcium channel blocking drugs will ultimately play a part in the treatment of achalasia.

Calcium channel blockade may also be a useful means of controlling the symptoms of the irritable bowel syndrome (IBS). Peppermint oil is widely used in the treatment of IBS [244]. The major active component in peppermint oil with regard to relaxation of gastrointestinal smooth muscle is menthol [245] and it has been shown that menthol inhibits gastrointestinal smooth muscle activity by blocking calcium channels [246–248]. It has also been shown that nifedipine reduces the colonic motor response to eating in IBS patients [249]. It would therefore appear that other calcium channel blockers may be potentially useful in controlling the symptoms associated with the irritable bowel syndrome.

Relaxation of myometrial smooth muscle is another activity of the calcium channel blocking drugs which may be exploited for therapeutic purposes. For instance, nifedipine has been shown to be effective in controlling the pain of dysmenorrhoea and in delaying premature labour [250]. These examples of possible new indications for the calcium channel blocking drugs illustrate the increasing importance of this group of drugs as therapeutic agents.

CONCLUSIONS

Calcium channel blocking drugs have been the subject of extensive investigation since 1964, when Fleckenstein first presented evidence that excitation-contraction coupling of the mammalian myocardium can be blocked *in vivo* and *in vitro* by verapamil or prenylamine [1]. However, while we now know a great deal more about how this class of drug interacts with the calcium channel, full elucidation of the nature of the calcium channel remains unresolved. Development of new methodology for measuring single channel currents has been useful in determining the functioning of calcium channels, particularly with respect to mechanisms involved in their activation and inactivation (see review by Tsien [251]). Many other aspects of the nature of the calcium channels are not yet clearly understood. Indeed, there is now strong evidence that different subtypes of calcium channel exist [53, 54, 118, 252, 253]. Receptor-operated calcium channels have been proposed in smooth muscle, although their specific properties are largely unknown. In contrast, the voltage-operated channels have been well characterized [252].

One major question still to be answered satisfactorily is whether or not the binding sites for calcium channel blocking drugs represent physiologically relevant receptor sites associated with the calcium channel. Solubilization of the high-affinity 1,4-dihydropyridine binding site in skeletal muscle has provided some evidence that the binding site is part of the calcium channel. For example, the detergent-solubilized material (a) shows allosteric regulation of 1,4-dihydropyridine binding by diltiazem or D600, (b) has a metalloprotein nature, and (c) has a molecular weight which indicates a structure large enough to span the cell membrane [254]. There is also some similarity between the solubilized 1,4-dihydropyridine-binding site and the solubilized sodium channel [254]. The molecular weights of various subunits of the putative channel have been determined using either target size analysis [255, 256] or gel electrophoresis of the solubilized 1,4-dihydropyridine-binding site [97, 147, 257]. From this data, the calcium channel would appear to consist of three subunits with molecular weights of 130,000, 50,000 to 60,000 and 30,000 to 45,000 [97, 147, 255, 257]. However, further evidence is required before concluding that the drug-binding sites may form an integral part of the calcium channel.

Other areas where there is a lack of general agreement about the nature of the binding sites for calcium channel blocking drugs include: the subcellular distribution of binding sites in cardiac membrane fractions; whether low-affinity binding sites exist in all tissues; how allosteric interactions occur between different types of binding site; and the precise effects of metal cations on binding. Continued investigation of these areas of disagreement will hopefully provide more information on how the calcium channel blocking drugs interact with their receptors on cell membranes. Furthermore, it is important to know whether the calcium channel blocking drugs bind to the channels in their activated or inactivated states and to elucidate the mechanisms by which the drugs affect the gating of the channels.

Without doubt, the calcium channel blocking drugs have proved to be most useful therapeutic agents for cardiovascular disease and as a result many compounds are currently under development. New indications for this group of drugs such as the treatment of achalasia or migraine have been proposed and doubtless the calcium channel blockers will be found to possess a broad spectrum of therapeutic activity. As a result, the calcium channel blocking drugs are likely to become established as one of the most useful groups of therapeutic agents of this decade.

REFERENCES

1 A. Fleckenstein, Annu. Rev. Pharmacol. Toxicol. 17 (1977) 149.

2 C.J. Cohen, R.A. Janis, D.G. Taylor and A. Scriabine, in Calcium Antagonists and Cardiovascular Disease, ed. L.H. Opie (Raven Press, New York, 1984) pp. 151–163.

2a P.M. Vanhoutte and R. Paoletti, Trends Pharmacol. Sci. 8 (1987) 4.

3 J.C. Stoclet, C. Lugnier, A. Follenius, J.M. Scheftel and D. Gerard, in Calcium Entry Blockers and Tissue Protection, eds. T. Godfraind, P.M. Vanhoutte, S. Govoni and R. Paoletti (Raven Press, New York, 1985) pp. 31–40.

4 C. Lugnier, A. Follenius, D. Gerard and J.C. Stoclet, Eur. J. Pharmacol., 98 (1984) 157.

5 M. Spedding, Br. J. Pharmacol., 79 (1983) 225.

6 D.A. Lathrop, A.M. Murphy, T. Humphreys and A. Schwartz, Eur. J. Pharmacol., 118 (1985) 283.

7 J. Brown, R.J. Marshall and E. Winslow, Br. J. Pharmacol., 86 (1985) 7.

8 H.M. Einwachter, R. Kern and J. Herb, Eur. J. Pharmacol., 55 (1979) 225.

9 T. Anno, T. Furuta, M. Itho, I. Kodama, J. Toyama and K. Yamada, Br. J. Pharmacol., 81 (1984) 589.

10 R.G. Leonard and R.L. Talbert, Clin. Pharm., 1 (1982) 17.

11 P.H. Stone, E.M. Antman, J.E. Muller and E. Braunwald, Ann. Intern. Med., 93 (1980) 886.

12 A.M. Katz, Am. J. Cardiol., 55 (1985) 2B.
13 J.R. Guerrero and S.S. Martin, Med. Res. Rev., 4 (1984) 87.
14 R.A. Murphy and W.T. Gerthoffer, in ref. 2, pp. 75–84.
15 W. Gevers, in ref. 2, pp. 67–74.
16 R.K. Saini, in Cardiovascular Pharmacology, ed. M.J. Antonaccio (Raven Press, New York) 2nd Edn. (1984) pp. 415–452.
17 W.G. Nayler and J. Szeto, Cardiovasc. Res., 6 (1972) 120.
18 W.H. Barry, J.D. Horowitz and T.W. Smith, Br. J. Pharmacol., 85 (1985) 51.
19 R. Bayer, R. Kaufmann and R. Mannhold, Naunyn-Schmiedebergs Arch. Pharmacol., 290 (1975) 69.
20 R. Bayer, R. Hennekes, R. Kaufmann and R. Mannhold, Naunyn-Schmiedebergs Arch. Pharmacol., 290 (1975) 49.
21 T. Godfraind, M. Finet, J.S. Lima and R.C. Miller, J. Pharmacol. Exp. Ther., 230 (1984) 514.
22 C. Cauvin, S. Lukeman, J. Cameron, O. Hwang, K. Meisheri, H. Yamamoto and C. van Breemen, J. Cardiovasc. Pharmacol., 6 (Suppl. 4) (1984) S630.
23 A. Scriabine, C.L. Anderson, R.A. Janis, K. Kojima, H. Rasmussen, S. Lee and U. Michal, J. Cardiovasc. Pharmacol., 6 (Suppl. 7) (1984) S937.
24 C. Cauvin, R. Loutzenhiser and C. Van Breemen, Annu. Rev. Pharmacol. Toxicol., 23 (1983) 373.
25 G. Haeusler, in IUPHAR 9th International Congress of Pharmacology London 1984 Proceedings, eds. W. Paton, J. Mitchell and P. Turner (Macmillan, London) Vol. 2 (1984) pp. 337–342.
26 G.T. Bolger, P. Gengo, R. Klockowski, E. Luchowski, H. Siegel, R.A. Janis, A.M. Triggle and D.J. Triggle, J. Pharmacol. Exp. Ther., 225 (1983) 291.
27 G.B. Frank, J. Pharmacol. Exp. Ther., 236 (1986) 505.
28 K. Jim, A. Harris, L.B. Rosenberger and D.J. Triggle, Eur. J. Pharmacol., 76 (1981) 67.
29 J. Church and T.T. Zsoter, Can. J. Physiol. Pharmacol., 58 (1980) 254.
30 F.B. Yousif and D.J. Triggle, Can. J. Physiol. Pharmacol., 64 (1986) 273.
31 D.W.P. Hay and R.M. Wadsworth, Br. J. Pharmacol., 76 (1982) 103.
32 Y. Ito, H. Kuriyama and H. Suzuki, Br. J. Pharmacol., 64 (1978) 503.
33 C. van Breemen, O. Hwang and K.D. Meisheri, J. Pharmacol. Exp. Ther., 218 (1981) 459.
34 R.J. Miller and S.B. Freedman, Life Sci., 34 (1984) 1205.
35 M. Schramm and R. Towart, Life Sci., 37 (1985) 1843.
36 M. Spedding and I. Cavero, Life Sci., 35 (1984) 575.
37 T. Godfraind, in ref. 3, pp. 1–19.
38 R.A. Janis and A. Scriabine, Biochem. Pharmacol., 32 (1983) 3499.
39 N.J. Lodge, A.L. Bassett and H. Gelband, Pflügers Arch., 403 (1985) 222.
40 K. Shigenobu, J.A. Schneider and N. Sperelakis, J. Pharmacol. Exp. Ther., 190 (1974) 280.
41 H. Nawrath, R.E. Ten Eick, T.F. McDonald and W. Trautwein, Circ. Res., 40 (1977) 408.
42 S. Kanaya and B.G. Katzung, J. Pharmacol. Exp. Ther., 228 (1984) 245.
43 R.S. Kass, J. Pharmacol. Exp. Ther., 223 (1982) 446.
43a P.A. Molyvdas and N. Sperelakis, J. Cardiovasc. Pharmacol., 8 (1986) 449.
44 R. Bayer, D. Kalusche, R. Kaufmann and R. Mannhold, Naunyn-Schmiedebergs Arch. Pharmacol., 290 (1975) 81.
45 T. Itjima, T. Morita and N. Taira, Arch. Int. Pharmacodyn. Ther., 277 (1985) 77.
46 T. Li and N. Sperelakis, Can. J. Physiol. Pharmacol., 61 (1983) 957.
47 M.C. Sanguinetti and R.S. Kass, Circ. Res., 55 (1984) 336.

48 T. Ehara and R. Kaufmann, J. Pharmacol. Exp. Ther., 207 (1978) 49.
49 S. Kanaya, P. Arlock, B.G. Katzung and L.M. Hondeghem, J. Mol. Cell. Cardiol., 15 (1983) 145.
50 M.D. Payet, O.F. Schanne and E. Ruiz-Ceretti, J. Mol. Cell. Cardiol., 12 (1980) 635.
51 M.D. Payet, O.F. Schanne, E. Ruiz-Ceretti and J.M. Demers, J. Mol. Cell. Cardiol., 12 (1980) 187.
52 K.S. Lee and R.W. Tsien, Nature (London), 302 (1983) 790.
53 B. Nilius, P. Hess, J.B. Lansman and R.W. Tsien, Nature (London), 316 (1985) 443.
54 B.P. Bean, J. Gen. Physiol., 86 (1985) 1.
55 P. Hess and R.W. Tsien, Nature (London), 309 (1984) 453.
56 H. Reuter, Nature (London), 301 (1983) 569.
57 P. Hess, J.B. Lansman and R.W. Tsien, Nature (London), 311 (1984) 538.
58 D. Noble, The Initiation of the Heart Beat (Clarendon Press, Oxford, 1975) pp. 39–48.
59 J.M. Nerbonne, S. Richard and J. Nargeot, J. Mol. Cell. Cardiol., 17 (1985) 511.
60 L.M. Hondeghem and B.G. Katzung, Annu. Rev. Pharmacol. Toxicol., 24 (1984) 387.
61 N. Akaike, A.M. Brown, K. Nishi and Y. Tsuda, Br. J. Pharmacol., 74 (1981) 87.
62 R. Mannhold, R. Steiner, W. Haas and R. Kaufmann, Naunyn-Schmiedebergs Arch. Pharmacol., 302 (1978) 217.
63 H. Meyer, Annu. Rep. Med. Chem., 17 (1982) 71.
63a A. Goll, R. Glossmann and R. Mannhold, Naunyn-Schmiedebergs Arch. Pharmacol., 334 (1986) 303.
64 C.J. Biswas, P.A. Molyvdas, N. Sperelakis and T.B. Rogers, Eur. J. Pharmacol., 104 (1984) 267.
65 W. Kobinger and C. Lillie, Eur. J. Pharmacol., 104 (1984) 9.
66 P.W. Sprague and J.R. Powell, Annu. Rep. Med. Chem., 20 (1985) 61.
67 S.U. Sys, N.M. DeClerk and D.L. Brutsaert, Eur. J. Pharmacol., 103 (1984) 33.
68 D.R. Ferry, A. Goll, C. Gadow and H. Glossmann, Naunyn-Schmiedebergs Arch. Pharmacol., 327 (1984) 183.
69 C. Ludwig and H. Nawrath, Br. J. Pharmacol., 59 (1977) 411.
70 D.A. Langs and D.J. Triggle, in ref. 25, pp. 323–328.
71 B. Loev, M.M. Goodman, K.M. Snader, R. Tedeschi and E. Macko, J. Med. Chem., 17 (1974) 956.
72 D. Rampe, C.M. Su, F. Yousif and D.J. Triggle, Br. J. Clin. Pharmacol., 20 (Suppl. 2) (1985) 247S.
73 R.A. Janis and D.J. Triggle, J. Med. Chem., 26 (1983) 775.
74 P. Sprague and J.R. Powell, Annu. Rep. Med. Chem., 19 (1984) 61.
75 H. Meyer, in ref. 2, pp. 165–173.
76 R. Rodenkirchen, R. Bayer, R. Steiner, F. Bossert, H. Meyer and E. Moller, Naunyn-Schmiedebergs Arch. Pharmacol., 310 (1979) 69.
77 R. Fossheim, K. Svarteng, A. Mostad, C. Romming, E. Shefter and D.J. Triggle, J. Med. Chem., 25 (1982) 126.
77a M. Mahmoudian and W.G. Richards, J. Pharm. Pharmacol., 38 (1986) 272.
78 S. Kongsamut, T.J. Kamp, R.J. Miller and M.C. Sanguinetti, Biochem. Biophys. Res. Commun., 130 (1985) 141.
79 G. Franckowiak, M. Bechem, M. Schramm and G. Thomas, Eur. J. Pharmacol., 114 (1985) 223.
80 D.A. Langs and D.J. Triggle, Mol. Pharmacol., 27 (1985) 544.
81 K.C. Preuss, G.J. Gross, H.L. Brooks and D.C. Warltier, Life Sci., 37 (1985) 1271.

82 G. Thomas, M. Chung and C.J. Cohen, Circ. Res., 56 (1985) 87.
83 G. Thomas, R. Gross and M. Schramm, J. Cardiovasc. Pharmacol., 6 (1984) 1170.
84 J.S. Williams, I.L. Grupp, G. Grupp, P.L. Vaghy, L. Dumont, A. Schwartz, A. Yatani, S. Hamilton and A.M. Brown, Biochem. Biophys. Res. Commun., 131 (1985) 13.
85 D.J. Triggle and R.A. Janis, J. Cardiovasc. Pharmacol., 6 (Suppl. 7) (1984) S949.
86 T. Morita, T. Fukuda, T. Sukamoto, S. Tajima, I. Sato, Y. Ikeda, T. Kanazawa, Y. Hamada, Y. Morimoto, H. Kitamura, K. Kawakami, K. Ito and T. Nose, Arzneim.-Forsch., 32 (1982) 1060.
87 P. Bellemann, D. Ferry, F. Lubbecke and H. Glossmann, Arzneim.-Forsch., 31 (1981) 2064.
88 A. Schwartz, I.L. Grupp, G. Grupp, J.S. Williams and P.L. Vaghy, Biochem. Biophys. Res. Commun., 125 (1984) 387.
89 P.L. Vaghy, I.L. Grupp, G. Grupp and A. Schwartz, Circ. Res., 55 (1984) 549.
90 A.B. Fawzi and J.H. McNeill, Eur. J. Pharmacol., 104 (1984) 357.
91 R.J. Gould, K.M. Murphy and S.H. Snyder, Mol. Pharmacol., 25 (1984) 235.
92 E.M. Luchowski, F. Yousif, D.J. Triggle, S.C. Maurer, J.G. Sarmiento and R.A. Janis, J. Pharmacol. Exp. Ther., 230 (1984) 607.
93 M.R. Bristow, R. Ginsburg, J.A. Laser, B.J. McAuley and W. Minobe, Br. J. Pharmacol., 82 (1984) 309.
94 K.M.M. Murphy and S.H. Snyder, Eur. J. Pharmacol., 77 (1982) 201.
95 M.L. Garcia, M.J. Trumble, J.P. Reuben and G.J. Kaczorowski, J. Biol. Chem., 259 (1984) 15013.
96 J.D. Marsh, E. Loh, D. Lachance, W.H. Barry and T.W. Smith, Circ. Res., 53 (1983) 539.
97 A. Rengasamy, J. Ptasienski and M.M. Hosey, Biochem. Biophys. Res. Commun., 126 (1985) 1.
98 W. McBride, A. Mukherjee, Z. Haghani, E. Wheeler-Clark, J. Brady, T. Gandler, L. Bush, L.M. Buja and J.T. Willerson, Am. J. Physiol., 247 (1984) H775.
99 J. Scott Hayes, N. Bowling, B.G. Conery and R.F. Kaufmann, Biochim. Biophys. Acta, 812 (1985) 313.
100 W.G. Nayler, J.S. Dillon, J.S. Elz and M. McKelvie, Eur. J. Pharmacol., 115 (1985) 81.
101 A. De Pover, M.A. Matlib, S.W. Lee, G.P. Dube, I.L. Grupp, G. Grupp and A. Schwartz, Biochem. Biophys. Res. Commun., 108 (1982) 110.
102 F.J. Ehlert, E. Itoga, W.R. Roeske and H.I. Yamamura, Biochem. Biophys. Res. Commun., 104 (1982) 937.
103 J.G. Sarmiento, R.A. Janis, R.A. Colvin, D.J. Triggle and A.M. Katz, J. Mol. Cell. Cardiol., 15 (1983) 135.
104 G.T. Bolger, P.J. Gengo, E.M. Luchowski, H. Siegel, D.J. Triggle and R.A. Janis, Biochem. Biophys. Res. Commun., 104 (1982) 1604.
105 P.J. Marangos, J. Patel, C. Miller and A.M. Martino, Life Sci., 31 (1982) 1575.
106 H.R. Lee, W.R. Roeske and H.I. Yamamura, Life Sci., 35 (1984) 721.
107 A.S. Fairhurst, S.A. Thayer, J.E. Colker and D.A. Beatty, Life Sci., 32 (1983) 1331.
108 J.F. Renaud, T. Kazazoglou, A. Schmid, G. Romey and M. Lazdunski, Eur. J. Biochem., 139 (1984) 673.
109 T. Godfraind and M. Wibo, Br. J. Pharmacol., 85 (1985) 335.
110 M.E. Goldman and J.J. Pisano, Life Sci., 37 (1985) 1301.
111 A.K. Grover, C.Y. Kwan, E. Luchowski, E.E. Daniel and D.J. Triggle, J. Biol. Chem., 259 (1984) 2223.
112 M. Fosset, E. Jaimovich, E. Delpont and M. Lazdunski, J. Biol. Chem., 258 (1983) 6086.

113 H.I. Yamamura, H. Schoemaker, R.G. Boles and W.R. Roeske, Biochem. Biophys. Res. Commun., 108 (1982) 640.
114 M.S. Finkel, G. Aguilera, K.J. Catt and H.R. Keiser, Life Sci., 35 (1984) 905.
115 H. Glossmann and D.R. Ferry, Drug Dev. Eval., 9 (1983) 63.
116 R.A. Janis, S.C. Maurer, J.G. Sarmiento, G.T. Bolger and D.J. Triggle, Eur. J. Pharmacol., 82 (1982) 191.
117 D.R. Ferry, H. Glossmann and A.J. Kaumann, Br. J. Pharmacol., 84 (1985) 811.
118 H. Glossmann, D.R. Ferry, A. Goll and M. Rombusch, J. Cardiovasc. Pharmacol., 6 (Suppl. 4) (1984) S608.
119 D.R. Ferry and H. Glossmann, Naunyn-Schmiedebergs Arch. Pharmacol., 321 (1982) 80.
120 P. Ruth, V. Flockerzi, E. von Nettelbladt, J. Oeken and F. Hofmann, Eur. J. Biochem., 150 (1985) 313.
121 J. Striessnig, G. Zernig and H. Glossmann, Eur. J. Pharmacol., 108 (1985) 329.
122 D.R. Ferry, A. Goll, M. Rombusch and H. Glossmann, Br. J. Clin. Pharmacol., 20 (Suppl. 2) (1985) 233S.
123 H. Glossmann, D.R. Ferry and C.B. Boschek, Naunyn-Schmiedebergs Arch. Pharmacol., 323 (1983) 1.
124 M. Holck, S. Thorens and G. Haeusler, Eur. J. Pharmacol., 85 (1982) 305.
125 J. Ptasienski, K.K. McMahon and M.M. Hosey, Biochem. Biophys. Res. Commun., 129 (1985) 910.
126 H. Glossmann, D.R. Ferry, A. Goll, J. Striessnig and M. Schober, J. Cardiovasc. Pharmacol., 7 (Suppl. 6) (1985) S20.
127 R.A. Janis, J.G. Sarmiento, S.C. Maurer, G.T. Bolger and D.J. Triggle, J. Pharmacol. Exp. Ther., 231 (1984) 8.
128 L.T. Williams and L.R. Jones, J. Biol. Chem., 258 (1983) 5344.
129 A. De Pover, S.W. Lee, M.A. Matlib, K. Whitmer, B.A. Davis, T. Powell and A. Schwartz, Biochem. Biophys. Res. Commun., 113 (1983) 185.
130 L.T. Williams and P. Tremble, J. Clin. Invest., 70 (1982) 209.
131 M. Holck, W. Fischli and U. Hengartner, J. Receptor Res., 4 (1984) 557.
132 F.J. Ehlert, W.R. Roeske, E. Itoga and H.I. Yamamura, Life Sci., 30 (1982) 2191.
133 H. Rogg, L. Criscione, A. Truog and M. Meier, J. Cardiovasc. Pharmacol., 7 (Suppl. 6) (1985) S31.
134 I.J. Reynolds, R.J. Gould and S.H. Snyder, Eur. J. Pharmacol., 95 (1983) 319.
135 J.P. Galizzi, M. Fosset and M. Lazdunski, Biochem. Biophys. Res. Commun., 118 (1984) 239.
136 P. Bellemann, J. Receptor Res., 4 (1984) 571.
137 H.J. Oeken, M. Zimmer and B. Jung, Naunyn-Schmiedebergs Arch. Pharmacol., 332 (Suppl.) (1986) R43.
138 J.P. Galizzi, M. Fosset and M. Lazdunski, Biochem. Biophys. Res. Commun., 132 (1985) 49.
139 A. Goll, D.R. Ferry, J. Striessnig, M. Schober and H. Glossmann, FEBS Lett., 176 (1984) 371.
140 I.J. Reynolds, A.M. Snowman and S.H. Snyder, J. Pharmacol. Exp. Ther., 237 (1986) 731.
141 J.P. Galizzi, M. Fosset and M. Lazdunski, Eur. J. Biochem., 144 (1984) 211.
142 H. Glossmann, T. Linn, M. Rombusch and D.R. Ferry, FEBS Lett., 160 (1983) 226.
143 H. Glossmann, D.R. Ferry and A. Goll, in Ref. 25, pp. 329–335.
144 F.J. Green, B.B. Farmer, G.L. Wiseman, M.J. Jose and A.M. Watanabe, Circ. Res., 56 (1985) 576.
145 K.M.M. Murphy, R.J. Gould, B.L. Largent and S.H. Snyder, Proc. Natl. Acad. Sci. USA, 80 (1983) 860.

146 A. DePover, I.L. Grupp, G. Grupp and A. Schwartz, Biochem. Biophys. Res. Commun., 114 (1983) 922.

147 T.L. Kirley and A. Schwartz, Biochem. Biophys. Res. Commun., 123 (1984) 41.

148 E. Carafoli, in Ref. 2, pp. 29–41.

149 M. Spedding and D.N. Middlemiss, Trends Pharmacol. Sci., 6 (1985) 309.

150 V. Ramkumar and E.E. El-Fakahany, Trends Pharmacol. Sci., 7 (1986) 171.

151 E.M. Gallant and V.M. Goettl, Eur. J. Pharmacol., 117 (1985) 259.

152 P.L. Vaghy, I.L. Grupp, G. Grupp, J.L. Balwierczak, J.S. Williams and A. Schwartz, Eur. J. Pharmacol., 102 (1984) 373.

153 J.O. Bustamante, Pflügers Arch., 403 (1985) 225.

154 J.B. Galper and W.A. Catterall, Mol. Pharmacol., 15 (1979) 174.

155 A. Yatani and A.M. Brown, Circ. Res., 56 (1985) 868.

156 M.R. Rosen, J.P. Ilvento, H. Gelband and C. Merker, J. Pharmacol. Exp. Ther., 189 (1974) 414.

157 C. Chen and L.S. Gettes, J. Pharmacol. Exp. Ther., 209 (1979) 415.

158 H. Reuter, in Ref. 2, pp. 43–51.

159 S.A. Siegelbaum, R.W. Tsien and R.S. Kass, Nature (London), 269 (1977) 611.

160 J.R. Hume, J. Pharmacol. Exp. Ther., 234 (1985) 134.

161 P.L. Vaghy, J.D. Johnson, M.A. Matlib, T. Wang and A. Schwartz, J. Biol. Chem., 257 (1982) 6000.

162 M.A. Matlib and A. Schwartz, Life Sci., 32 (1983) 2837.

163 M. Pan and R.A. Janis, Biochem. Pharmacol., 33 (1984) 787.

164 S. Takeo, K. Adachi and M. Sakanashi, Biochem. Pharmacol., 34 (1985) 2303.

165 A. Erdreich and H. Rahamimoff, Biochem. Pharmacol., 33 (1984) 2315.

165a F. van Amsterdam and J. Zaagsma, Eur. J. Pharmacol., 123 (1986) 441.

166 D.C. Pang and N. Sperelakis, Eur. J. Pharmacol., 81 (1982) 403.

167 J. Mas-Oliva and W.G. Nayler, Br. J. Pharmacol., 70 (1980) 617.

168 M. David Dufilho, M.A. Devynck, S. Kazda and P. Meyer, Eur. J. Pharmacol., 97 (1984) 121.

169 J.D. Johnson, Biophys. J., 45 (1984) 134.

170 J.D. Johnson and L.A. Wittenauer, Biochem. J., 211 (1983) 473.

171 P.M. Epstein, K. Fiss, R. Hachisu and D.M. Andrenyak, Biochem. Biophys. Res. Commun., 105 (1982) 1142.

172 J.D. Johnson, Biochem. Biophys. Res. Commun., 112 (1983) 787.

173 S. Bostrom, B. Ljung, S. Mardh, S. Forsen and E. Thulin, Nature (London), 292 (1981) 777.

174 M.A. Movsesian, A.L. Swain and R.S. Adelstein, Biochem. Pharmacol., 33 (1984) 3759.

175 J.M.J. Lamers, K.J. Cysouw and P.D. Verdouw, Biochem. Pharmacol., 34 (1985) 3837.

176 J.A. Norman, J. Ansell and M.A. Phillipps, Eur. J. Pharmacol., 93 (1983) 107.

177 K. Kubo, Y. Matsuda, H. Kase and K. Yamada, Biochem. Biophys. Res. Commun., 124 (1984) 315.

178 M.J. Daly, S. Perry and W.G. Nayler, Eur. J. Pharmacol., 90 (1983) 103.

179 P.J. Silver, J.M. Ambrose, R.J. Michalak and J. Dachiw, Eur. J. Pharmacol., 102 (1984) 417.

180 M.A. Movsesian, I.S. Ambudkar, R.S. Adelstein and A.E. Shamoo, Biochem. Pharmacol., 34 (1985) 195.

181 T. Wang, L.I. Tsai and A. Schwartz, Eur. J. Pharmacol., 100 (1984) 253.

182 M.L. Entman, J.C. Allen, E.P. Bornet, P.C. Gillette, E.T. Wallick and A. Schwartz, J. Mol. Cell. Cardiol. 4 (1972) 681.

183 A.M. Watanabe and H.R. Besch, J. Pharmacol. Exp. Ther., 191 (1974) 241.

183a R.A. Colvin, N. Pearson, F.C. Messineo and A.M. Katz, J. Cardiovasc. Pharmacol., 4 (1982) 935.

184 A.M. Katz, in Ref. 2, pp. 53–66.

185 D.J. De Vries and P.M. Beart, Eur. J. Pharmacol., 106 (1985) 133.

186 W.G. Nayler, J.E. Thompson and B. Jarrott, J. Mol. Cell. Cardiol., 14 (1982) 185.

187 J. Nishimura, H. Kanaide and M. Nakamura, J. Pharmacol. Exp. Ther., 236 (1986) 789.

188 M. Waelbroek, R. Robberecht, P. de Neef and J. Christophe, Biochem. Biophys. Res. Commun., 121 (1984) 340.

189 W.G. Nayler and J.S. Dillon, Br. J. Clin. Pharmacol., 21 (Suppl. 2) (1986) 97S.

190 C. Kawai, T. Konishi, E. Maisuyama and H. Okazaki, Circulation, 63 (1981) 1035.

191 H.C. Stern, J.H. Matthews and G.G. Belz, Eur. J. Clin. Pharmacol., 29 (1986) 541.

192 M.D. Winniford, J.T. Willerson and L.D. Hillis, Am. J. Cardiol., 55 (1985) 116B.

193 G. Ellrodt, C.Y.C. Chew and B.N. Singh, Circulation, 62 (1980) 669.

194 A.H. Meltzer, K.R. Kafka and W.H. Frishman, Hosp. Formulary, 21 (1986) 299.

195 M.F. Rousseau, M.F. Vincent, F. van Hoof, G. van den Berghe, A.A. Charlier and H. Pouleur, Am. J. Cardiol., 54 (1984) 1189.

196 H. Emanuelsson, A. Hjalmarson, S. Holmberg and F. Waagstein, Eur. Heart J., 5 (1984) 308.

197 C.E. Handler and E. Sowton, Eur. J. Clin. Pharmacol., 27 (1984) 415.

198 N.S. Khurmi, M.J. O'Hara, M.J. Bowles, V.B. Subramanian and E.B. Raftery, Am. J. Cardiol., 53 (1984) 684.

199 N. Danchin, J.P. Clozel, K. Khalife, F. Elkik, F. Boulay, J.L. Neimann and F. Cherrier, Am. Heart J., 109 (1985) 764.

200 W.W. Klein, Am. J. Med., 77 (1984) 143.

201 B.F. Robinson, Am. J. Cardiol., 55 (1985) 102B.

202 I.K. Inouye, B.M. Massie, N. Benowitz, P. Simpson and D. Loge, Am. J. Cardiol., 53 (1984) 1588.

203 B. Trimarco, N. DeLuca, B. Ricciardelli, M. Volpe, A. Veniero, A. Cuocolo and M. Cicala, J. Clin. Pharmacol., 24 (1984) 218.

204 D. Elmfeldt and T. Hedner, Drugs, 29 (Suppl. 2) (1985) 109.

205 H. Aberg, M. Lindsjo and B. Morlin, Drugs, 29 (Suppl. 2) (1985) 117.

206 L.A. Ferrara, M.L. Fasano, G. de Simone, S. Soro and R. Gagliardi, Clin. Pharmacol. Ther., 38 (1985) 434.

207 P. Balansard, F. Elkik, J.A. Levenson, M. Ciampi and P. Sans, Br. J. Clin. Pharmacol., 18 (1984) 823.

208 D.M. Krikler, Br. J. Clin. Pharmacol., 21 (Suppl. 2) (1981b) 183S.

209 L. Schamroth, Am. Heart J., 100 (1980) 1070.

210 R.F. Gilmour and D.P. Zipes, Am. J. Cardiol., 55 (1985) 89B.

211 A. Betriu, B.R. Chaitman, M.G. Bourassa, G. Brevers, J.M. Scholl, P. Bruneau, P. Gagne and M. Chabot, Circulation, 67 (1983) 88.

212 E. Rowland, W.J. McKenna, H. Gulker and D.M. Krikler, Circ. Res., 52 (1983) I-163.

213 C.R. Smith and R.J. Rodeheffer, Am. J. Cardiol., 55 (1985) 154B.

214 C.J. White, W.A. Phillips, L.A. Abrahams, T.D. Watson and P.T. Singleton, Am. J. Med., 80 (1986) 623.

215 T. Gjorup, H. Kelbaek, O.J. Hartling and S.L. Nielsen, Am. Heart J. 111 (1986) 742.

216 M.A. Creager, K.M. Pariser, E.M. Winston, H.M. Rasmussen, K.B. Miller and J.D. Coffman, A. Heart J., 108 (1984) 370.

217 W.H. Frishman, P. Weinberg, H.B. Peled, B. Kimmel, S. Charlap and N. Beer, Am. J. Med., 77 (Suppl. 2B) (1984) 35.
218 O. Bertel and L.D. Conen, Am. J. Med., 79 (Suppl. 4A) (1985) 31.
219 M.C. Houston, Am. Heart J., 111 (1986) 963.
220 J. de Leiris, V. Richard and S. Pestre, in Ref. 2, pp. 105–115.
221 W.D. Bussmann, W. Seher and M. Gruengras, Am. J. Cardiol., 54 (1984) 1224.
222 J. Heikkila and M.S. Nieminen, Am. Heart J., 107 (1984) 241.
223 The Danish study group on verapamil in myocardial infarction, Br. J. Clin. Pharmacol., 21 (Suppl. 2) (1986) 197S.
224 F. Crea, J. Deanfield, P. Crean, M. Sharom, G. Davies and A. Maseri, Am. J. Cardiol., 55 (1985) 900.
225 P.A. Sirnes, K. Overskeid, T.R. Pedersen, J. Bathen, A. Drivenes, G.S. Froland, J.K. Kiekshun, K. Landmark, R. Rokseth, K.B. Sirnes, A. Sundoy, B.R. Torjussen, K.M. Westlund and B.A. Wik, Circulation, 70 (1984) 638.
226 R.O. Bonow, Am. J. Cardiol., 55 (1985) 172B.
227 B.H. Lorell, Am. J. Med., 78 (Suppl. 2B) (1985) 43.
228 D.R. Rosing, U. Idanpaan-Heikkila, B.J. Maron, R.O. Bonow and S.E. Epstein, Am. J. Cardiol., 55 (1985) 185B.
229 W.S. Colucci, M.A. Fifer, B.H. Lorell and J. Wynne, Am. J. Med., 78 (Suppl. 2B) (1985) 9.
230 U. Elkayam, L. Weber, C.R. McKay and S.H. Rahimtoola, Am. J. Cardiol., 54 (1984) 126.
231 M.A. Joesphson and B.N. Singh, Am. J. Cardiol., 55 (1985) 81B.
232 J. Ferlinz and P.D. Citron, Am. J. Cardiol., 51 (1983) 1339.
233 M. Packer, Am. J. Cardiol., 55 (1985) 196B.
234 L.J. Rubin, Chest, 88 (Suppl.) (1985) 257S.
235 H.J. Gelmers, Am. J. Cardiol., 55 (1985) 144B.
236 G.S. Allen, Am. J. Cardiol., 55 (1985) 149B.
237 J.S. Meyer, Ration. Drug. Ther., 19 (1985) 1.
238 H.J. Gelmers, Am. J. Cardiol., 55 (1985) 139B.
239 P.J. Barnes, Br. J. Clin. Pharmacol., 20 (Suppl. 2) (1985) 289S.
240 M. Traube and R.W. McCallum, Am. J. Gastroenterol., 79 (1984) 892.
241 M. Traube, M. Hongo, L. Magyar and R.W. McCallum, J. Am. Med. Assoc., 252 (1984) 1733.
242 M. Traube, S. DuBovik, L. Magyar and R.W. McCallum, Gastroenterology, 90 (1986) 1670.
243 B.S. Becker and R. Burakoff, Am. J. Gastroenterol., 78 (1983) 773.
244 W.D.W. Rees, B.K. Evans and J. Rhodes, Br. Med. J., 2 (1979) 835.
245 B.A. Taylor, D.K. Luscombe and H.L. Duthie, Br. J. Surg., 71 (1984) 385.
246 B.A. Taylor, P. Collins, D.K. Luscombe and H.L. Duthie, Br. J. Surg., 71 (1984) 902.
247 B.A. Taylor, H.L. Duthie and D.K. Luscombe, Br. J. Clin. Pharmacol., 20 (1985) 293P.
248 B.A. Taylor, D.K. Luscombe and H.L. Duthie, Gut, 25 (1984) 1168.
249 F. Narducci, G. Bassotti, M. Gaburri, F. Farroni and A. Morelli, Am. J. Gastroenterol., 80 (1985) 317.
250 M.L. Schwartz, H.M. Rotmensch, P.H. Vlasses and R.K. Ferguson, Arch. Intern. Med., 144 (1984) 1425.
251 R.W. Tsien, Annu. Rev. Physiol., 45 (1983) 341.
252 L. Hurwitz, Annu. Rev. Pharmacol. Toxicol., 26 (1986) 225.
253 M.C. Nowycky, A.P. Fox and R.W. Tsien, Nature (London), 316 (1985) 440.

254 H. Glossmann and D.R. Ferry, Naunyn-Schmiedebergs Arch. Pharmacol. 323 (1983) 279.
255 J.C. Venter, C.M. Fraser, J.S. Schaber, C.Y. Jung, G. Bolger and D.J. Triggle, J. Biol. Chem., 258 (1983) 9344.
256 D.R. Ferry, A. Goll and H. Glossmann, Naunyn-Schmiedebergs Arch. Pharmacol., 323 (1983) 292.
257 B.M. Curtis and W.A. Catterall, Biochemistry, 23 (1984) 2113.

Progress in Medicinal Chemistry – Vol. 24, edited by G.P. Ellis and G.B. West

8 The Medicinal Chemistry of Aldose Reductase Inhibitors

LESLIE G. HUMBER, Ph.D.

Chemistry Department, Ayerst Laboratories Research, Inc., CN 8000, Princeton, NJ 08540, U.S.A.

INTRODUCTION

Insulin therapy has been dramatically effective in eliminating keto-acidotic coma as a cause of death in diabetics. However, while prolonging life, insulin therapy does not prevent the occurrence of disabling complications of chronic diabetes, such as neuropathy, nephropathy, retinopathy and cataracts [1]. Insulin therapy is only partially effective in normalizing glucose levels, and the occurrence of diabetic complications appears to be related to the severity and duration of diabetic hyperglycaemia.

The tissues involved (nerve, kidney, retina and lens) do not require insulin for glucose uptake and consequently are exposed to elevated glucose levels. Under normal conditions, glucose is metabolized *via* the energy-producing glycolytic pathway involving an initial phosphorylation by the enzyme hexokinase. Evidence is accumulating which shows that under conditions of diabetic hyperglycaemia, hexokinase becomes saturated and excess glucose is metabolized by the polyol pathway which was first identified in seminal vesicle in 1956 [2], and has since been found to be functional in a variety of tissues, including those involved with the manifestation of diabetic complications [3–5]. The polyol pathway, for which there is no known physiological rôle [6], consists of the enzymes aldose reductase (AR), which reduces glucose to sorbitol, and sorbitol dehydrogenase, which converts sorbitol to fructose.

THE POLYOL PATHWAY

$$\text{glucose} + \text{NADPH} \xrightarrow{\text{aldose reductase}} \text{sorbitol} + \text{NADP}^+$$

$$\text{sorbitol} + \text{NAD}^+ \xrightarrow{\text{sorbitol dehydrogenase}} \text{fructose} + \text{NADH}$$

While glucose has a high affinity for hexokinase [7], the affinity for AR is low [8]. As a result, in the presence of high glucose concentrations, hexokinase becomes saturated and the polyol pathway becomes activated, resulting in the intracellular production of sorbitol and fructose in tissues where this pathway exists. Since the rate of formation of sorbitol is faster than its conversion to fructose, and because the polarity of the polyol prevents it from exiting the cell, sorbitol selectively accumulates in affected tissues of animals with experimentally induced diabetes [9, 10]. The increased flux of glucose through the polyol pathway is believed to result in altered cellular function, and ultimately, the production of the pathology associated with diabetic complications [11, 12].

Numerous other investigations have produced evidence which provides support for a major rôle for AR in the manifestation of various diabetic complications (for recent reviews see [13–16]), and it has become apparent that inhibitors of AR may be able to prevent, retard, or reverse the complications of chronic diabetes.

The search for inhibitors of AR was started after the enzyme was isolated from bovine lens and characterized by Hayman and Kinoshita in 1965 [8]. *In vitro* studies have also utilized enzyme from other species and tissues, including rabbit lens, human lens, rat lens and human placenta [14]. Several animal models are available for the study of AR inhibitors (for reviews see [14, 15, 17]); the most widely used models are the galactosaemic rat and the streptozotocin-induced diabetic rat, where the accumulation of galactitol and sorbitol, respectively, can be measured in target tissues.

Considerable effort has been spent in the attempts to develop clinically useful AR inhibitors and this chapter reviews the classes of compounds that have been investigated in the past 20 years.

THE DEVELOPMENT OF ALRESTATIN

Alrestatin (AY-22,284, 1), the first AR inhibitor to be clinically investigated, was developed at the Ayerst Research Laboratories. In 1966, work on AR inhibitors was initiated at Ayerst, after Hayman and Kinoshita [8], observed that, *in vitro*, the enzyme was inhibited by a variety of aliphatic and keto acids, octanoic acid being one of the most potent, although cytotoxic, causing 59% inhibition at 1×10^{-4} M. The programme at Ayerst, using a partially purified enzyme preparation isolated from calf lens, involved the systematic screening of various straight-chain carboxylic acids. For aliphatic carboxylic acids containing one to 23 carbon atoms (*Table 8.1*), there was a relationship between chain length and inhibitory activity; maximal activity was associated with a chain length of 8 to 12 carbon atoms. Straight-chain α,ω-dicarboxylic acids (*Table 8.2*) showed a similar relationship, with inhibitory activity being maximal between the C_7 and C_{12} acids.

A series of dicarboxylic acids was synthesized in which the chain separating the carboxylic acid groups was substituted by a ring (*Table 8.3*). The most

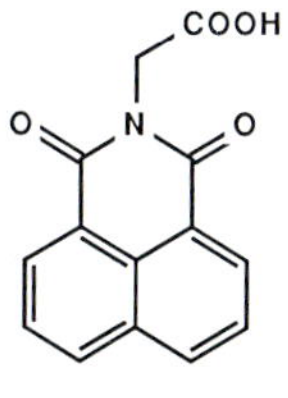

COOH
O N O

(1)

Table 8.1. AR INHIBITORY ACTIVITY OF ALIPHATIC CARBOXYLIC ACIDS: R – COOH

R	*Acid*	*% inhibition 1×10^{-4} M*	*R*	*Acid*	*% inhibition 1×10^{-4} M*
C_1	acetic	0	C_{11}	lauric	56
C_2	propionic	0	C_{12}	tridecanoic	52
C_3	butyric	1	C_{13}	myristic	44
C_4	valeric	11	C_{15}	palmitic	31
C_5	caproic	27	C_{16}	margaric	28
C_6	enanthic	35	C_{17}	stearic	7
C_7	caprylic	48	C_{18}	nonadecanoic	0
C_8	pelargonic	56	C_{19}	icosanoic	0
C_9	capric	47	C_{20}	henicosanoic	4
C_{10}	undecanoic	56	C_{23}	lignoceric	0

Table 8.2. AR INHIBITORY ACTIVITY OF α,ω-DICARBOXYLIC ACIDS: HOOC – X – COOH

X	*Acid*	*% inhibition 1×10^{-4} M*	*X*	*Acid*	*% inhibition 1×10^{-4} M*
–	oxalic	0	C_7	azelaic	40
C_1	malonic	0	C_8	sebacic	56
C_2	succinic	0	C_9	undecanedioic	48
C_3	glutaric	2	C_{10}	dodecanedioic	55
C_4	adipic	2	C_{11}	tridecanedioic	63
C_5	pimelic	7	C_{12}	tetradecanedioic	63
C_6	suberic	34			

active was found to be AY-20,037, tetramethyleneglutaric acid (TMG), which inhibited calf lens AR by 30% at 1×10^{-6} M (*Table 8.3*, No. 1). TMG was used to validate, *in vitro*, the hypothesis that AR was involved in the initiation and development of galactose cataracts [18], although it was inactive when administered orally [19, 20].

A series of pyridylanilines was next investigated; *N*-(3-nitro-2-pyridyl)-3-trifluoromethylaniline, (2), (AY-20,263), was found to inhibit calf lens AR by 56% at 1×10^{-5} M, and was the first AR inhibitor demonstrated to be active upon injection into the vitreous cavity of the eye of rats fed a high-galactose diet, by reducing galactitol levels in the lens by 53%, and delaying the onset of cataract [21].

Table 8.3. AR INHIBITORY ACTIVITY OF GLUTARIC ACID DERIVATIVES

No.	Compound	% inhibition 1×10^{-4} M	1×10^{-5} M	1×10^{-6} M
1 (TMG)	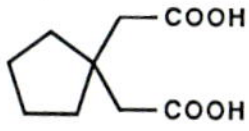	90	60	30
2	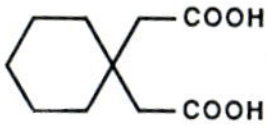	65	30	–
3	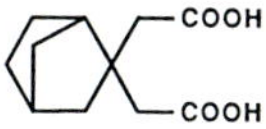	55	33	–
4		51	–	–
5	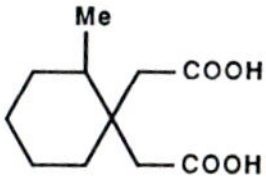	11	–	–
6	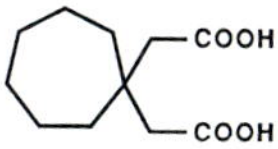	55	28	–
7	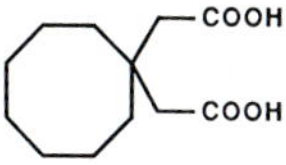	47	–	–

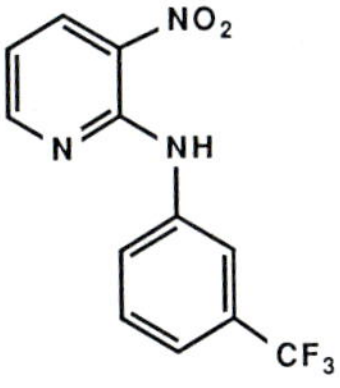

(2)

Hundreds of compounds were subsequently tested. Phthaloylglycine (3) was found to be potent *in vitro* ($IC_{50} = 2.5 \times 10^{-6}$ M) but was inactive upon oral administration. This was attributed to a facile opening of the five-membered ring, giving the corresponding phthalamic acid (4), which had virtually no effect

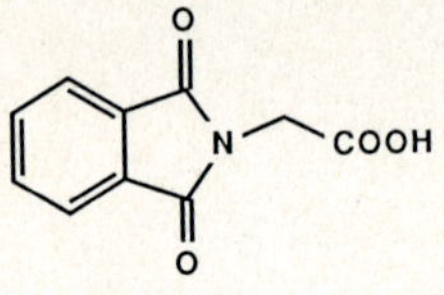

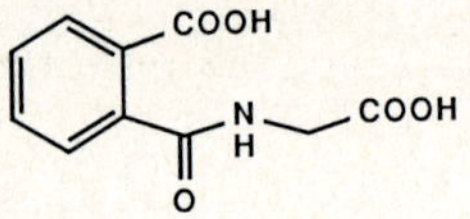

(3)

(4)

on AR *in vitro*. In order to retain structural features of phthaloylglycine, but in a molecular framework which would be resistant to hydrolysis, alrestatin (AY-22,284) (1), a compound first described in 1931 [22], was synthesized [23], tested and found to be active *in vitro* ($IC_{50} = 3.5 \times 10^{-6}$ M). It was also the first AR inhibitor to be effective upon oral administration, decreasing polyol accumulation in both streptozotocin-induced diabetic rats and in rats fed galactose, and delaying cataract formation in the latter model [19].

Alrestatin was remarkably stable *in vivo*, being excreted unchanged in several species, and after extensive pharmacological studies, alrestatin was tested in man at high dosage in diabetic patients with clinical signs of neuropathy [24] and impaired nerve conduction velocity [25].

FLAVONOIDS

Flavonoids were first reported to be AR inhibitors in 1975 [26]. With partially purified rat lens AR, eight of them were tested and three were found to retain activity at 1×10^{-7} M, the most potent being quercitrin, the 3-L-rhamnoside of quercetin (*Table 8.4*, No. 3), which inhibited AR by 55% at 1×10^{-7} M. At 1×10^{-4} M, it decreased xylitol accumulation by 80% in rat lens incubated in a high-xylose medium [26].

Oral administration of quercitrin, in the diet, to the streptozotocin-induced diabetic degu (*Octodon degus*) decreased lens sorbitol accumulation by 50% and delayed the onset of cataracts [27]. The degu, a rodent native to the South American Andes, was selected for the study because its lens AR activity is 3- to 4-times higher than that of the rat, and as a consequence it develops lens opacities within 10–12 days, rather than in the 3–4 months required for the rat

Table 8.4. AR INHIBITION BY FLAVONOIDS HAVING IC_{50} VALUES UNDER 1×10^{-7} M

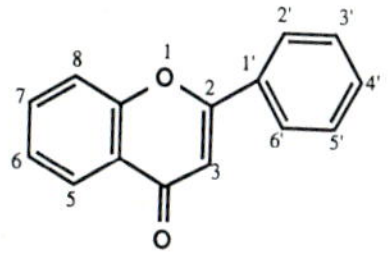

A purified rat lens AR was used except where indicated.

No.		*Substitution*	*IC_{50} (M)*	*Ref.*
1	ascillarin	3′,4′,5,7-$(OH)_4$-3,6-$(OMe)_2$	2.6×10^{-8}	29
			3×10^{-8}	30
2		3′,4′,6-$(OH)_3$-5,7,8-$(OMe)_3$	3.6×10^{-8}	29
			3.4×10^{-8}	30
3	quercitrin	3′,4′,3,5,7-$(OH)_5$-3-rhamnoside	4.9×10^{-7}	29
			1×10^{-7}	28
			1×10^{-6} [a]	37
4	quercitrin 2″-acetate	3′,4′,3,5,7-$(OH)_5$-3-rhamnose-2″-OAc	4×10^{-8}	28
5	quercetin 3′,4′,3,7-tetrasulphate	5-OH-3′,4′,3,7-$(OSO_3H)_4$	$\sim 10^{-8}$ [b]	35
6	quercetin 3-acetate-3′,4′,7-trisulphate	5-OH-3-OAc-3′,4′,7-$(OSO_3H)_3$	$\sim 10^{-9}$ [b]	35
			1×10^{-7} [a]	37
7		3′-4′-$(OH)_2$-5,6,7-$(OMe)_3$	9×10^{-8}	30
8		3′,4′,5,6,7-$(OH)_5$-3-(OMe)	5.8×10^{-8}	30
9		3′,4′,5-$(OH)_3$-7,8-$(OMe)_2$	7.8×10^{-8}	30
10		3′,4′,7-$(OH)_3$-5,8-$(OMe)_2$	7.4×10^{-8}	30
11		3′,4′-$(OH)_2$-5,7,8-$(OMe)_3$	4.5×10^{-8}	30
12		3′,4′,5,7-$(OH)_4$-6-(OMe); 8-CH_2Ph	3.4×10^{-8}	30
13		3′,4′,5-$(OH)_3$-6,7,8-$(OMe)_3$	3.9×10^{-8}	30
14		3′,4′-$(OH)_2$-5,6,7,8-$(OMe)_4$	3.2×10^{-8}	30

[a] Human lens AR.
[b] The source of enzyme and its purity were not indicated (see [35]).

after induction of diabetes. Based on an estimated food intake of 10 g/day, the total administered dose of quercitrin was between 3.2 and 3.5 g/kg per day. This high oral dose of quercitrin required to elicit an effect *in vivo*, compared with its high *in vitro* potency as an AR inhibitor, reflects the poor systemic availability of orally administered quercitrin.

The most potent inhibitor among a group of flavonoids studied by Varma and Kinoshita [28], was quercitrin 2″-acetate (*Table 8.4*, No. 4) (50% at 4×10^{-8} M), which decreased polyol accumulation by 58% at 1×10^{-5} M in the intact rat lens. In contrast to most flavonoids, it is water-soluble and was considered for the topical treatment of diabetic cataracts. Another series of 60 flavonoids has been investigated [29, 30] and structure-activity relationships for this class of compounds have been detailed.

Several other groups have reported on the AR inhibitory activity of flavonoids *in vitro*. Three flavone glycosides isolated from *Cannabis sativa* L had moderate activity [31]. Kador and Sharpless [32] reported the activity of 12 flavonoids; three flavonoids newly isolated from *Brickella arguta* were studied [33], and another series of 73 flavonoids were tested [34].

The AR inhibiting activities of flavonoids that exhibit an IC_{50} value of 1×10^{-7} M or lower are listed in *Table 8.4*. The most potent described to date is quercetin 3-acetate-3′,4′,7-trisulphate (*Table 8.4*, No. 6) [35] (IC_{50} = 1×10^{-9} M). Of the many flavonoids shown to be potent inhibitors of AR *in vitro*, only quercitrin has been investigated *in vivo* [27].

The activity of quercitrin and rutin [28] prompted the study of a series of acetal derivatives involving the 2- and 3-hydroxyls of the rhamnose moiety; the most potent, the acetal of quercitrin with cyclohexanone, had an IC_{50} value of 8.1×10^{-8} M. No *in vivo* results were described [36].

An early report by Bjeldanes and Chang [38] showed that a number of flavonoids, including quercetin, were mutagenic to certain strains of *Salmonella*; however, in a more recent study, in rats fed quercetin for up to 850 days, no significant difference between the incidence of tumours in the treated and control animals was observed [39].

A novel drug delivery to a specific anatomical site has been described [40] using a quercitrin derivative, NAP-HEX-Q, containing the light- or X-ray-activatable 2-nitro-4-azidophenyl group. NAP-HEX-Q was reported to have structure (5), although no proof-of-structure was presented [40a]. NAP-HEX-Q was injected into the aqueous humour of rats maintained on a high-galactose diet. Irradiation with light or X-rays was expected to cause irreversible inacti-

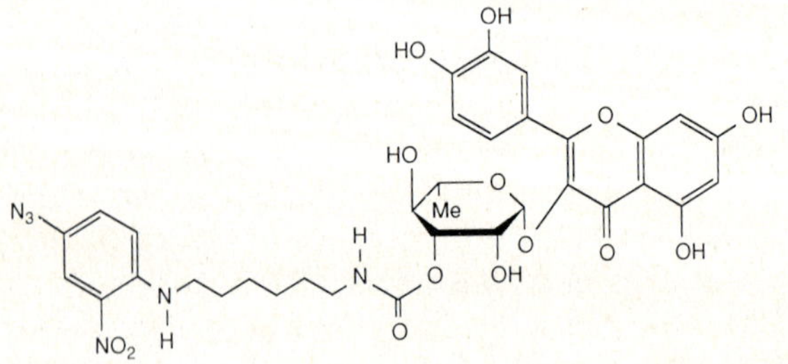

OH HO O OH HO O Me N_3 H N O O OH O OH N H NO_2 O

(5)

vation of AR *via* the formation of a drug-enzyme covalent bond. Irradiated animals were protected from cataract formation for 15 days, while in the non-irradiated group, cataracts appeared after 4 days. In cultured rat lens, complete protection from galactose-induced cataracts was observed after treatment with (5), followed by irradiation [40].

In conclusion, the great interest in the naturally occurring flavonoids as AR inhibitors over the past decade arose because of the ready availability of these products and because various biological rôles in man have been proposed for them since 1936, when it was suggested that they had an effect on capillary permeability [41].

A series of synthetic flavones have been described [42] of which 7-methylsulphonamido-3-propylflavone-4-carboxylic acid, (6), was the most potent, having an IC_{50} value of 1.3×10^{-7} M with calf lens AR. At 100 mg/kg per day for 2 days in the streptozotocin-induced diabetic rat, (6) lowered sorbitol levels in the sciatic nerve by 57%, and inositol levels by 28%. The route of administration was not specified. The same group have prepared the 3-methanesulphonyl derivative of quercitin, (7), and found it to inhibit calf lens AR with an IC_{50} value of 1.6×10^{-7} M, and in the same *in vivo* rat model used above [42], to lower sorbitol levels in the sciatic nerve by 41.2% at a dose of 100 mg/kg, again by an unspecified route of administration [43].

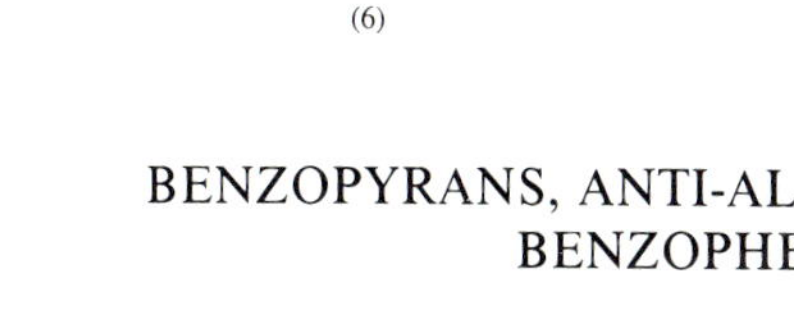

MeSO$_2$HN, O, COOH, Pr, O

(6)

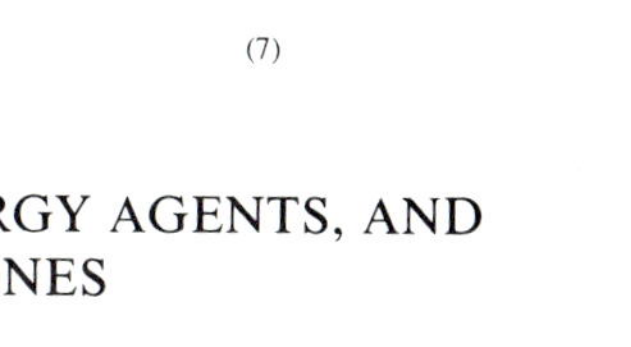

HO, O, OH, OH, OSO$_2$Me, OH, O

(7)

BENZOPYRANS, ANTI-ALLERGY AGENTS, AND BENZOPHENONES

Kador and Sharpless [32] have tested a series of compounds containing the 4*H*-chromen-4-one (4*H*-1-benzopyran-4-one) system found in flavonoids; only weak activity was observed, the most potent, (8), inhibiting AR by 33% at 1×10^{-6} M. No studies were done *in vivo*. Replacing the carboxyl group of (8) by a tetrazole ring gave a compound with comparable activity. Importantly, it was observed that there was a significant correlation between AR inhibition and the energy of the lowest empty molecular orbital (LEMO) of the

4*H*-chromen-4-one system in various flavonoids as well as in a series of 4-oxo-4*H*-chromene-2-carboxylic acid derivatives. Based on this observation, they proposed that a charge-transfer interaction occurs at the reactive carbonyl group *via* a reversible nucleophilic attack by the receptor. A similar type of receptor interaction was proposed [44] to account for the activity of a series of anti-allergy agents comprising oxanilic acids, 1,4-dihydro-4-oxoquinaldic acids, and 4-oxo-4*H*-1-benzopyran-2-carboxylic acids. Over 60 compounds which had been prepared as potential anti-allergy agents have been tested [45, 46]. These included cromoglycate (9) and analogues, as well as a series of

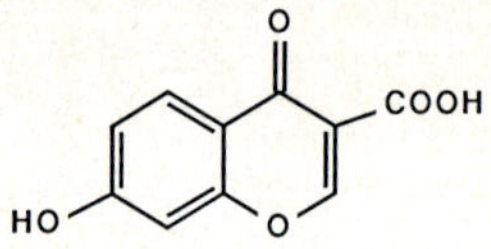

(8)

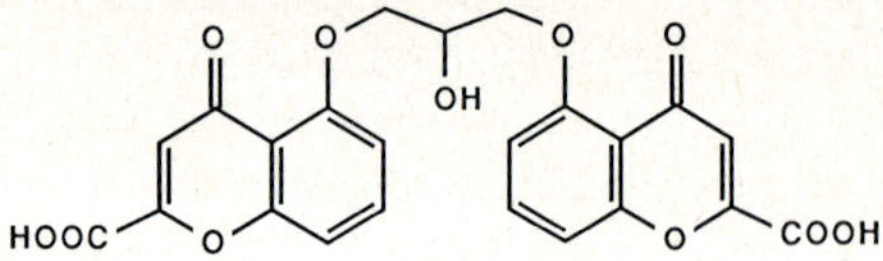

(9)

4-oxo-4*H*-quinoline-2-carboxylic acids (10), oxanilic acids (11; X = N, CH), 3,4-dihydro-4-oxothieno[2,3-*d*]pyrimidine carboxylic acids and analogues (12; X = O, NH), 4-hydroxycoumarins (13), xanthone-2-carboxylic acids (14), 11-oxo-11*H*-pyrido[2,1-*b*]quinazoline-2-carboxylic acids and analogues (15), and 1,6-dihydro-6-oxo-2-phenylpyrimidine-5-carboxylic acids (16). The most potent was the oxanilic acid ester (17), which had an IC_{50} value of 1×10^{-7} M with rat lens AR. No reports of *in vivo* studies with this compound have appeared.

A series of 29 substituted xanthone-2-carboxylic acids, which had also been prepared as anti-allergy agents, were studied as AR inhibitors [47]. The dimethylaminosulphamoyl derivative (18), a water insoluble compound, inhibited rabbit lens AR by 67% at 1×10^{-6} M and was reported to be active both topically and orally in delaying the appearance of cataracts in rats receiving a diet containing 35% galactose. Replacing an *N*-methyl group by 2-hydroxyethyl in an attempt to improve water solubility gave (19), which inhibited the enzyme by 83% at 1×10^{-6} M. The authors suggested that, since the xanthonecarboxylic acids are colourless compounds, they would be more appropriate for topical eye application than the yellow-coloured flavonoids.

A series of benzophenones, which retain some features of the xanthones and which are polyhydroxylated like the flavonoids, have been investigated [48]. EISAI 70-A-196 (20), 2,2′,4,4′-tetrahydroxybenzophenone, inhibited rat lens AR with an IC_{50} value of 1×10^{-7} M; in streptozotocin-induced diabetic rats

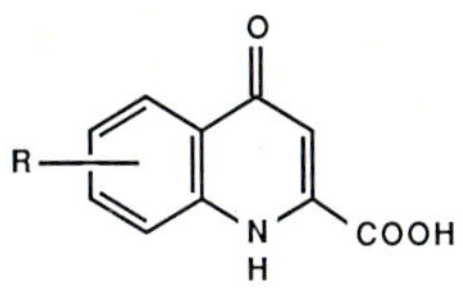

(10) 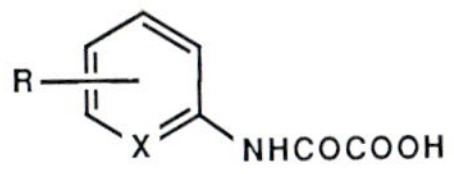 (11) 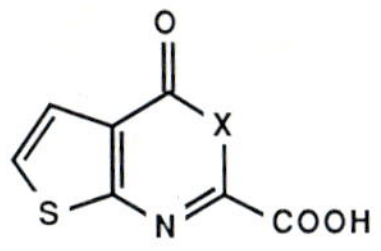 (12)

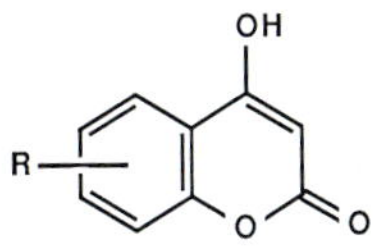

(13) 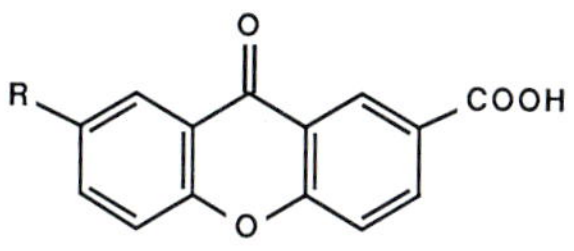 (14) 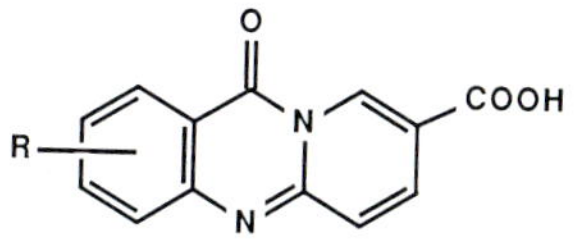 (15)

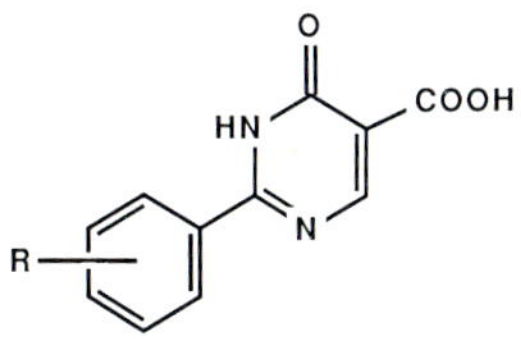

(16) 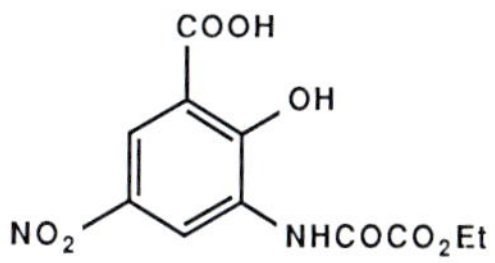 (17)

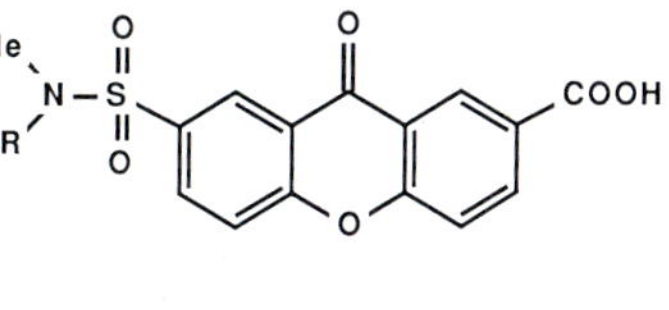

(18) R = Me

(19) R = CH_2CH_2OH

given a 35% xylose diet, it also inhibited cataract formation upon oral administration [49].

Two groups have investigated compounds containing the coumarin-4-acetic acid system (21). The 3,6,7-trimethyl derivative inhibited calf lens AR by 81% at 1×10^{-4} M and caused a 21% inhibition of sciatic nerve sorbitol accumulation in the streptozotocin-induced diabetic rat at 25 mg/kg given three times daily [50]. The 5,6-cyclohexano derivative had an IC_{50} value of 2×10^{-8} M for inhibition of rat lens AR but was not studied *in vivo* [51].

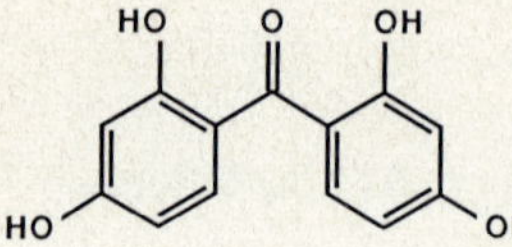

(20) 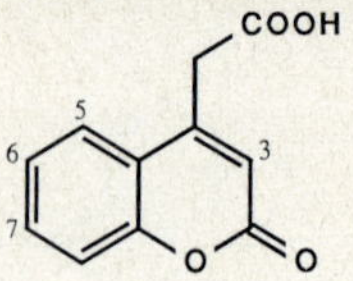 (21) 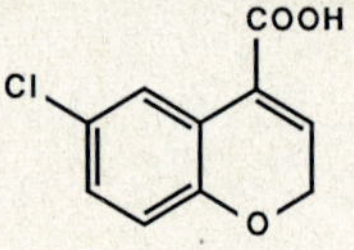 (22)

A series of benzopyran- and benzothiopyrancarboxylic acids were studied; compound (22) was found to inhibit AR by 80% at 1×10^{-4} M and to have moderate activity in the streptozotocin-induced diabetic rat model [52].

Brazilin, (23), and haematoxylin, (24), pigments from a plant used in Korean folk medicine for the treatment of diabetic complications, showed weak activity (48% and 57%, respectively, at 1×10^{-4} M) in inhibiting calf lens AR [53].

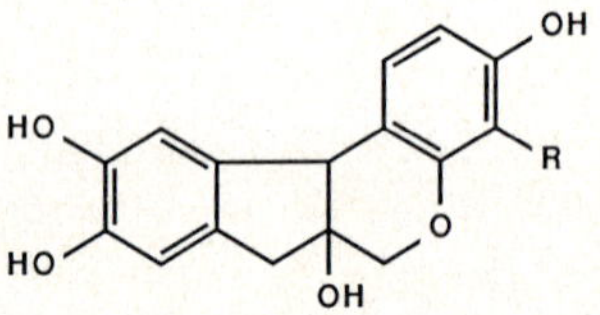

(23) R = H

(24) R = OH

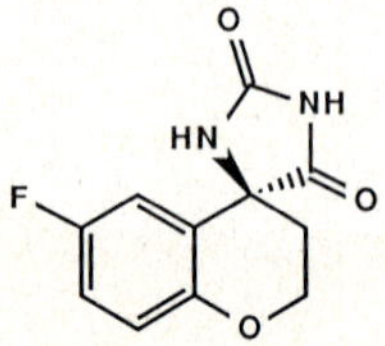

(25)(SORBINIL)

SPIROHYDANTOINS AND RELATED STRUCTURES

AR inhibitors containing a spirohydantoin or related system have been developed by Pfizer. Sorbinil (CP-45,634; (*S*)(+)-6-fluoro-2,3-dihydrospiro-[4*H*-1-benzopyran-4,4′-imidazolidine]-2′,5′-dione, (25), was first disclosed in 1978 [54] and is the best known member of this chemical class. Its absolute configuration was elucidated by X-ray crystallographic studies [55, 56] which show that in the crystal, the pyran ring of sorbinil adopts a pseudochair conformation (*Figure 8.1*) such that the N3′ nitrogen is attached pseudoequatorially. The alternative pseudochair, with the N3′ nitrogen pseudoaxially attached, was found, by molecular mechanics calculations, to be only 0.57 kcal/mol less stable than the solid-state conformation. NMR studies indicate the existence of both conformers in solution [56], therefore these studies do not provide a basis for identifying the conformation which interacts with the enzyme.

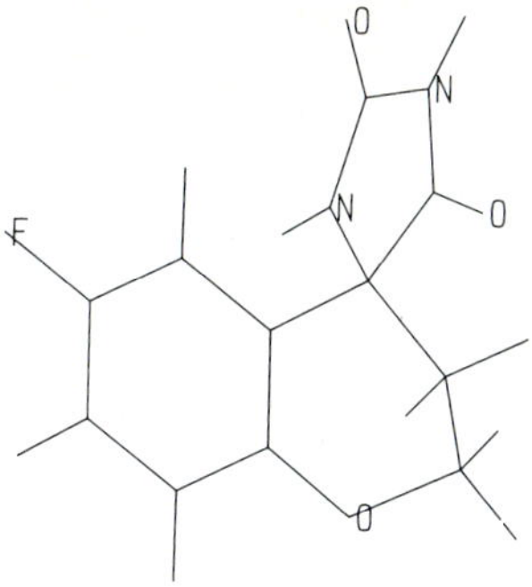

Figure 8.1. X-ray structure of sorbinil.

The effects of sorbinil, its enantiomer, and the racemic form on calf lens AR *in vitro*, and *in vivo* on sorbitol accumulation in the sciatic nerve of diabetic rats, are shown in *Table 8.5*. At 1×10^{-6} M, (−)-sorbinil was only marginally active (23%), while (+)-sorbinil caused 98% inhibition. A limited number of sorbinil analogues with different substitution patterns have been disclosed; they are shown in *Table 8.5* together with their activities *in vitro* and *in vivo*. The racemic 6,7-dichloro, 6,8-dichloro and 6-chloro analogues (Nos. 5, 6 and 7) are more potent *in vitro* than racemic sorbinil (No. 1); this is also the case *in vivo* for the 6,8-dichloro and the 6-chloro analogues, which inhibit sorbitol accumulation by 82 and 64%, compared with 45% for (±)-sorbinil, at a dose of 0.75 mg/kg p.o. in the streptozotocin-induced diabetic rat. An account of the development of sorbinil, and additional structure-activity relationships in that series, has recently appeared [60].

EISAI M-79,175, methosorbinil, the 2-methyl analogue of sorbinil (*Table 8.5*, No. 17), of unknown stereochemistry, was reported to have an IC_{50} value of 6.5×10^{-8} M using a rabbit lens AR [61]. Sorbinil, tested under the same conditions, had an IC_{50} of 1.1×10^{-7} M [61]. Methosorbinil was also reported to markedly reduce sorbitol levels in the lens of streptozotocin-induced diabetic rats upon oral administration of doses of 0.1 to 1.0 mg/kg [62]. An analogue of methosorbinil bearing a dimethylaminoethyl group on the 1′-nitrogen of the spirohydantoin moiety (*Table 8.5*, No. 22) was tested and found to be devoid of AR-inhibiting activity at 1×10^{-5} M [61].

Although a number of patents disclose analogues claimed to be active, in addition to those shown in *Table 8.5*, only compounds supported by experimental data are included in this review.

The 2,3-dihydro[4*H*]-1-benzopyran nucleus of sorbinil has been replaced

Table 8.5. *IN VITRO* AND *IN VIVO* AR INHIBITORY ACTIVITIES OF 2,3-DIHYDRO[4*H*]-1-BENZOPYRAN-4,4′-IMIDAZOLIDINE-2′,5′-DIONES

			% inhib. of AR (concn., M)[a]				*% decrease of sorbitol accumulation* in vivo *(dose, mg/kg)*[b]					
No.	*Ref.*	*Substitution*	*10^{-4}*	*10^{-5}*	*10^{-6}*	*10^{-7}*	*0.25*	*0.75*	*1.5*	*2.5*	*5.0*	*10.0*
1 (±)	54	6-F	85	58	52	3	19	45	72			
2 (+)[c]	54	6-F	100	100	98	39	47	78				
3 (−)	54	6-F	88	63	23			19		6	22	
4	57	6-OMe	100	92	35	7				24		
5	57	6,7-Cl_2	59	96	91	84			64			
6	57	6,8-Cl_2	85	90	78	81		82				
7	57	6-Cl	73	81	77	64		64	84			
8	57	8-Cl	87	85	52	6			30			
9	57	6-Br	74									
10	57	6,8-Me_2	71	84	54	17						
11	58	6-Ph		$IC_{50} < 1 \times 10^{-4}$ M					0			
12	58	8-Ph		$IC_{50} = 1 \times 10^{-6}$ M								54
13	58	6-OPh		$IC_{50} = 1 \times 10^{-5}$ M					0			
14	58	6-Ph-8-Cl		$IC_{50} = 1 \times 10^{-5}$ M								
15	59	6-Cl-8-NH_2			58[d]		17					
16	59	6-F-8-NH_2			40[d]		40					
17[e]	61	2-Me-6-F		$IC_{50} = 6.5 \times 10^{-8}$ M								
18	61	2-Me-6-F-8-NO_2		$IC_{50} = 5.0 \times 10^{-8}$ M								

Table 8.5. continued

No.	Ref.	Substitution	% inhib. of AR (concn., M)[a] 10^{-4}	10^{-5}	10^{-6}	10^{-7}	% decrease of sorbitol accumulation in vivo (dose, mg/kg)[b] 0.25	0.75	1.5	2.5	5.0	10.0
19	61	2-Me-6-F-8-NH_2		$IC_{50} = 1.7 \times 10^{-8}$ M								
20	61	2-Me-6-Cl		$IC_{50} = 4.7 \times 10^{-8}$ M								
21	61	2-Me-6-Cl-8-NH_2		$IC_{50} = 8.3 \times 10^{-8}$ M								
22	61	2-Me-6-F-1′-$(CH_2)_2NMe_2$		0								
23	63	6-SMe		$IC_{50} = 6.6 \times 10^{-7}$ M[d]								
24	63	6-SOMe		$IC_{50} = 1.2 \times 10^{-5}$ M[d]								

[a] A partially purified calf lens AR was used unless indicated otherwise.

[b] Rats were made diabetic with streptozotocin and compounds were administered orally at 4, 8 and 24 h following streptozotocin administration. Sorbitol levels in the sciatic nerve were measured and compared to controls.

[c] Sorbinil.

[d] Human placental AR was used.

[e] Methosorbinil.

Table 8.6. *IN VITRO* AND *IN VIVO* AR INHIBITORY ACTIVITIES OF MISCELLANEOUS SPIRO-IMIDAZOLIDINE-2′,5′-DIONES

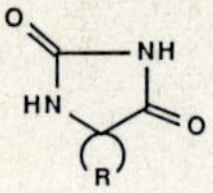

No.	*Ref.*	*R*	*% inhibition of A.R. (concn.)*[a]	*% inhibition of sorbitol accumulation* in vivo *(dose)*[b]
1	57	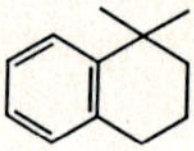	34% (1×10^{-5} M)	3% (5 mg/kg)
2	57	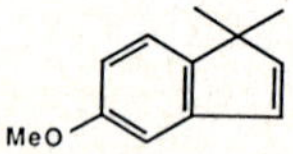 MeO	81% (1×10^{-5} M)	33% (5 mg/kg)
3	54	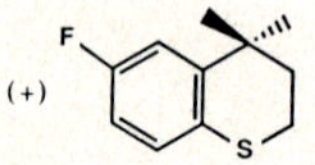 (+) F, S	74% (1×10^{-7} M)	55% (0.75 mg/kg)
4	57	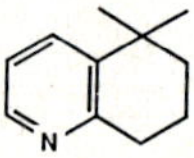 N	54% (1×10^{-5} M)	5% (5 mg/kg)
5	57	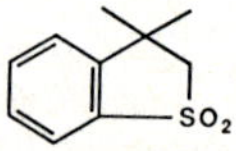 SO_2	64% (1×10^{-5} M)	5% (5 mg/kg)
6	64	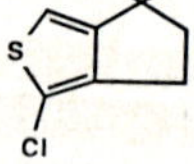 S, Cl	67% (1×10^{-5} M)	–
7	64	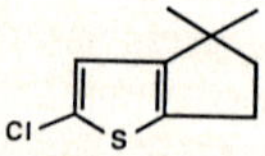 Cl, S	55% (1×10^{-5} M)	20% (2.5 mg/kg)
8	64	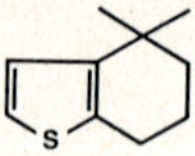 S	52% (1×10^{-4} M)	56% (25 mg/kg)

Table 8.6. continued

No.	*Ref.*	*R*	*% inhibition of A.R. (concn.)*[a]	*% inhibition of sorbitol accumulation* in vivo *(dose)*[b]
9	65	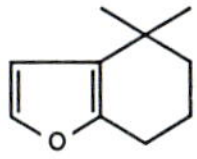 O	50% (1×10^{-4} M)	46% (25 mg/kg)
10	66	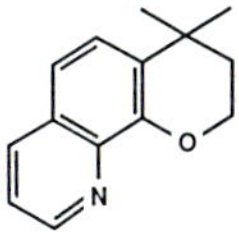 O N	75% (1×10^{-4} M)	27% (1.5 mg/kg)
11	66	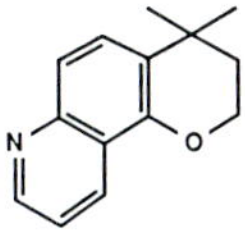 N O	94% (1×10^{-4} M)	92% (10 mg/kg)
12	67	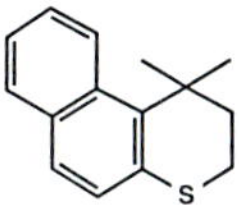 S	59% (1×10^{-4} M)	–
13	67	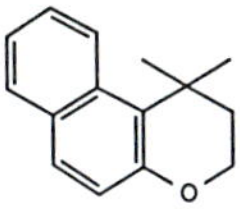 O	78% (1×10^{-4} M)	inactive at 1.5 mg/kg
14	67	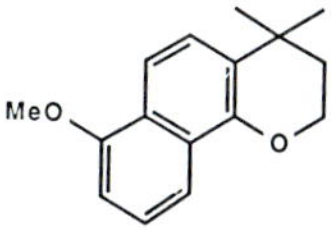 MeO O	94% (1×10^{-4} M)	inactive at 1.5 mg/kg
15	67	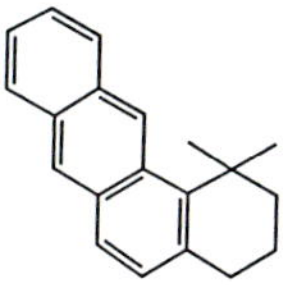	26% (1×10^{-4} M)	–
16	67	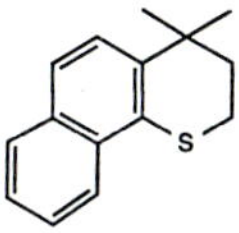 S	100% (1×10^{-4} M)	32% (1.5 mg/kg)

continued

Table 8.6. continued

No.	*Ref.*	*R*	*% inhibition of A.R. (concn.)*[a]	*% inhibition of sorbitol accumulation* in vivo *(dose)*[b]
17	67	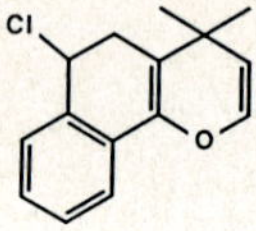 Cl, O	78% (1×10^{-4} M)	–
18	67	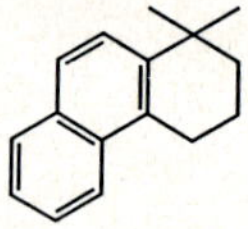	86% (1×10^{-4} M)	inactive at 1.5 mg/kg
19	67	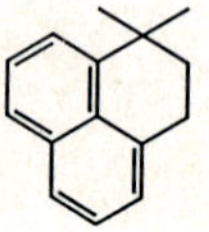	71% (1×10^{-4} M)	28% (5 mg/kg)
20	67	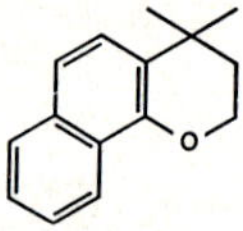 O	72% (1×10^{-4} M)	30% (1.5 mg/kg)
21	67	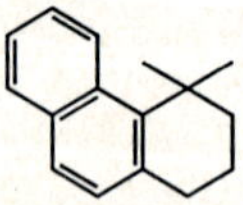	60% (1×10^{-4} M)	–
22	68	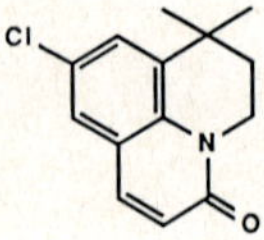 Cl, N, O	94% (1×10^{-6} M)	32% (1.5 mg/kg)

Table 8.6. continued

No.	*Ref.*	*R*	*% inhibition of A.R. (concn.)*[a]	*% inhibition of sorbitol accumulation* in vivo *(dose)*[b]
23	69	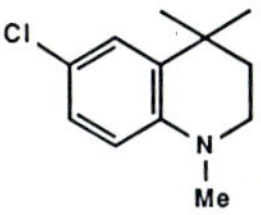	$IC_{50} = 1 \times 10^{-6}$ M	80% (10 mg/kg)
24	69	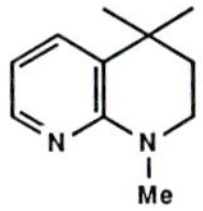	$IC_{50} = 1 \times 10^{-5}$ M	–
25	70	(+) - 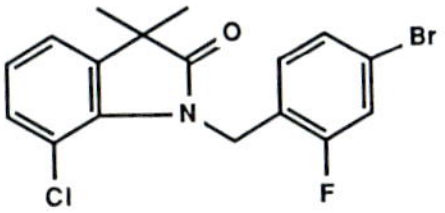	$IC_{50} = 9.8 \times 10^{-9}$ M	ED_{50}: 0.62 mg/kg
26	71	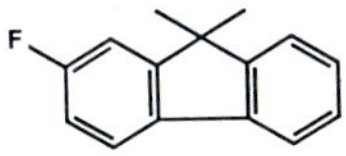 (AL-1567)	$IC_{50} = 4.3 \times 10^{-8}$ M[c]	–

[a] A calf lens AR was used unless otherwise indicated.
[b] See footnote b, *Table 8.5.*, for details.
[c] Rat lens AR.

by a wide variety of other rings; the most potent representatives are shown in *Table 8.6* with reported AR inhibitory activities *in vitro* and *in vivo*. Of 26 systems reported, three have IC_{50} values of less than 1×10^{-7} M; the thiochroman, the spiroindolinone, and the fluorenone derivatives, Nos. 3, 25, and 26 (alconil, AL-1567, Alcon Laboratories). AR inhibitory activities of twenty analogues of alconil were recently reported [72]. Alconil has been extensively studied *in vivo*; on p.o. administration it reduced sciatic nerve sorbitol levels in the streptozotocin-induced diabetic rat and prevented cataract formation [71, 73].

Table 8.7. AR INHIBITORS CONTAINING ACIDIC HETEROCYCLE OTHER THAN A SPIROHYDANTOIN

No.	*Ref.*	*Structure*	*% inhib. of AR*[a] *(concn. M)*	*% decrease of sorbitol accumulation* in vivo[b] *(dosage)*
1	74		$IC_{50} = 1 \times 10^{-7}$ M	ED_{50} = 1.5–2.5 mg/kg
2	75		$IC_{50} = 8 \times 10^{-9}$ M	–
3	76		–	46% (5 mg/kg)

Table 8.7 continued

No.	*Ref.*	*Structure*	*% inhib. of AR*[a] *(concn. M)*	*% decrease of sorbitol accumulation* in vivo[b] *(dosage)*
4	77	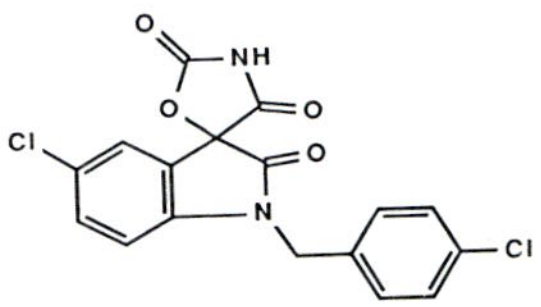	60% (1×10^{-7} M)	–
5	78	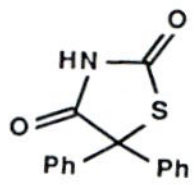	40% (1×10^{-7} M)[d]	–
6	79 (diphenylhydantoin)	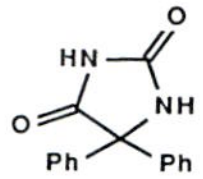	71% (5×10^{-4} M)[e]	–
7	79 (ethosuccinimid)	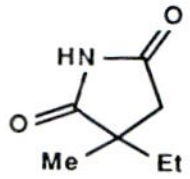	15% (2.5×10^{-4} M)[e]	–

continued

Table 8.7 continued

No.	*Ref.*	*Structure*	*% inhib. of AR*[a] *(concn. M)*	*% decrease of sorbitol accumulation* in vivo[b] *(dosage)*
8	80 (phenobarbital)	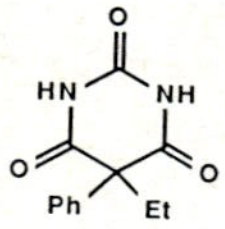	73% (1×10^{-3} M)[f]	–
9	81 (ciglitazone)	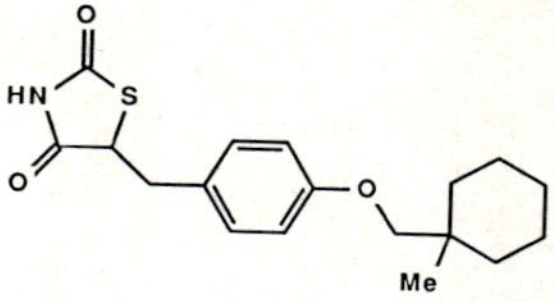	39% (1×10^{-4} M)[d]	–
10	83	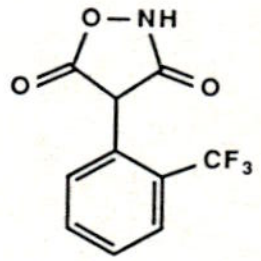	80% (5×10^{-5} M)[c]	–
11	84	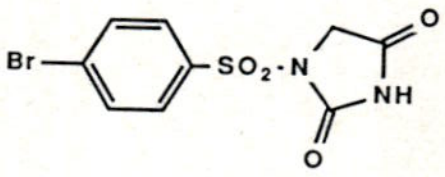	$IC_{50} = 3.7 \times 10^{-7}$ M	–

Table 8.7. continued

No.	*Ref.*	*Structure*	*% inhib. of AR*[a] *(concn. M)*	*% decrease of sorbitol accumulation* in vivo[b] *(dosage)*
12	86	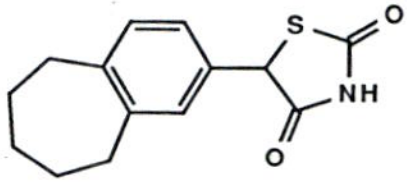	34% (1×10^{-6} M)[d]	83% (25 mg/kg)
13	78	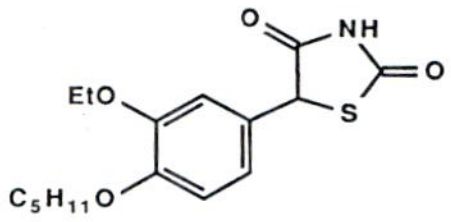	35% (1×10^{-6} M)[d]	–
14	87	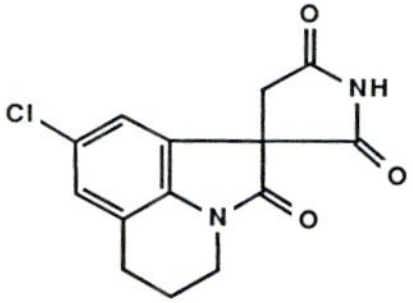	$IC_{50} = 4 \times 10^{-7}$ M[c]	–

[a] Partially purified calf lens AR was used except where indicated.
[b] See footnote b, *Table 8.5.*, for details.
[c] Rat lens AR.
[d] Human placenta AR.
[e] Human brain AR.
[f] Rabbit lens AR.

The spirohydantoin moiety, in turn, can be replaced by other systems containing an acidic hydrogen; representative examples with their AR inhibitory activities are collected in *Table 8.7*. The spiro-oxazolidinedione, No. 1 [74], had an IC_{50} value of 1×10^{-7} M (calf lens AR), while the spirothiazolidinedione, No. 2 [75], was more potent, with an IC_{50} value of 8×10^{-9} M. *In vivo* data have been reported for the spirosuccinimide, No. 3 [76]. A series of thiazolidine-2,4-diones spiro-fused to a diphenylmethane group was investigated [78]; the most potent, No. 5, inhibited human placental AR by 40% at 1×10^{-7} M. It is a thio analogue of the anticonvulsant, phenytoin, No. 6, which was shown by O'Brien and Schofield [79] to inhibit human brain AR by 71% at 5×10^{-4} M. These latter authors have also tested the anticonvulsant ethosuccinimid [79], No. 7, and have shown that it is a very weak AR inhibitor. A number of hypnotic barbiturates have been tested [79, 80], using rabbit lens AR. Both groups found that the barbiturates are very weak inhibitors of AR, phenobarbital, No. 8, being relatively the most potent, showing 73% inhibition at 1×10^{-3} M. An analogue of the thiazolidine-2,4-dione, No. 5, is the hypoglycaemic and hypolipidaemic agent, ciglitazone (*Table 8.7*, No. 9), (ADD-3878; U-63287); it exhibited marginal AR inhibitory activity of 39% at 1×10^{-4} M [81, 82].

Five-membered heterocycles which contain an acidic hydrogen but which are not spiro-fused were also investigated. Compound 10 (*Table 8.7*) was the most potent of a series of isoxazolidinediones [83], while the phenylsulphonylhydantoin, No. 11 [84], was the best of the 54 analogues tested. No. 11 reduced galactitol accumulation in the galactosemic rat model at 50 mg/kg p.o., and decreased cataract formation in rats at the same dose [85]. In a series of phenyl-substituted thiazolidinediones, No. 12 [86] was the most effective, inhibiting sorbitol accumulation in sciatic nerve by 83% at a dose of 25 mg/kg. The 3-ethoxy-4-pentyloxythiazolidinedione, No. 13 (*Table 8.7*) (CT-112) [78], was reported to be in development by the Takeda Company. *In vitro*, CT-112 inhibited human placental AR by 35% at 1×10^{-6} M. The spiro[pyrrolidine-3,1′[4*H*]-pyrrolo[3,2,1-*ij*]quinoline]-2,2′,5(1′*H*)-trione (14) (*Table 8.7*) had an IC_{50} value of 4×10^{-7} M for inhibition of rat lens AR [87]. After p.o. administration to rats at 25 mg/kg, blood was taken and 0.5 ml of serum was added to the *in vitro* incubation mixture containing rat lens AR; after 5 h, AR was inhibited by 68.5% [87]. No. 14, phenytoin and sorbinil were evaluated p.o. for inhibition of pentylenetetrazole-induced convulsions in mice; the results (see *Figure 8.2*) show that, in contrast to phenytoin and sorbinil, No. 14 is devoid of anticonvulsant activity [87].

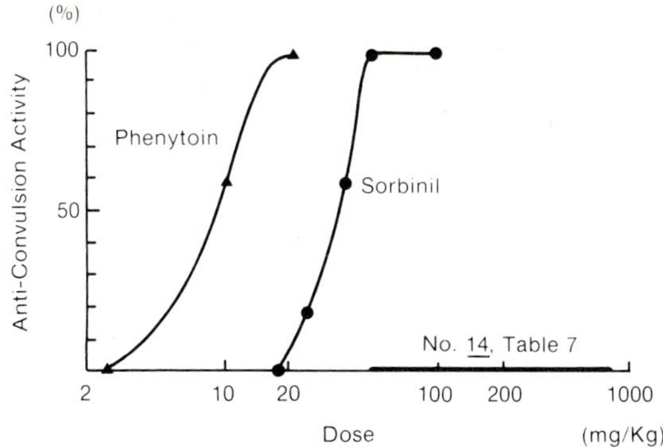

Figure 8.2. Anticonvulsant activities of phenytoin, sorbinil and No. 14, Table 8.7

HETEROCYCLIC ALKANOIC ACIDS

Several potent AR inhibitors have been developed which, structurally, are heterocyclic alkanoic acids. Of a large series of rhodanine derivatives [88, 89], (*E*)-3-carboxymethyl-5-[(2*E*)-methyl-3-phenylpropenylidene]rhodanine (ONO-62235; epalrestat, 26) has been selected for clinical development. Epalrestat has an IC_{50} value of 1×10^{-8} M with AR from rat lens and an IC_{50} of 2.6×10^{-6} M with human placental AR [90]. Another series of rhodanines has been reported, the most potent of which was the aminomethylidene derivative (27), which had an IC_{50} value of 2.3×10^{-8} M with rat lens AR [91]. No *in vivo* results were presented.

The quinolineacetic acid (ICI-105,552) (28) [92] and the phthalazinoneacetic acid (statil; ICI-128,436) (29) [93] have been studied in laboratory animals. *In vitro*, at 5×10^{-5} M, ICI-105,552 prevented sorbitol accumulation in monkey kidney epithelial cells cultured in a glucose-containing medium [94]; *in vivo*, it caused an 86% reduction of sorbitol levels in the sciatic nerve of streptozotocin-induced diabetic rats upon oral administration of 50 mg/kg per day [92]. Statil inhibited human lens AR with an IC_{50} value of 2×10^{-8} M [95], and on treatment of streptozotocin-induced diabetic rats for 3 weeks at 25 mg/kg per day, p.o., elevated sciatic nerve sorbitol levels were reduced to levels found in non-diabetic controls [96]. A review of structure-activity relationships in compounds related to (28) and (29) was recently presented [97].

The furanone propionic acid (FR-51785) (30), isolated from cultures of *Chaetomella raphigera*, inhibited rabbit lens AR with an IC_{50} value of

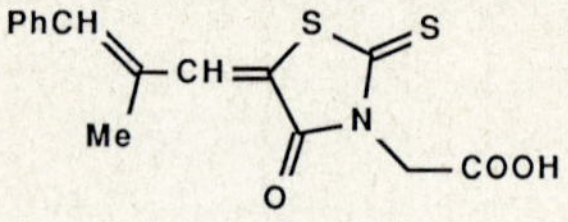

(26)

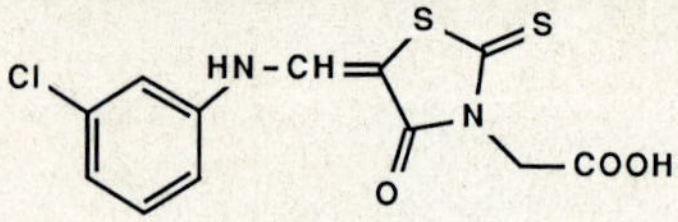

(27)

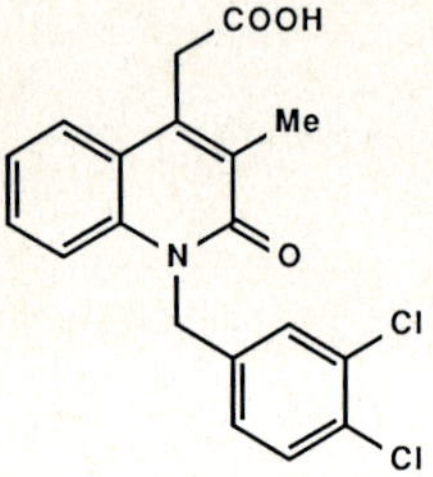

(28)

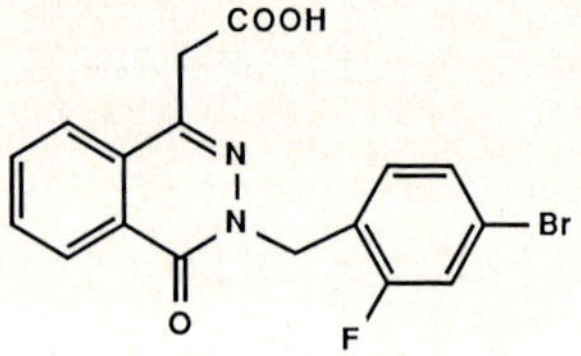

(29)

1.6×10^{-7} M [98]. A number of synthetic analogues of FR-51785 were investigated, one of the most potent being the 3,4-dichlorophenyl analogue, which had an IC_{50} value of 3.1×10^{-7} M and which was active *in vivo*, inhibiting sorbitol accumulation in the streptozotocin-induced diabetic rat by 41% at 10 mg/kg p.o. [99].

A series of benzothiazolinylalkanoic acids was tested using rabbit lens AR; the most potent member, (31), had an IC_{50} value of 2.9×10^{-8} M [100]. The benzimidazoline derivative (32), the best of a series, inhibited calf lens AR by 92% at 1×10^{-4} M [101], and the isoquinolineacetic acid, (33), showed 97% inhibition at 1×10^{-5} M [102].

A series of 3-oxo- and 3-thioxobenzothiazine derivatives were studied as inhibitors of rat lens AR; one of the most potent compounds, (34), had an IC_{50} value of 4.2×10^{-8} M [103]. No *in vivo* results were reported.

Several reports have appeared on the AR inhibitory activity of a series of *N*-acylthiazolidinecarboxylic acids, using a partially purified rat lens enzyme [104, 106]. (Some of the compounds [104] also inhibit angiotensin-converting enzyme.) Of over sixty analogues studied, compound (35) [104] was the most potent, having an IC_{50} value of 1.6×10^{-10} M. Other structural variations with high AR inhibitory activity were the symmetrical disulphide (36) [105], compound (37) [106], and the unsymmetrical disulphide (38) [105], which had IC_{50} values of 5.8×10^{-9}, 3.7×10^{-10} and 1.4×10^{-8} M, respectively. No *in vivo* data were reported for any of these compounds. The AR inhibition

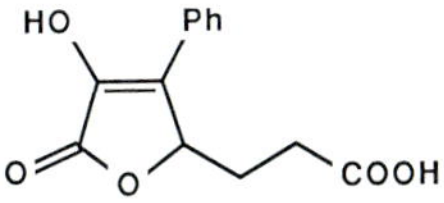

(30) 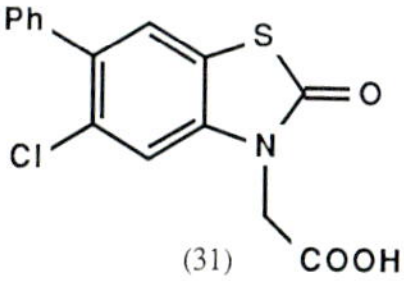 (31) 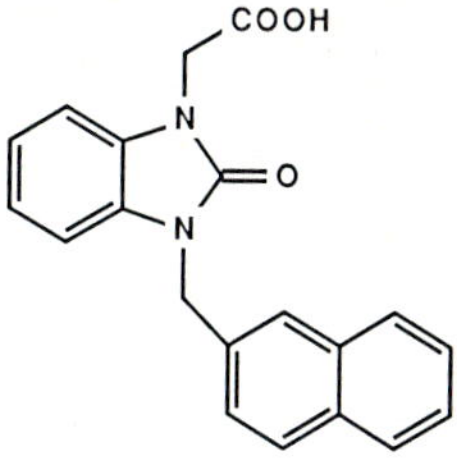 (32)

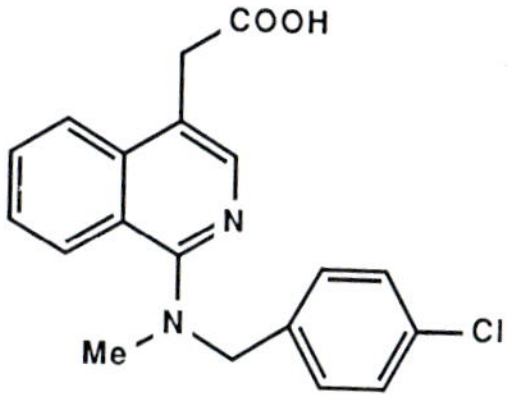

(33) 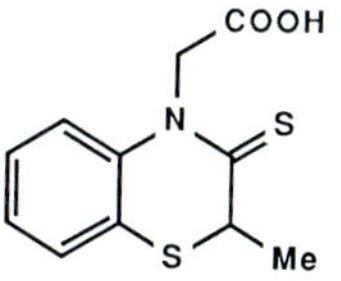 (34)

reported for (35) could not be repeated in our laboratories [107], where it was found to inhibit calf lens AR by only 50% at 1×10^{-7} M, and to be inactive in the 4-day galactosemic rat model [108] at 52 mg/kg per day p.o. While rat lens AR is somewhat more susceptible to inhibition than the enzyme from calf lens (for example, tolrestat had IC_{50} values of 5.7×10^{-9} and 3.4×10^{-8} M with rat and calf lens AR, respectively [107]), the greater than 600-fold difference between the literature IC_{50} value for (35) [104] and the value obtained in our laboratories [107] remains unexplained.

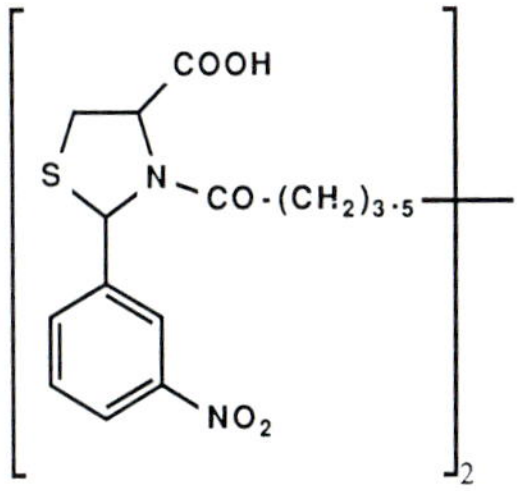

(35) 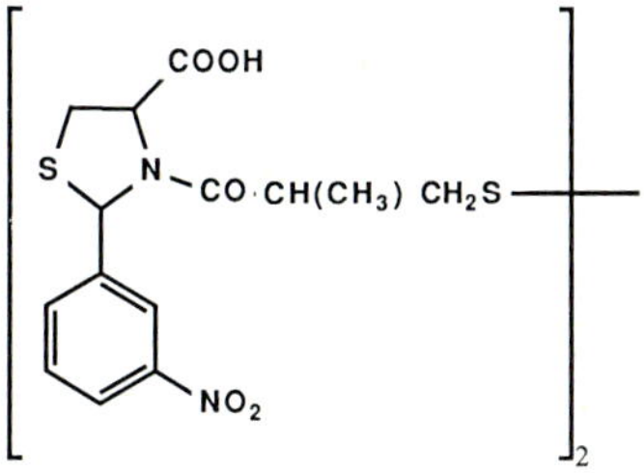 (36)

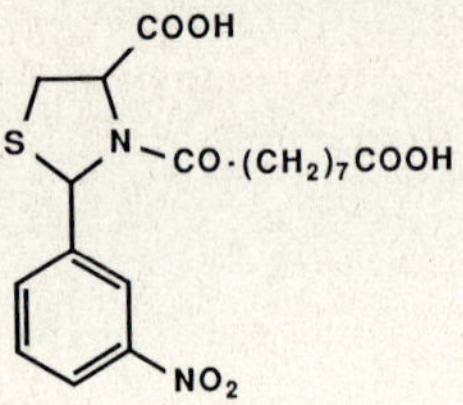

(37)

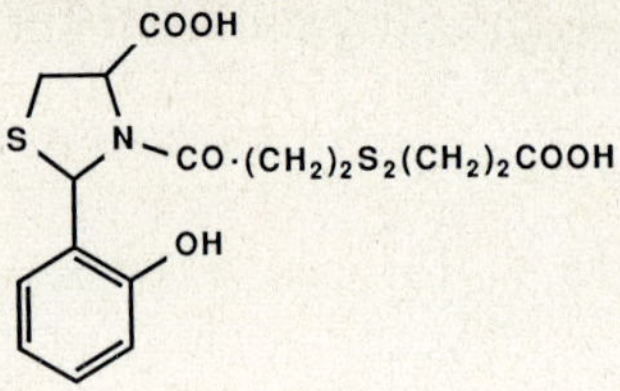

(38)

NON-STEROIDAL ANTI-INFLAMMATORY AGENTS

The fact that the majority of AR inhibitors are acidic compounds prompted the testing of a series of non-steroidal anti-inflammatory-analgesic agents at the Ayerst Research Laboratories. The compounds tested included aspirin, salicylic acid, phenylbutazone, indomethacin, ibuprofen, naproxen, tolmetin, mefenamic acid and piroxicam. Sulindac and its metabolites, sulindac sulphide and sulindac sulphone, were also tested, as well as flurbiprofen, suprofen, benoxaprofen and the anti-inflammatory-analgesics furobufen, prodolic acid and etodolac. The results are shown in *Table 8.8* along with those reported by other investigators. Sulindac was the most potent AR inhibitor, with an IC_{50} value of 1×10^{-7} to 4×10^{-7} M, depending on the source of the enzyme. Sulindac sulphide, the metabolite which is responsible for the anti-inflammatory activity of sulindac, and sulindac sulphone were about as potent *in vitro* as sulindac. In the 4-day galactosemic rat model [108], sulindac was weakly active, preventing galactitol accumulation in the lens and in the sciatic nerve at 62 mg/kg per day, the highest dose tested [109].

While some of the non-steroidal anti-inflammatory agents tested were found to inhibit AR *in vitro*, only weak activity was found *in vivo*, hence this class of drugs is not a likely source of clinically useful AR inhibitors. That some diabetics with rheumatoid arthritis treated with aspirin had a reduced incidence of cataracts [110] cannot be explained by inhibition of AR by aspirin.

ALKALOIDS

A series of alkaloids has been tested as inhibitors of AR, using a rabbit lens preparation [61]. Nandazurine (39), an aporphine, had an IC_{50} value of 2.0×10^{-4} M, and the benzylisoquinoline, papaverine (40), and berberine (41), a bisisoquinoline, had IC_{50} values of 1.4×10^{-4} and 5.2×10^{-5} M, respec-

Table 8.8. EFFECTS OF NONSTEROIDAL ANTI-INFLAMMATORY DRUGS ON AR[a]

No.	*Name*	*Structure*	*Inhibition of AR*	*Ref.*
1	aspirin	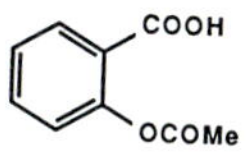	2% (1×10^{-4} M) 18% (1×10^{-5} M)[b]	107 37
2	salicylic acid	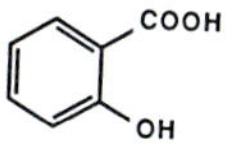	16% (1×10^{-4} M)	107
3	phenylbutazone	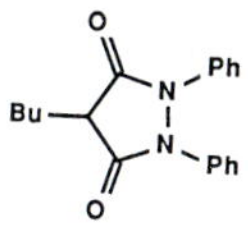	14% (1×10^{-4} M)	107
4	indomethacin	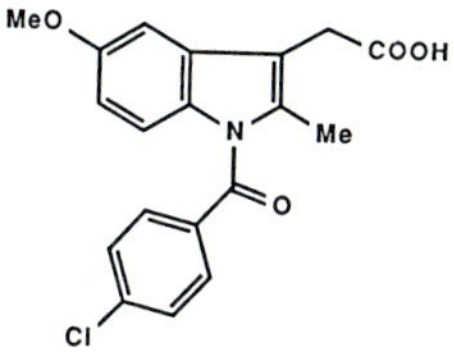	52% (1×10^{-5} M) $IC_{50} = 5 \times 10^{-4}$ M[c]	107 111
5	ibuprofen	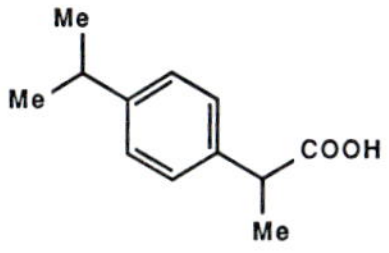	6% (1×10^{-5} M) $K_i = (1 \times 10^{-4}$ M)	107 112
6	naproxen	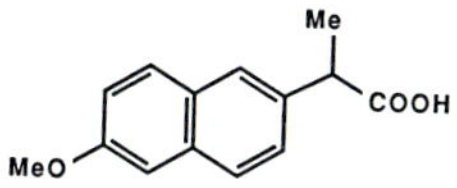	0% (1×10^{-5} M)	107
7	tolmetin	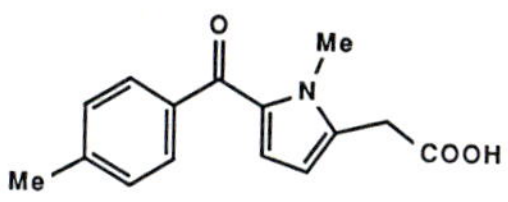	44% (1×10^{-5} M)	107

continued

Table 8.8 continued

No.	*Name*	*Structure*	*Inhibition of AR*	*Ref.*
8	mefenamic acid	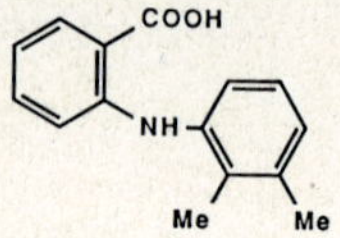	1% (1×10^{-5} M)	107
9	piroxicam	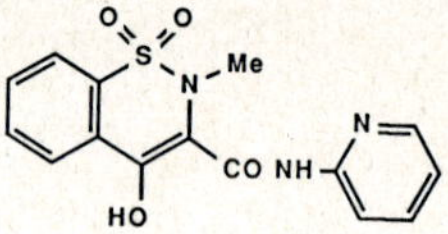	0% (1×10^{-5} M)	107
		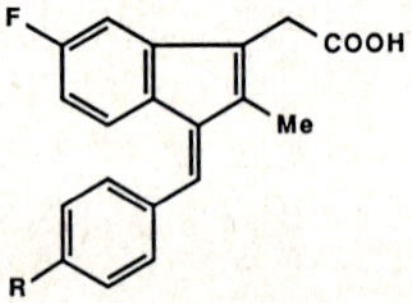 R		
10	sulindac	SOMe	$IC_{50} = 4 \times 10^{-7}$ M[b]	37
			$IC_{50} = 1 \times 10^{-7}$ M	109
11	sulindac sulphide	SMe	50% (1×10^{-6} M)	109
12	sulindac sulphone	SO_2Me	50% (1×10^{-7} M)	109
13	flurbiprofen	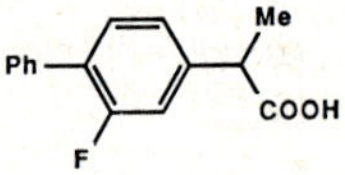	27% (1×10^{-5} M)	107
14	suprofen	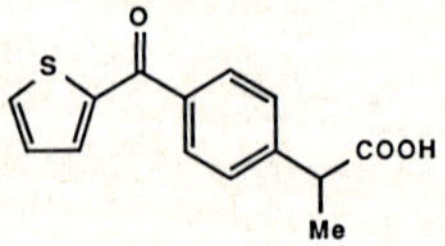	8% (1×10^{-5} M)	107
15	benoxaprofen	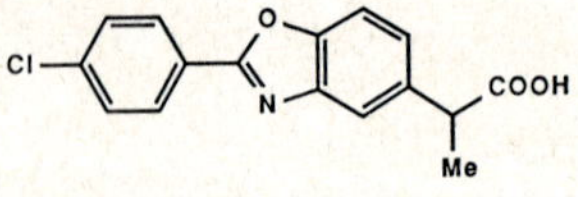	32% (1×10^{-5} M)	107

Table 8.8. continued

No.	*Name*	*Structure*	*Inhibition of AR*	*Ref.*
16	furobufen	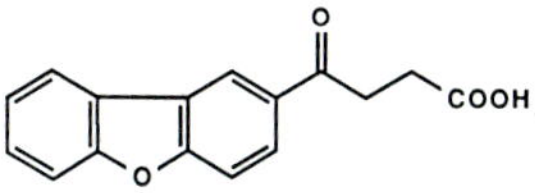	46% (1×10^{-5} M)	107
17	prodolic acid	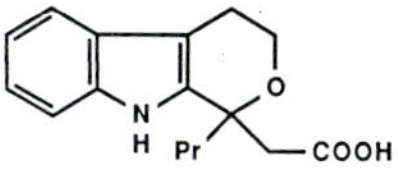	4% (1×10^{-5} M)	107
18	etodolac	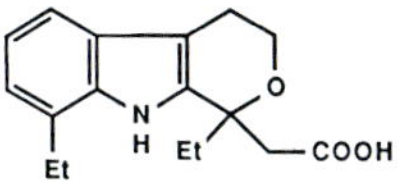	3% (1×10^{-5} M)	107

[a] Calf lens AR was used unless otherwise indicated.
[b] Human lens AR.
[c] Human cataract AR.

tively [61]. No *in vivo* studies were reported. Kinetic studies of AR inhibition by papaverine and berberine were carried out with DL-glyceraldehyde as substrate, and analysis of the data showed that both alkaloids were competitive inhibitors of AR [61]. All other inhibitors of AR whose kinetics have been studied have been found to be noncompetitive or uncompetitive inhibitors [28, 29, 32, 84, 90, 95, 111–116]. Finding competitive inhibition with compounds whose structures bear no resemblance to either substrates or products of AR is unexpected and confirmation would be desirable.

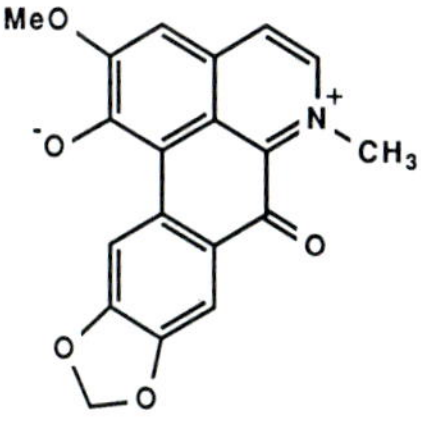

(39) 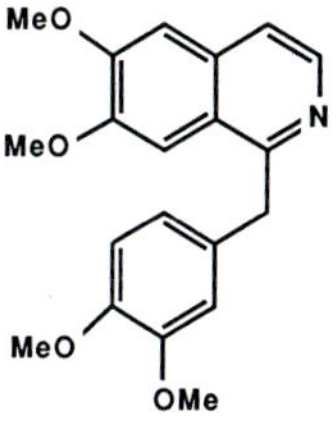 (40) 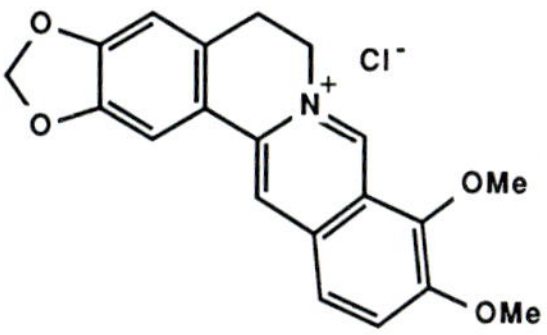 (41)

An alkaloid-like compound, GPA-1734, described as 8,9-dihydroxy-7-methylbenzo[*b*]quinolizinium bromide [117] (probable structure, (42)) was reported to inhibit a partially purified calf lens AR and to have an IC_{50} value of 1×10^{-5} to 7×10^{-6} M [117]. No *in vivo* results were reported.

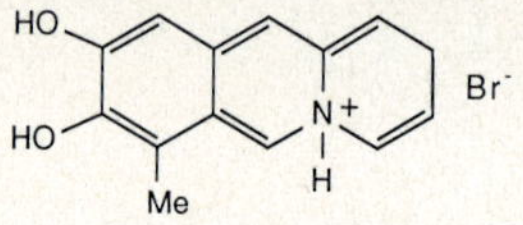

(42)

MISCELLANEOUS ALDOSE REDUCTASE INHIBITORS

Various compounds not in the previous categories have been reported to inhibit AR. The pyrroloquinolinedione (43) (AHR-5191) had an IC_{50} value of 8×10^{-5} M with calf lens AR and caused a 43% decrease in rat lens galactitol levels when administered at 100 mg/kg per day, p.o. for 3 days [118]. Of a series of oxazoles studied [119], the carbamate (44) was the most potent, having an IC_{50} value of 1.5×10^{-5} M with rabbit lens AR.

Prompted by a possible rôle for AR in the pathogenesis of muscular dystrophy [120], Murphy and Davidson tested menadione, (45) [121], (vitamin K_3) and menadione bisulphite, (46) [122], as inhibitors of AR. Working with a purified chicken breast AR and using pyridine-3-carboxaldehyde as sub-

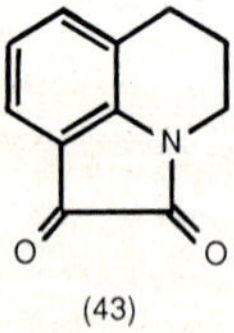

(43)

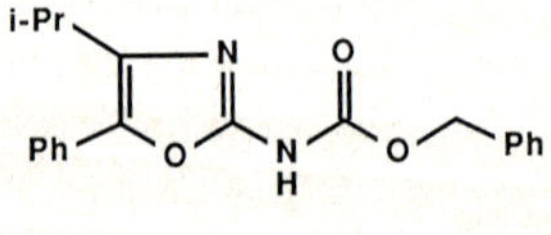

(44)

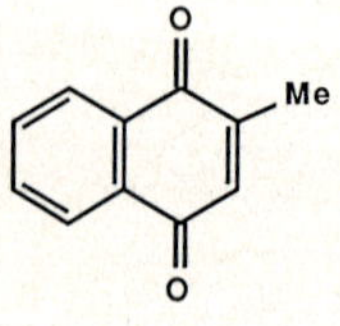

(45)

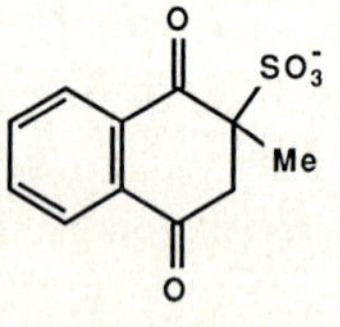

(46)

strate, menadione had a K_i of 4.25×10^{-2} M and the water-soluble menadione bisulphite had a K_i of 2.5×10^{-5} M. In our laboratories, menadione inhibited calf lens AR by 27% at 1×10^{-5} M and by 17% at 1×10^{-6} M [107]. The kinetics of the inhibitions of chicken breast AR by menadione and menadione bisulphite were studied; in both cases the inhibitions were competitive with respect to substrate. Along with papaverine and berberine [61], these are the only competitive inhibitors of AR that have been described.

A MODEL OF THE ALDOSE REDUCTASE INHIBITOR SITE

The kinetics of AR inhibition by several inhibitors have been studied; the flavonoids quercitrin [28] and axillarin [29], as well as sulindac [111], alrestatin [28], indomethacin [111], and *p*-bromophenylsulphonylhydantoin [84], have been shown to be noncompetitive inhibitors. In addition, epalrestat [90], sorbinil [113, 114], TMG [115], 7-hydroxy-4-oxo-4*H*-chromene-6-carboxylic acid [32] and statil [95] were found to exhibit mixed uncompetitive-noncompetitive inhibition. Hence, these and other AR inhibitors [116] do not compete for the substrate-binding site on the enzyme. Furthermore, other studies show that various AR inhibitors do not compete for the nucleotide-cofactor-binding site [114, 116].

Kador and Sharpless [116] have suggested that a wide range of AR inhibitors bind at a common inhibitor site on the enzyme. Based on this, plus the observed correlation between their LEMOs and inhibitory potencies, and the identification of a polarizable carbonyl group in each inhibitor, these investigators have proposed a mode for the superpositioning of these inhibitors in which the polarizable carbonyl groups lie in a common region. This led further to a proposal of a model of the inhibitor binding site which comprises planar lipophilic regions, hydrogen bonding sites, and a sterically constrained charge-transfer pocket. They have also presented evidence supporting the presence of a tyrosine residue in close proximity to the inhibitor-binding site, and the hydroxyl group of the tyrosine is presumed to be the nucleophile which reversibly participates in an interaction with a polarizable carbonyl group of the inhibitors. Kador and Sharpless [116] believe they have described the minimal structural features required for AR inhibition and have suggested that their proposal may permit the rational design of more potent and specific inhibitors of AR. But while such a model can be of value, it should be pointed out that, of the six AR inhibitors which are currently being studied clinically (see *Table 8.9*), only the spirohydantoin type has been considered in the development of the model [116]. If all the compounds in *Table 8.9* should indeed act

Table 8.9. AR INHIBITORS UNDERGOING CLINICAL TRIALS

No.	*Structure*	*Name*	*Company*
1	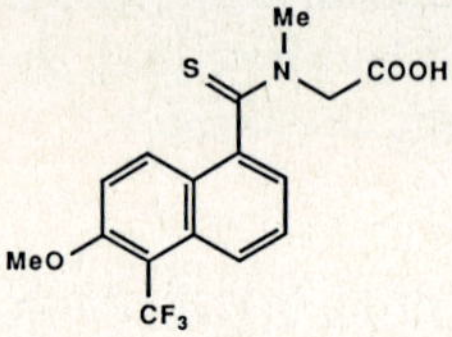 Me, S, N, COOH, MeO, CF_3	AY-27,773 Tolrestat Alredase®	Ayerst
2	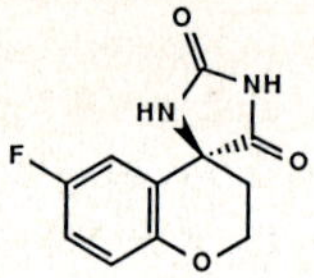 O, NH, HN, O, F, O	CP-45,634 Sorbinil	Pfizer
3	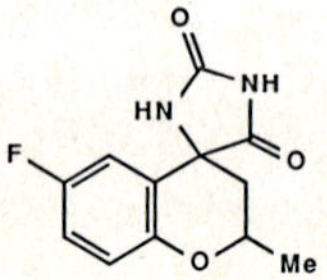 O, NH, HN, O, F, O, Me	M-79,175 Methosorbinil	Eisai
4	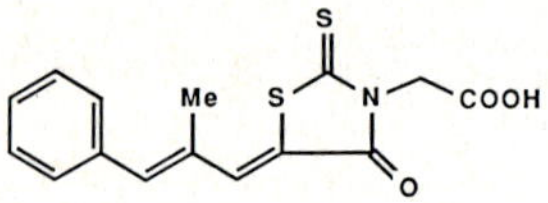 S, Me, S, N, COOH, O	ONO-2235 Epalrestat	Ono
5	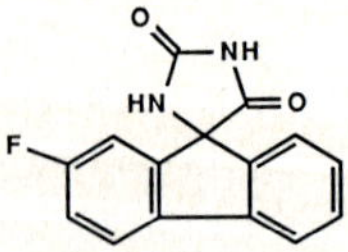 O, NH, HN, O, F	AL-1567 Alconil	Alcon
6	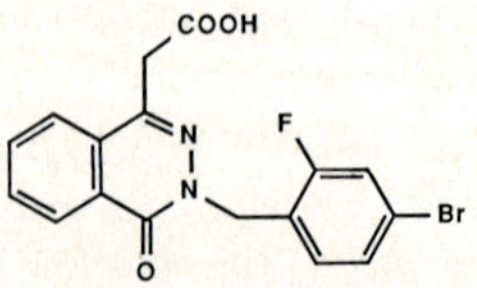 COOH, N, N, F, Br, O	ICI-128,436 Statil	I.C.I.

at the same site on the enzyme, incorporation of features derived from an analysis of the structures of tolrestat, epalrestat, alconil and statil could generate a modified model of the AR inhibitory site which might be of even greater value. At the time of writing, one series of compounds, 2-(arylamino)-4(3*H*)-quinazolinones, have been designed as AR inhibitors based exclusively on requirements of the model; however, *in vitro* results were disappointing [123].

THE DEVELOPMENT OF TOLRESTAT

Since alrestatin, albeit at relatively high doses, showed clinical efficacy in the treatment of chronic complications of diabetes [24, 25], Ayerst Research Laboratories continued the search for another, more potent, orally active AR inhibitor. Initial attempts were directed towards analogues of alrestatin, and a series was synthesized [124] (*Table 8.10*). Substitution at the 6-position generally increased potency, the highest *in vitro* being obtained with the readily

Table 8.10 AR INHIBITORY ACTIVITY OF ALRESTATIN ANALOGUES

No.	*Y*	*X*	*% inhibition* 1×10^{-5} *M*	1×10^{-6} *M*	1×10^{-7} *M*
1[a]	O	H	74	32	–
2	O	5-NO_2	64	46	–
3	O	5-NH_2	62	20	–
4	O	6-Cl	86	42	–
5	O	6-Br	70	40	–
6	O	6-COPh	85	65	22
7	O	6-SPh	92	71	22
8	O	6-(2-COOH) – C_6H_4S –	91	61	17
9	O	6-Me$(CH_2)_9$S –	60	33	26
10	S	H	89	75	28

[a] Alrestatin.

accessible thio analogue (*Table 8.10*, No. 10), which, upon oral administration, was about 6-times more potent than alrestatin in lowering galactitol levels in the lens and sciatic nerve of galactosemic rats [124]. Since the thio analogue tends to undergo hydrolysis to alrestatin, it served only as a prototype for further molecular modifications. One type involved the deletion of the oxo function to generate thionaphthostyrilacetic acids (*Table 8.11*) [125]. Halogen derivatives of this system (*Table 8.11*) showed high activity both *in vitro* and *in vivo*, and were considerably more potent than the thio analogue of alrestatin. Thus, the 6-bromo derivative (*Table 8.11*, No. 2) inhibited AR by 79% at 1×10^{-7} M, and *in vivo*, at a dose of 131 mg/kg p.o., lowered nerve galactitol levels in the galactosemic rat by 80% [125]. However, compounds containing the naphthostyril nucleus were coloured and produced staining of animal tissues.

Further molecular modifications, generated by cleaving the five-membered ring, afforded a series of colourless naphthoylglycine derivatives [126–129] (*Table 8.12*), some of which were highly potent, both *in vitro* and *in vivo*. The unsubstituted naphthoylglycine derivative (No. 1, *Table 8.12*) had moderate activity *in vitro* (51% at 1×10^{-6} M) which was enhanced by a variety of single substituents located at positions 3 to 8 of the naphthalene ring (see Nos. 2–15, *Table 8.12*). Maximal inhibitory activity (47% at 1×10^{-7} M) was obtained with the 5-bromo derivative, No. 4. The effect of changing the nature of the

Table 8.11. AR INHIBITORY ACTIVITY OF NAPHTHOSTYRIL DERIVATIVES

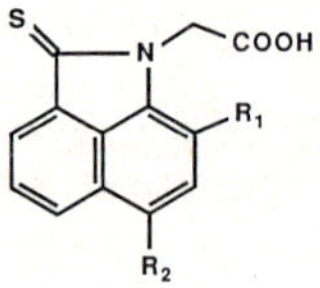

S, N, COOH, R_1, R_2

No.	R_1	R_2	% inhibition 1×10^{-5} M	1×10^{-6} M	1×10^{-7} M
1	H	H	93	87	61
2	H	Br	95	91	79
3	H	Cl	96	91	66
4	H	F	94	88	56
5	H	I	93	91	81
6	Br	Br	89	56	13
7	Cl	Cl	94	76	30

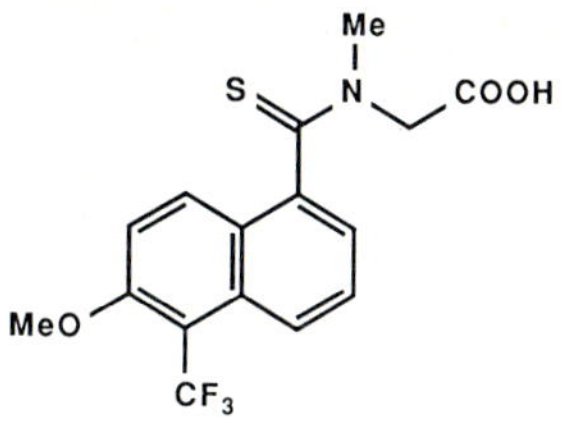

(47) TOLRESTAT

N-methyl group in this compound was investigated; when replaced by hydrogen (No. 26), or by hydrocarbon groups (Nos. 27 to 31), a substantial reduction in potency was consistently observed, whereas disubstitution of the naphthalene nucleus enhanced activity, and several derivatives (Nos. 17–19, 21, 22, 24, 25) showed more than 50% inhibition of AR at 1×10^{-7} M. Of these, No. 24, tolrestat (47), (AY-27,773) was selected for further development. Tolrestat (Aldredase®) inhibited calf lens AR with an IC_{50} value of 3.5×10^{-8} [130]. Extensive pharmacological and biochemical studies were conducted with tolrestat [108, 130–137].

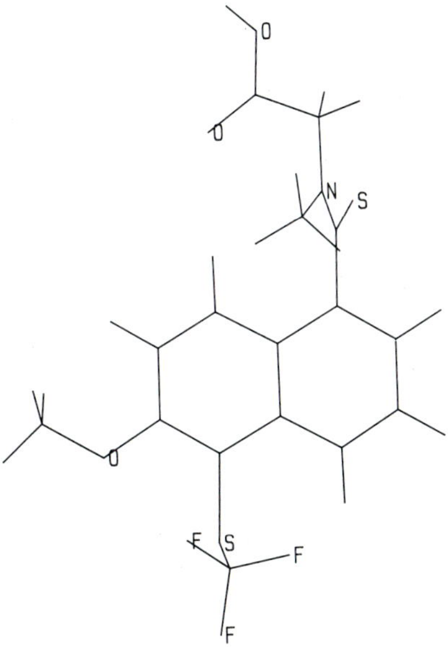

Figure 8.3. X-ray structure of trifluoromethylthio analogue of tolrestat.

Table 8.12. AR INHIBITORY ACTIVITY OF NAPHTHOYLGLYCINE DERIVATIVES

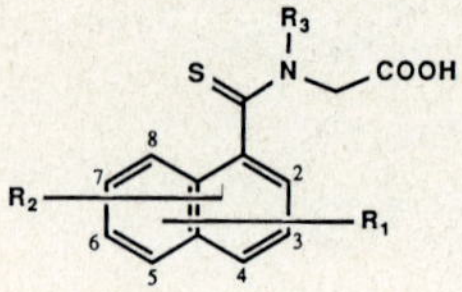

No.	R_1	R_2	R_3	*% inhibition* 1×10^{-5} *M*	1×10^{-6} *M*	1×10^{-7} *M*
1	H	H	Me	85	51	13
2	3-Br	H	Me	90	74	28
3	4-Br	H	Me	91	77	32
4	5-Br	H	Me	93	87	47
5	6-Br	H	Me	91	75	18
6	7-Br	H	Me	86	58	15
7	8-Br	H	Me	88	75	24
8	5-OMe	H	Me	83	64	17
9	5-Me	H	Me	89	74	26
10	5-CN	H	Me	89	79	32
11	5-NO_2	H	Me	91	83	43
12	5-Cl	H	Me	91	83	40
13	5-COMe	H	Me	92	74	19
14	5-Pr^i	H	Me	91	72	21
15	5-CF_3	H	Me	93	84	33
16	3-Cl	4-OMe	Me	85	78	33
17	5-Br	6-$O(CH_2)_4Me$	Me	93	91	55
18	5-Br	6-OMe	Me	99	91	72
19	5-Br	6-Me	Me	92	88	55
20	5-Cl	7-Cl	Me	88	75	29
21	5-I	6-OMe	Me	98	95	72
22	5-CN	6-OMe	Me	98	93	74
23	5-Br	6-$O(CH_2)_3OMe$	Me	92	87	38
24	5-CF_3	6-OMe	Me	98	94	65
25	5-SCF_3	6-OMe	Me	–	94	73
26	5-Br	H	H	54	14	–
27	5-Br	H	Pr^n	91	70	19
28	5-Br	H	$CH_2CH=CH_2$	92	77	27
29	5-Br	H	Et	85	72	24
30	5-Br	H	Bu^n	86	65	19
31	5-Br	H	$C_6H_5CH_2$	86	69	20

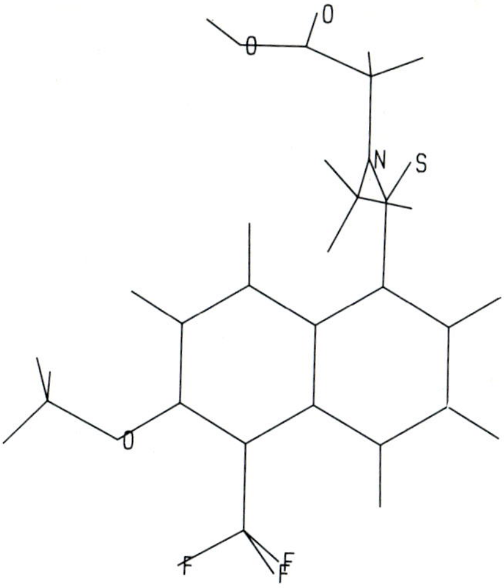

Figure 8.4. Energy-minimized computer-generated conformation of tolrestat.

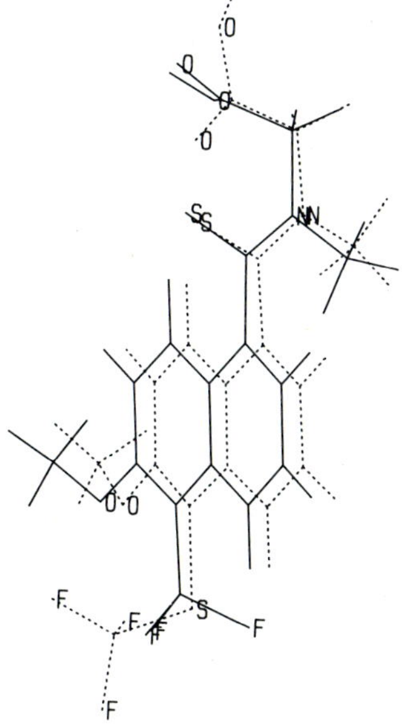

Figure 8.5. Superimposition of energy-minimized structure of tolrestat and X-ray structure of trifluoromethylthio analogue of tolrestat.

Although tolrestat is a highly crystalline compound, crystals suitable for crystallographic analysis could not be obtained. Suitable crystals, however, were obtained for the trifluoromethylthio analogue No. 25 (*Table 8.12*), which is equipotent to tolrestat *in vitro*. *Figure 8.3* shows a computer-generated projection of this compound derived from its crystallographic coordinates [138]. The energy-minimized conformation of tolrestat obtained by molecular mechanics calculations using the SIMPLEX algorithm is shown in *Figure 8.4*, while *Figure 8.5* is a superpositioning of *Figures 8.3 and 8.4*. The conformations of the side-chains on both molecules are virtually identical in the crystal and as obtained by minimization, and it is probable that the conformation seen in *Figure 8.5*, or a very similar low-energy conformation, may represent the conformations of these compounds at the receptor site.

CONCLUSION

Since the partial purification of AR in 1965 [8] and the elaboration of a rôle for this enzyme in the pathogenesis of some of the complications of chronic diabetes, thousands of compounds, both natural products and synthetic molecules, have been tested as AR inhibitors in various laboratories. Many compounds have been found to be active *in vitro*, and the present survey shows that *in vitro* AR inhibitory can be readily found. High *in vitro* activity is, however, not the only prerequisite for oral activity *in vivo*. Not only must the drug be absorbed, but it must also reach sufficiently high levels in the target organ or tissue to effectively inhibit AR. Clearly, the structural and physicochemical requirements for effective *in vivo* inhibition of AR at the desired sites are unique and are met by relatively few compounds. The discovery of clinically useful AR inhibitors has depended less on the finding of high *in vitro* activity than on the identification of compounds with suitable pharmacokinetics and tissue distribution patterns.

ACKNOWLEDGEMENT

The author would like to express his thanks to Dr. D. Dvornik for valuable discussions during the preparation of this article.

REFERENCES

1 M. Ellenberg, N.Y. State J. Med., (1979) 2005.
2 H.G. Hers, Biochim. Biophys. Acta, 13 (1974) 713.

3 R. Van Heyningen, Nature (London), 184 (1959) 194.
4 S. Hayman, M.F. Lous, L.O. Merola and J.H. Kinoshita, Biochim. Biophys. Acta, 128 (1966) 474.
5 K.H. Gabbay. Isr. J. Med. Sci., 8 (1972) 1626.
6 H. Weiner and B. Wermuth, in Enzymology of Carbonyl Metabolism: Aldehyde Dehydrogenase and Aldo/Keto Reductase (Alan R. Liss, New York, 1982) pp. 398–400.
7 P.K. Pottinger, Biochem. J., 104 (1967) 663.
8 S. Hayman and J.H. Kinoshita, J. Biol. Chem., 240 (1965) 877.
9 K.H. Gabbay, L.O. Merola and R.A. Field, Science (Washington, DC), 151 (1966) 209.
10 J.H. Kinoshita, P. Kador and M. Catiles, J. Am. Med. Assoc., 246 (1981) 257.
11 K.H. Gabbay and J.B. O'Sullivan, Diabetes, 17 (1968) 239.
12 J.H. Kinoshita, L.O. Merola, K. Satoh and E. Dikmak, Nature (London), 194 (1962) 1085.
13 C.A. Lipinski and N.J. Hutson, Annu. Rep. Med. Chem., 19 (1984) 169.
14 P.F. Kador, J.H. Kinoshita and N.E. Sharpless, J. Med. Chem., 28 (1985) 841.
15 P.F. Kador, W.G. Robinson, Jr. and J.H. Kinoshita, Annu. Rev. Pharmacol., Toxicol. 25 (1985) 691.
16 D. Dvornik, Annu. Rep. Med. Chem., 13 (1978) 159.
17 N. Canal and G. Comi, Trends Pharmacol. Sci., (1985) 328.
18 J.H. Kinoshita, D. Dvornik, M. Kraml and K.H. Gabbay, Biochim. Biophys. Acta, 158 (1968) 472.
19 D. Dvornik, N. Simard-Duquesne, M. Kraml, K. Sestanj, K.H. Gabbay, J.H. Kinoshita, S.D. Varma and L. Merola, Science (Washington, DC), 182 (1973) 1146.
20 J.C. Hutton, P.J. Schofield, J.F. Williams and F.C. Hollows, Biochem. Pharmacol., 23 (1974) 2991.
21 J.H. Kinoshita, Ophthalmology, 13 (1974) 713.
22 H. Stephen, J. Chem. Soc., (1931) 871.
23 K. Sestanj, N. Simard-Duquesne and D. Dvornik, U.S. Pat., 3,821,383; Chem. Abstr., 81 (1974) 176158.
24 J. Fagius and S. Jameson, J. Neurol. Neurosurg. Psychiatr., 44 (1981) 991.
25 A. Culebras, J. Alio, J.L. Herrera and I.P. Lopez-Fraile, Arch. Neurol., 38 (1981) 133.
26 S.D. Varma, I. Mikuni and J.H. Kinoshita, Science (Washington, DC), 188 (1975) 1215.
27 S.D. Varma, A. Mizuno and J.H. Kinoshita, Science (Washington, DC), 195 (1977) 205.
28 S.D. Varma and J.H. Kinoshita, Biochem. Pharmacol., 25 (1976) 2505.
29 J. Okuda, I. Miwa, K. Inagaki, T. Horie and M. Nakayama, Biochem. Pharmacol., 31 (1982) 3807.
30 J. Okuda, I. Miwa, K. Inagaki, T. Horie and M. Nakayama, Chem. Pharm. Bull., 32 (1984) 767.
31 A.B. Segelman, F.P. Segelman, S.D. Varma, H. Wagner and O. Seligmann, J. Pharm. Sci., 66 (1977) 1358.
32 P.F. Kador and N.E. Sharpless, Biophys. Chem., 8 (1978) 81.
33 K-H.A. Rosler, R.S. Goodwin, T.J. Mabry, S.D. Varma and J. Norris, J. Nat. Prod., 47 (1984) 316.
34 M. Shimizu, T. Ito, S. Terashima, T. Hayashi, M. Arisawa, N. Morita, S. Kurokawa, K. Ito and Y. Hashimoto, Phytochemistry, 23 (1984) 1885.
35 S.D. Varma, Curr. Top. Eye Res., 3 (1980) 91.
36 F. Fauran, C. Feniou, J. Mosser and G. Prat, Eur. J. Med. Chem., 13 (1978) 503.
37 P.S. Chaudhry, J. Cabrera, H.R. Juliani and S.D. Varma, Biochem. Pharmacol., 32 (1983) 1995.

38 L.F. Bjeldanes and G.W. Chang, Science, (Washington, DC), 197 (1977) 577.

39 I. Hirono, I. Ueno, S. Hosaka, H. Takanashi, T. Matsushima, T. Sugimura and S. Natori, Cancer Lett., 13 (1981) 15.

40 W. Cohen and S. Chau, Fed. Proc., 43 (1984) 563.

40a W. Cohen, S.V. Chau, P.R. Kastl and D.R. Caldwell, J. Ocular Pharmacol., 2 (1986) 151.

41 A. Bentsath, St. Rusznyak and A. Szent-Gyorgyi, Nature, 138 (1936) 798.

42 M.H. Creuset, C. Feniou, F. Guichard, J. Mosser, H. Pontagnier and G. Prat, Fr. Pat., 2, 543, 140; Chem. Abstr., 102 (1984) 149115.

43 M-H. Crueset, C. Feniou, F. Guichard, J. Mosser, G. Prat and H. Pontagnier, Eur. Pat., 121, 489; Chem. Abstr., 102 (1984) 78724.

44 B.V. Cheney, J.B. Wright, C.M. Hall, H.G. Johnson and R.E. Christoffersen, J. Med. Chem., 21 (1978) 936.

45 P.F. Kador, N.E. Sharpless and J.D. Goosey, in Progress in Clinical and Biological Research. *114*. Workshop on Enzymology of Aldehyde Dehydrogenase and Aldehyde Ketone Reductase; Weiner, H. and Wermuth, B., eds (Alan R. Liss, New York, 1982) pp. 243–259.

46 P.F. Kador, J.H. Kinoshita, W.H. Tung and L.T. Chylack Jr., Invest. Ophthalmol. Vis. Sci., 19 (1980) 980.

47 J.R. Pfister, W.E. Wymann, J.M. Mahoney and L.D. Waterbury, J. Med. Chem., 23 (1980) 1264.

48 H. Ono and S. Hayano, Acta Soc. Ophthalmol. Jpn., 86 (1982) 353.

49 H. Ono, I. Ohtsuka, M. Kawase, T. Shoji, T. Wakabayashi, S. Hayano and Y. Nozawa, J. Ophthalmol. Jpn., 26 (1982) 114.

50 R. Sarges, J.L. Belletire and R.C. Schnur, U.S. Pat., 4, 210, 667; Chem. Abstr., 93 (1980) 245470.

51 A.N. Brubaker, J. DeRuiter and W.M. Whitmer, J. Med. Chem., 29 (1986) 1094.

52 J.L. Belletire, Eur. Pat., 11, 426; Chem. Abstr., 94 (1981) 121235; equivalent to U.S. Pat., 4, 210, 663.

53 C.K. Moon, Y.P. Yun, J.H. Lee, H. Wagner and Y.S. Shin, Planta Med., (1985) 66.

54 R. Sarges, U.S. Pat., 4, 130, 714; Chem. Abstr., 92 (1980) 94401.

55 C.R. Kissinger, E.T. Adman, J.I. Clark and R.E. Stenkamp, Acta Crystallogr., 41 (1985) 988.

56 R. Sarges, J. Bordner, B.W. Dominy, M.J. Peterson and E.B. Whipple, J. Med. Chem., 28 (1985) 1716.

57 R. Sarges, W. Ger. Pat 2, 746, 244; Chem. Abstr., 89 (1978) 24308; equivalent to U.S. Pat. 4, 117, 230.

58 R. Sarges, U.S. Pat., 4, 181, 729; Chem. Abstr., 92 (1980) 163972.

59 R. Sarges and R.C. Schnur, U.S. Pat. 4, 248, 882; Chem. Abstr., 94 (1981) 192339.

60 R. Sarges and M.J. Peterson, Metabolism, 35 (1986) 101.

61 N. Nakai, Y. Fujii, K. Kobashi and K. Nomura, Arch. Biochem. Biophys., 239 (1985) 491.

62 H. Ono, Y. Nozawa and S. Hayano, Acta Soc. Ophthalmol. Jpn., 86 (1982) 1343.

63 B.M. York Jr., Pat. Coop. Treat. Intl. Appl. 8, 303, 542; Chem. Abstr., 100 (1984) 185790; equivalent to U.S. Pat. 4, 464, 385.

64 R. Sarges and R.C. Schnur, U.S. Pat., 4, 127, 665; Chem. Abstr., 90 (1979) 87464.

65 P.R. Kelbaugh and R. Sarges, U.S. Pat., 4, 147, 797; Chem. Abstr., 91 (1979) 20511.

66 R.C. Schnur, U.S. Pat., 4, 176, 185; Chem. Abstr., 92 (1980) 111015.

67 R. Sarges and J.L. Belletire, U.S. Pat., 4, 193,996; Chem. Abstr., 93 (1980) 46449.

68 R.C. Schnur, U.S. Pat., 4, 193, 996; Chem. Abstr., 93 (1980) 46449.

69 R. Sarges, U.S. Pat., 4,235,911; Chem. Abstr., 94 (1981) 121540.
70 D.R. Brittain and R. Wood, Eur. Pat., 125,090; Chem. Abstr., 102 (1985) 113493.
71 B.W. Griffin, M.L. Chandler and L. DeSantis, Invest. Ophthalmol. Vis. Sci., 25 (Suppl. 2) (1984) 136.
72 B. York, Pat. Coop. Treat. Intl. Appl., 8, 303, 543; Chem. Abstr., 100 (1984) 139108; equivalent to U.S. Pat. 4, 540, 700.
73 B.W. Griffin, M.L. Chandler, W.A. Shannon and L. DeSantis, Invest. Ophthalmol. Vis. Sci., 25 (Suppl) (1984) 159.
74 R.C. Schnur, R. Sarges and M.J. Peterson, J. Med. Chem., 25 (1982) 1451.
75 H. Hegler, F. Schneider and F. Lederer, Eur. Pat., 129, 747; Chem. Abstr., 102 (1985) 166740.
76 J.L. Belletire and R. Sarges, U.S. Pat., 4, 307, 108; Chem. Abstr., 96 (1982) 181165.
77 A.J. Hutchinson, Eur. Pat., 79, 675; Chem. Abstr., 99 (1983) 194944.
78 T. Sohda, K. Mizuno, E. Imamiya, H. Tawada, K. Meguro, Y. Kawamatsu and Y. Yamamoto, Chem. Pharm. Bull., 30 (1982) 3601.
79 M.M. O'Brien and P.J. Schofield, Biochem. J., 187 (1980) 21.
80 T. Tanimoto, H. Fukuda and J. Kawamura, Chem. Pharm. Bull., 31 (1983) 2395.
81 T. Sohda, K. Mizuno, E. Imamiya, Y. Sugiyama, T. Fukita and Y. Kawamatsu, Chem. Pharm. Bull., 30 (1982) 3580.
82 A.Y. Chang, B.M. Wyse and B.J. Gilchrist, Diabetes, 32 (1983) 839.
83 A.B. Richon, M.E. Maragoudakis and J.S. Wasvary, J. Med. Chem., 25 (1982) 745.
84 K. Inagaki, I. Miwa, T. Yashiro and J. Okuda, Chem. Pharm. Bull., 30 (1982) 3244.
85 J. Okuda, K. Yashima, K. Inagaki and I. Miwa, Chem. Pharm. Bull., 33 (1985) 2990.
86 Y. Kawamatsu, T. Fugita and Y. Yamamoto, Eur. Pat., 91, 761; Chem. Abstr. 100 (1984) 103326; equivalent to U.S. Pat., 4, 486, 594.
87 T. Irikura, K. Takagi, S. Fujimori, H.Y. Hirata, Eur. Pat., 147, 805; Chem. Abstr., 104 (1986) 5884.
88 T. Tadeo, S. Satoghi, K. Masanori, H. Masaki, T. Hiroshi and F. Hirata, Eur. Pat., 45, 165; Chem. Abstr., 96 (1982) 217830.
89 T. Tadeo, K. Masanori, A. Akio, M. Tetsuya, H. Masaki, T. Hiroshi, H. Fumio and M. Takashi, Eur. Pat., 47, 109; Chem. Abstr., 97 (1982) 23781.
90 H. Terashima, K. Hama, R. Yamamoto, M. Tsuboshima, R. Kikkawa, I. Hatanaka and Y. Shigeta, J. Pharmacol. Exp. Ther., 229 (1984) 226.
91 Y. Ohishi, M. Nagahara, Y. Takehisa, M. Yajima, S. Kurokawa, N. Kallkawa, A. Itoh and K. Nogimori, Eur. Pat., 143, 461; Chem. Abstr., 103 (1985) 196070.
92 R. Poulsom and H. Heath, Biochem. Pharmacol., 32 (1983) 1495.
93 D. Stribling, D.J. Mirrlees and D.C.N. Earl, Diabetologia, 25 (1983) 196.
94 R. Boot-Hanford and H. Heath, Biochem. Pharmacol., 30 (1981) 3065.
95 D. Stribling, D.J. Mirrlees, H.E. Harrison and C.N. Earl, Metabolism, 34 (1985) 336.
96 P. Fretten, D.R. Tomlinson and J. Townsend, Br. J. Pharmacol., 82 (1984) 265P.
97 D.R. Brittain, in Proceedings of the Third SCI-RSC Medicinal Chemistry Symposium, ed. R.W. Lambert (Royal Society of Chemistry, London, 1986), p. 210.
98 M. Okamoto, I. Uchida, K. Umehara, M. Kohsaka and H. Imanaka, Eur. Pat., 99, 692; Chem. Abstr., 100 (1984) 191723.
99 Y. Ito, Y. Kitaura, T. Oku, K. Sawada, Y. Suzuki and S. Hashimoto, Jap. Pat. 60/178879; Chem. Abstr., 104 (1986) 207131.
100 I. Udea, M. Matsumo, S. Satoh and T. Watanabe, Eur. Pat., 22, 317; Chem. Abstr., 95 (1981) 7226; equivalent to U.S. Pat., 4, 370, 340.

101 R. Sarges, U.S. Pat., 4, 209, 527; Chem. Abstr., 93 (1980) 204655.
102 R.C. Schnur, Eur. Pat., 30, 861; Chem. Abstr., 95 (1981) 169009; equivalent to U.S. Pat., 4, 283, 539.
103 J.M. Teulon, Eur. Pat., 161, 776; Chem. Abstr., 104 (1986) 109668.
104 M. Oya and T. Iso, Jap. Pat., 57/106, 658; Chem. Abstr., 97 (1982) 182401; equivalent to U.S. Pat., 4, 499, 102.
105 E. Kato, K. Yamamoto, T. Baba, T. Watanabe, Y. Kawashima, H. Masuda, M. Horiuchi, M. Oya, T. Iso and J.I. Iwao, Chem. Pharm. Bull., 33 (1985) 74.
106 E. Kato, K. Yamamoto, T. Baba, T. Watanabe, Y. Kawashima, M. Horiuchi, M. Oya, T. Iso and J.I. Iwao, Chem. Pharm. Bull., 33 (1985) 5341.
107 Unpublished observations, Ayerst Laboratories Research, Inc.
108 N. Simard-Duquesne, E. Greselin, R. Gonzalez and D. Dvornik, Proc. Soc. Exp. Biol. Med., 178 (1985) 599.
109 D. Dvornik and N. Simard-Duquesne, Belg. Pat., 885, 368; Chem. Abstr., 95 (1981) 103336; equivalent to U.S. Pat. 4, 307,114.
110 E. Cotlier and Y.R. Sharma, Lancet, i (1981) 338.
111 Y.R. Sharma and E. Cotlier, Exp. Eye Res., 35 (1982) 21.
112 R.C. Hunt and A.D. Morrison, Clin. Res., 32 (1984) 398A.
113 M.J. Peterson, R. Sarges, C.E. Aldinger and D.P. MacDonald, Metabolism, 28 (Suppl. 1) (1979) 456.
114 P.F. Kador, J.D. Goosey, N.E. Sharpless, J. Kolish and D.D. Miller, Eur. J. Med. Chem., 16 (1981) 293.
115 J.A. Jedziniak and J.H. Kinoshita, Invest. Ophthalmol., 10 (1971) 357.
116 P.F. Kador and N.E. Sharpless, Mol. Pharmacol., 24 (1983) 521.
117 M.E. Maragoudakis, H. Kalinsky, J. Wasvary and H. Hankin, Fed. Proc., 35 (1976) 679.
118 H.R. Munson Jr., U.S. Pat., 4, 198, 414; Chem Abstr., 93 (1980) 71574.
119 T. Tanimoto, H. Fukuda, J. Kawamura, M. Nakao, U. Shimada, A. Yamada and C. Tanaka, Chem. Pharm. Bull., 34 (1986) 2501.
120 D. Ellis, Br. Med. Bull., 36 (1980) 165.
121 W.S. Davidson and D.G. Murphy, in Enzymology of Carbonyl Metabolism 2: Aldehyde Dehydrogenase, Aldo-Keto Reductase, and Alcohol Dehydrogenase (Alan R. Liss, New York, 1985) p. 251.
122 D.G. Murphy and W.S. Davidson, Biochem. Pharmacol., 34 (1985) 2961.
123 J. DeRuiter, A.N. Brubaker, J. Millen and T.N. Riley, J. Med. Chem. 29 (1986) 627.
124 K. Sestanj, U.S Pat., 4, 254, 108; Chem. Abstr., 95 (1981) 24852.
125 K. Sestanj, U.S Pat., 4, 369, 188; Chem. Abstr., 98 (1983) 179217.
126 K. Sestanj, N.A. Abraham, F. Bellini and A. Treasurywala, Can Pat., 1, 147, 739; Chem. Abstr., 100 (1984) 7158; equivalent to U.S. Pat., 4, 391, 816.
127 F. Bellini, K. Sestanj and L.G. Humber, Eur. Pat., 59, 596; Chem. Abstr., 98 (1983) 71727; equivalent to U.S. Pat., 4, 391, 825.
128 K. Sestanj, N.A. Abraham, F. Bellini and A. Treasurywala, Eur. Pat., 59, 596; Chem. Abstr., 98 (1983) 71727; equivalent to U.S. Pat., 4, 439, 617.
129 K. Sestanj, N.A. Abraham, F. Bellini and A. Treasurywala, Eur. Pat., 59, 596; Chem. Abstr., 98 (1983) 71727; equivalent to U.S. Pat., 4, 568, 693.
130 K. Sestanj, F. Bellini, S. Fung, N. Abraham, A. Treasurywala, L.G. Humber, N. Simard-Duquesne and D. Dvornik, J. Med. Chem., 27 (1984) 255.
131 D. Dvornik, N. Simard-Duquesne, E. Greselin, J. Dubuc, D. Hicks and M. Kraml, Diabetes, 32 (Suppl. 1) (1983) 164A.
132 W.G. Robinson Jr., P.F. Kador, Y. Akagi and J.H. Kinoshita, Invest. Ophthalmol. Vis. Sci., 25 (Suppl. 2) (1984) 66.

133 D.R. Hicks, M. Kraml, M.N. Cayen, J. Dubuc, S. Ryder and D. Dvornik, Clin. Pharmacol., Ther., 36 (1984) 493.
134 M.N. Cayen, D.R. Hicks, E.S. Ferdinandi, M. Kraml, E. Greselin and D. Dvornik, Fed. Proc., 43 (1984) 750.
135 N. Simard-Duquesne, E. Greselin, J. Dubuc and D. Dvornik, Metabolism, 34 (1985) 885.
136 C. Kemper, L. Huslin and D. Dvornik, Fed. Proc., 44 (1985) 1392.
137 W.G. Robinson, P.F. Kador, Y. Akagi, J.H. Kinoshita, R. Gonzalez and D. Dvornik, Diabetes, 35 (1986) 295.
138 K.I. Varughese, M. Przybylska, K. Sestanj, F. Bellini and L.G. Humber, Can. J. Chem., 61 (1983) 2137.

Index

Cumulative Index of Authors for Volumes 1–24

The volume number, (year of publication) and page number are given in that order.

Cumulative Index of Subjects for Volumes 1–24

The volume number, (year of publication) and page number are given in that order.